Collecting the North

Edited by

Janicke S. Kaasa
Lene Liebe Delsett
Ulrike Spring

BRILL

LEIDEN | BOSTON

Chapter by Lene Liebe Delsett (CC BY-NC)
Chapter title: The Whales Are Leaving: a Dolphin Skeleton and Its History of Collection, Curation and Research.

Chapter by Grace Shephard (CC BY-NC)
G.E.S acknowledges support from the Research Council of Norway for 'POLARIS – Evolution of the Arctic in Deep Time', Project Number 326238, and its Centres of Excellence Scheme, Project Number 332523.

Chapter by Johannes Mattes (CC BY)
Chapter title: Expedition diary – Sensing, Observing, Measuring: József Pálffy's Diary and the Austro-Hungarian Jan Mayen Expedition of 1882/83.
This research was funded by the Marie Skłodowska-Curie actions (MSCA) grant agreement No 101066820 (Poles and People) from the European Commission.

Cover illustration: The Ephemerality of Collecting. Svalbard 2011. Photo: Ulrike Spring.

The Library of Congress Cataloging-in-Publication Data is available online at https://catalog.loc.gov
LC record available at https://lccn.loc.gov/2026930883

Typeface for the Latin, Greek, and Cyrillic scripts: "Brill". See and download: brill.com/brill-typeface.

ISSN 2772-9621
ISBN 978-90-04-74967-2 (hardback)
ISBN 978-90-04-74966-5 (e-book)
DOI 10.1163/9789004749665

Contents

Acknowledgements

This volume is a result of the international and interdisciplinary research group Collecting Norden (2021–2023, University of Oslo), financed by UiO: Nordic and Reimagining Norden in an Evolving World (ReNEW). We wish to thank the team of UiO:Nordic, under the leadership of Professor Tore Rem, for creating a supportive and intellectually inspiring environment for discussions about collecting and the North, the *Arctic Humanities* series editor Jan Borm and editorial board for their belief in this project, our research assistant Ruairi Timothy Long for his help with preparing the manuscript for publication, two anonymous peer-reviewers for their useful suggestions, and the team at Brill for their patience. Special thanks go to our contributors for generously sharing their ideas with us and for making this volume possible.

Figures and Tables

Figures

Tables

Notes on Contributors

Anna Heiða Baldursdóttir
is a research specialist at the Agricultural Museum of Iceland. Baldursdóttir's research focuses on material culture, everyday history, museum collections and cultural heritage from the late 18th century until the 21st century. She received her PhD degree in history from the University of Iceland in 2023. The doctoral research was a part of the project *My Favorite Things* which was a grant of excellence from the Icelandic Research Fund 2018–2022.

Sine Halkjelsvik Bjordal
is a postdoctoral researcher specializing in the cultural history of natural history, affiliated with *The Afterlives of Natural History* project at the University of Oslo. She previously completed a PhD as part of *The Printed and the Built – Architecture and Public Debate in Modern Europe* project at the Oslo School of Architecture and Design, where she investigated the history of knowledge of the Norwegian stave churches in the 18th and 19th centuries. Her research focuses on 18th- and 19th-century histories of knowledge, heritage studies, Scandinavian literature, and literary theory.

John Bodinger de Uriarte
is Professor of Anthropology at Susquehanna University, where he directs the Museum Studies and Diversity Studies Programs. He has been a Fulbright Fellow in Iceland and is currently on the roster of Fulbright Specialists. His research interests include museums; study abroad; contemporary Iceland; tourism; questions of identity, representation, and sovereignty; and how such issues are engaged in contemporary museum, casino, touristic, and photographic practice. He is the author of *Casino and Museum: Mashantucket Pequot Representation*, and lead editor of and contributor to *Study Abroad and the Quest for an Anti-Tourism Experience*.

Brita Brenna
is Professor of Museology and Cultural History at the University of Oslo, where she teaches in museum and heritage studies. Her research covers histories of collections, museums and landscapes in conversation with Natural History and Arctic humanities. Perspectives from STS, gender studies and the history of science are informing her work. She is currently PI of the research project *The Afterlives of Natural History*.

Elettra Carbone
is Associate Professor in Norwegian Studies in the UCL Department of Scandinavian Studies (School of European Languages, Culture and Society). Her main areas of research and teaching are nineteenth-century Norwegian and Nordic literatures, cultural mobility, sculpture, archival studies and object-based learning. In addition to several articles, she is the author of *Nordic Italies: Representations of Italy in Nordic Literature from the 1830s to the 1910s* (Nuova Cultura 2016) and of the practical grammar *Norwegian Tutor* (Hodder & Stoughton 2018, with G. Puzey). She has also co-edited volumes such as *The Public Sphere and Freedom of Expression in Northern Europe, 1814–1914* (Norvik 2020, with R. Hemstad), *Sculpture and the Nordic Region* (Routledge 2017, with S. Ayres) and *The Norwegian Constitution and Independence of 1814* (Norvik 2015, with R. Hemstad).

JoAnn Conrad
is a professor of folklore and cultural history. She has taught at the University of California, Berkeley, and the University of Missouri and currently teaches at Diablo Valley College. Her research deals with issues of nation-formation, children's literature, women artists and illustrators of the modern era, feminist theory, and narrative, particularly in Norway and Sweden, where she has been conducting research since the mid-1990s. Her dissertation was based on field work in Kautokeino during the 1990s, and in 2022 she was a Fulbright Scholar to the University of Iceland. She has published in *Ethnologia Europaea*, *Fabula*, *Journal of American Folklore*, *Narrative Culture*, and in many collected editions focusing on the North, and is currently working on a book project on Gustaf Tenggren.

Lene Liebe Delsett
is a palaeontologist studying marine vertebrates with ancestors on land, focusing on whales and extinct marine reptiles. She is a museum scientist that is interested in the history and biases in natural history collections. Currently, she holds a position at the Natural History Museum, University of Oslo where she leads the research project *ECHO – Evolutionary Convergences in Historical Oceans*. Her most recent publications are "Collecting whales: processes and biases in Nordic museum collections" and "Boreal waterways: An Early Cretaceous plesiosaur from Ellesmere Island, Nunavut, Canadian Arctic and its palaeobiogeography".

Jesper Hansen
is Associate Professor in UCL HEDS (Higher Education Development and Support Institute). His main areas of research are language teaching pedagogy,

queer pedagogy and student voice (with a particular focus on student-staff partnerships). He has published on education, co-authored the grammar book *Danish Tutor* (Hodder & Stoughton 2018, with A. Grydehøj) and is one of the founding editors of the journal Postgraduate Pedagogies Journal (LSE, Houghton St Press), which centres the voices of Graduate Teaching Assistants.

Heidi Hansson
is Professor of English Literature at Umeå University, Sweden. Her research has concerned Arctic and Northern studies, Irish women's literature, nineteenth-century periodical publishing and postmodern romance, and she has also published on education. Her most recent book is Heidi Hansson, Maria Lindgren Leavenworth and Anka Ryall, ed., *The Arctic in Literature for Children and Young Adults* (Routledge, 2020). At present, she is finalising an overview chapter about Lapland in nineteenth to twenty-first century literature by outsiders and involved in developing a project about relationships between activism and academic work.

Janicke S. Kaasa
is Associate Professor of Nordic literature at the Department of Linguistics and Scandinavian Studies, University of Oslo. Kaasa's fields of research include book history, the sociology of literature and literary representations of the North, and her objects of study range from eighteenth and nineteenth century periodical texts for children to twentieth century travel writing. Among her most recent publications are the articles "Arctic images in context: Rereading the Padleimiut photographs in Richard Harrington's *The Face of The Arctic*" (*Lychnos*, 2023) and "Collecting the Andrée expedition: Circulating Arctic objects in writing" (*Nordisk Museologi*, 2024).

Harald Lindbach
is Associate Professor (first archivist) at the National Archives, The regional state archive of Tromsø. His research area includes documentation strategies and archives in the 18th century, Sámi and Kven history and archival theory in a wider sense. He has worked with dissemination and exhibitions of archive material in various contexts, including at the indigenous festival Riddu Riđđu where he has also been a member of the board.

Sigurður Gylfi Magnússon
is Professor of Cultural History, Department of Philosophy, History and Archaeology at the University of Iceland. He is furthermore the chair of the Center for Microhistorical Research (www.microhistory.org). He has written thirty books and numerous articles published in Iceland and abroad. He is the founder and

one of the editors of a book series named *The Anthology from Icelandic Popular Culture* (Sýnisbók íslenskrar alþýðumenningar) which has so far published thirty-four books, mostly on the topics of egodocuments and everyday life history. He is also the co-editor with István M. Szijártó of *Microhistories*, an international book series published by Routledge.

Johannes Mattes
is a historian of knowledge and science specializing in the natural sciences (e.g., earth sciences, geography, cartography, archaeology, zoology, nuclear medicine) from the eighteenth to the twentieth century. He is currently a Marie Curie Fellow at the University of Oslo (Norway) and a Project Leader at the Austrian Academy of Sciences in Vienna. He is Secretary General of the International Union of Speleology (UIS) and Delegate to the International Science Council in Paris.

Davíð Ólafsson
is Associate Professor of Cultural studies at the Faculty of Icelandic and Comparative Cultural Studies, University of Iceland. He has studied Icelandic cultural history of nineteenth century for three decades, with emphasis on literacy practices, egodocuments, and microhistory. Among his latest publications are "Social Circularity of Books and Manuscripts," *Objects in the Archives: Modern Material Culture and Heritage in the North* (Routledge, 2025); "Many pages, much ink," *Paper Stories: Paper and Book History in Post-Medieval Europe*. (Gruyter, 2023), and (with Sigurður Gylfi Magnússon) *Minor Knowledge and Microhistory: Manuscript Culture in the Nineteenth Century* (Routledge, 2017).

Grace Shephard
is an interdisciplinary Earth Scientist with a background in geology and geophysics. Her research delves into what the planet looked like millions of years ago, including the past configurations of continents and oceans. She builds digital models that focus on the Arctic and allow a better understanding of processes including climate, volcanic eruptions, and mineral formation. She was lucky enough to get to the North Pole in 2016 onboard the Swedish Icebreaker, Oden. She received her PhD from the University of Sydney and currently holds appointments at the University of Oslo and the Australian National University.

Ulrike Spring
is Professor of Modern European History at the Department of Archaeology, Conservation and History, University of Oslo. Her research interests include the history of travel, literary museums, Arctic history and most recently, the relations between Norway and Japan in the nineteenth century. She was

project manager of the UiO:Nordic research group Collecting Norden (2021–2023) together with Janicke S. Kaasa. With Kaasa she has co-edited the thematic issue "Objects of the North" in *Nordisk Museologi* 2024.

Ryan R. Weber
is Associate Professor of Medical Humanities at the Geisinger Commonwealth School of Medicine (Pennsylvania, USA). He has published articles and book-length studies on topics ranging from the social constructions of pain and eugenics to disability studies, medical professionalism, and arts-based interventions for promoting health. He also serves on the editorial board of the *International Journal of Music, Health, and Wellbeing*. Prior to his appointment at GCSOM, Dr. Weber was Chair of the Department of Fine Arts and Program Director of Medical Humanities at Misericordia University.

CHAPTER 1

Collecting the North: Introduction

Ulrike Spring, Janicke S. Kaasa and Lene Liebe Delsett

Abstract

This introductory chapter explores the theoretical and methodological perspectives that underpin the *Collecting the North* volume. In its investigation of the narratives and images of the North through objects that were collected in the Nordic countries and the European Arctic during the nineteenth and early twentieth centuries, the chapter argues for a 'glocal' approach that acknowledges both the specific localities and global existence and distribution of northern objects. Drawing on insights from actor-network theory (ANT), assemblage thinking and new materialism, the chapter demonstrates the need for a 'nomadic' way of thinking about objects, as multiple, relational and inter-connected, an approach that is further emphasized by the radical cross-disciplinarity of the volume, represented by the fields of Cultural studies, Nordic literature, English literature, History, Cultural history, History of science, Scandinavian studies, Museology, Palaeontology, Musicology, Anthropology and Geology. The bringing together – the collecting – of such a wide range of scholarly discussions on such a wide selection of objects (human made, natural, sensory, material, immaterial, lost) is, the chapter contends, crucial to understand their relevance to the narratives, imaginings, and interpretations of the North in the past, present and future.

Keywords

objects – collecting – the North – 'glocal' – actor-network theory – assemblage

This project began with our discussions on what the term 'northern objects' entails. What do we mean by 'northern'? And what by 'object'? After all, natural phenomena such as polar bears, icebergs and northern lights, as well as human-made artefacts such as igloos, stave churches and Viking ships are in everyday speech often referred to as *northern objects*, whether or not they

 | DOI:10.1163/9789004749665_002

are material things or immaterial phenomena and whether or not they can be seen in regions beyond the North as well. Do objects necessarily have to belong somewhere? Are they inevitably linked to specific localities, such as the North? Certainly, the objects we discuss in this volume encompass a wide range of 'things', and they can indeed all be understood as 'northern'. Yet, during our work, we soon realised that most transgress such an interpretation, and that some of them destabilize established ideas about their 'northernness' and the North. The present volume, then, was sparked by our curiosity about what the various narratives that objects communicate, may tell us about the North.

While Timothy Morton's (2010; 2013, 1; 2) understanding of "hyperobjects", referring to delocalized "things that are massively distributed in time and space relative to humans", bearing "a significant impact on human social and psychic space", inspired us to think about objects and phenomena beyond borders, we decided in this volume to adopt a hands-on approach and to think about transgression and delocalization in relation to concrete material objects. The objects discussed here are physically stored in collections, or they are verbally and/or visually represented – and collected – in printed materials, such as books and maps, in and by themselves objects. The objects have their own local biographies (Kopytoff 1986) or, to avoid the limitations that the idea of a continuous line between 'life' and 'death' implies, "itineraries" (Fontijn 2013), that is, they follow various paths with detours, stops, and unexpected moves. At the same time, they belong to a shared space of experience that extends the localities of their storage place or itineraries, pervading our lives and beings, and thus are delocalized or rather non-local. Loosely drawing on Morton's concept: we argue that they belong to genres or specific media that cannot be tied merely to specific localities, but that are global in their existence and distribution. Thus, an Icelandic manuscript is a singular Icelandic object, but also part of a transcultural genre of artefacts. It has specific meaning as an object belonging to the Icelandic culture, but also as an object belonging to a format, which is global and communicates beyond its locality. The Icelandic manuscript, then, is an object that is both Icelandic, northern and global.

This leads us to another concept that inspired us: the 'glocal', the intersection of the local and the global. Some of the objects discussed in this volume were collected in the geographical North and others were collected outside of it; certain objects originate in the geographical North, others do not. Moreover, some objects do not lend themselves easily to a specific localization: different approaches to one and the same object may locate it within different geographical and cultural contexts, allowing for belonging to both the local and the global. The 'glocal', that is the global as occurring in the local, or, as Victor Roudometof (2016, 401; italics in original) defines it, "as experiencing the global locally or through local lenses", creates a central fulcrum for our investigation of how objects are made relevant to and shape the narratives, imaginings, and interpretations of the North in the past, present and future. These narratives, imaginings, and interpretations, we argue, cannot be understood merely through what is usually defined as a northern object, but through objects located in the fluid borderzone between the North and the world, between the local and the global – a borderzone that is constantly moving, producing new stories and narratives depending on from where and from which perspective we look at it.

Collecting the North originated in a conference organised by the interdisciplinary and interinstitutional research group 'Collecting Norden', based at the University of Oslo. With both the research group and the present volume we wish to contribute to studies of the North and to the field of Arctic humanities, through consistent crossdisciplinarity and, most of all, methodological and theoretical openness and curiosity (see also Spring & Kaasa 2024). The contributors to our volume share this curiosity and a keen interest in understanding how narratives and images of the North were produced during the nineteenth and early twentieth centuries through the prism of objects, collections and collecting practices. Such methodological and theoretical variation is crucial for our aim of bringing forth the manifold ways in which objects may be relevant to our understanding of the North. The contributors, like the editors, therefore come from a wide array of academic fields and professional backgrounds, including the humanities and the natural sciences.

1 The North

What, and above all, where is the North? The objects discussed in this volume refer primarily to the geopolitical North of the Nordic countries and the Arctic north of Europe. Such a demarcation, however, is not straightforward, nor is it a given. Definitions of the North vary with and within different disciplines. For example, there is more than one definition of the Arctic, depending on one's approach and position (Hamelin [1975] 1979; Plumb, Sanmark & Heddle 2020), although the most widespread (and strictest) definition locates the Arctic north of the Arctic Circle at approximately the 66th parallel north. The North is relational and might be somewhere else for people positioned in Southern Europe than for people living in the North (Spufford 1996; Schimanski & Spring 2020).

Canadian geographer Louis-Edmond Hamelin ([1975] 1979) has argued for an understanding of the North through the concept of nordicity (*nordicité*), which breaks with more traditional geopolitical approaches by referring to different degrees of northernness. Hamelin's concept allows for an identification of the North based on several overlapping, varying and even conflicting factors and criteria, opening the way for an understanding of not one unitary, but several geographical Norths. Despite the centring on the Nordics and the Arctic north of Europe in this volume, we approach the North as a flexible geography that is spatially, temporally, historically, and culturally variable. Indeed, we argue that the objects, collections and collecting practices discussed here serve to expand the North in a variety of ways, as the objects may be collected outside of the North, physically located in non-northern collections, or represented, verbally and/or visually, in printed and digital materials and archives that are not necessarily associated with the North. As such, we draw here on the notion of the global Arctic (Finger & Rekvig 2022), or to be precise, of a *glocal* Arctic, which recognizes the importance of approaching the global North from a consistently local perspective, in this volume made evident through the objects in question.

In his cultural history of the North, Peter Fjågesund (2014, 16) underlines the importance of the North as an orientation point, reflecting upon how the North, indeed, all the cardinal points, the South, East and West, represents "a mental matrix, an overriding metaphor, a kind of culturally defined and maintained structuring principle that helps us – literally like the compass itself – to orient ourselves in the world". The mental matrix that Fjågesund refers to applies to geography and space, but equally to our cultural imaginary of the North. Thus, our idea of *where* the North is, is intrinsically linked to our idea of *what* the North is.

Like the geographical North, the North as a cultural imaginary is manifold. In this volume, the North is a site of exploration, expansion, and conquest, but also of everyday life. The North is a space of scientific projects, commercial ventures, and self-realization. It is a destination for explorers, scholars, tourists, artists and hunters. It is home to many and *terra nullius*, Indigenous land and industrialized space. The North is material and immaterial, lost and found. It is contrast and difference, far and near, made meaningful through both similarity and difference. It should be clear, then, that we do not understand the North as one, unitary and static, but as multiple, dynamic and relational.

2 Objects

The objects collected in this volume are as varied as the Norths presented above, but all relate to the geographical North of Europe. They belong to different categories of objects: while most are tangible and material, they were used for different purposes, they are unique, and they are reproducible copies. They are created by humans and drawn from nature. Some exist only in fragments, others within the imagination. They range from a whale specimen (Lene Liebe Delsett) and a polar bear hide (Brita Brenna) to writing tools (Davíð Ólafsson), photographs (JoAnn Conrad, Ulrike Spring) and folk songs (Ryan R. Weber). They include everyday objects and treasures (Anna Heiða Baldursdóttir, John Bodinger de Uriarte), maps of imaginary islands (Grace Shephard), book illustrations (Sine Halkjelsvik Bjordal, Heidi Hansson, Janicke S. Kaasa), drawings, manuscripts (Elettra Carbone & Jesper Hansen, Sigurður Gylfi Magnússon, Johannes Mattes, Ólafsson) and archival files (Harald Lindbach). What they have in common is that they contribute to the production of knowledge about 'northernness'. This leads us to our use of the terms 'object' and 'northern objects'.

Our contributors acknowledge the different qualities objects possess, and the varied narratives objects convey. They apply different terms in their analyses besides object – such as item, thing, artefact, fragment, specimen – reflecting contemporary scholarly discussions on terminological challenges (Olsen, Burström, DeSilvey & Pétursdóttir 2020; Morton 2013; Sundqvist 2021). To make these variations manageable, we have decided to use 'object' as our umbrella term. The term carries the connotation of objectifying things, which has been a dominant Western scientific approach for at least the past 300 years, corresponding with our focus on nineteenth- and early twentieth-century European or, more generally Western, collecting practices, which often implied a colonial gaze. While we remain critical of the ontological distinction object-subject, we acknowledge that things traditionally are objectified in Western scientific thinking; they are literally made into objects.

The various chapters examine the objects as material things that bring forth specific narratives. While the materiality of the objects matters, their cultural embeddedness and potential for shaping specific perceptions and images of the North is central. This volume then takes more inspiration from material culture approaches than from studies of materiality (see McDonnell 2023, 199–200).

Objects have biographies or follow itineraries, but they are also nodes: objects, we presume, have agency and are interacting with other objects and subjects (human and non-human) in networks (Latour 2005; Law 1994; 2009; Harman 2017). Such a perspective implies an understanding of objects not as something in and by themselves, but as always acting and being in relation to something else. In this sense, we are inspired by Rosi Braidotti's (2013, 167) nomadic thought in our approach to objects as non-unitary, relational and inter-connected, always in their becoming. Centring our attention on single objects as nodes in networks allows us to explore the concept of the North from a multitude of perspectives, corresponding to Hamelin's understanding of the North through degrees of northernness. We could even say that once an object is inserted into a specific collection that is identified as northern, it becomes itself northern.

In some objects, the degrees of northernness are condensed, for instance with established 'northern objects' such as polar bears and stave churches. While these objects may be typical of a northern geographical locality, their relevance reaches far beyond this space. As Brenna demonstrates in her chapter, polar bears were turned into commodities in the nineteenth century and became part of European economic circulation; and Bjordal shows in her contribution how Norwegian stave churches were seen in relation to Chinese pagodas. Moreover, polar bears and stave churches are examples of objects that were mobile and transferred from the North to other parts of Europe. Similarly, some collections could become part of an imagined Northern space as far as in Australia and the US, as Weber argues in his chapter on the Nordic folk songs collected by composer and musician Percy Grainger. Conversely, objects that did not exclusively belong to a northern geographical locality could nevertheless become representative of that space. In the nineteenth century, whale specimens in Scandinavian museums, examined by Delsett, were often considered typical of the northern regions, despite whales having a global habitat, living in every ocean on the planet (see also Simon-Ekeland & Delsett 2024). Likewise, as Spring discusses in her chapter, a photograph of German tourists in Sámi costumes was perceived as an authentic Sámi photograph, becoming representative of the North.

Most objects explored in this volume were widely disseminated and actively shaped ideas of the North. Some were part of everyday life during the nineteenth century, and only later entered collections as typical northern objects. In their chapters, Baldursdóttir and Bodinger de Uriarte examine vernacular objects in Icelandic cultural collections and their transformation from the mundane into collectible 'treasures' or objects worthy of display in museums, thus becoming part of Icelandic national narratives. Also, several of the contributions point to how the idea of a northern object may be questioned and nuanced. In Ólafsson's chapter on the Icelandic peasant Sighvatur Grímsson's manuscripts, the emphasis on the writing tools that made the writing possible embeds the manuscript in a global network of consumerism, which complicates the manuscript's local belonging: the ink and pens needed came from outside of Iceland. Moreover, the diary of József Pálffy, participant in the Austro-Hungarian Jan Mayen Expedition of 1882–1883, discussed by Mattes, is today kept in a private Austrian collection. The northernness of this object may be confirmed by its place of origin – the Arctic – and its content: it was written in the North about the North. Yet, it can just as easily be read through the biography of its author and its present-day location and thus be identified as an Austrian object.

Some objects may be lost to us in their physical form but may still be relevant to our interpretation of the North. As Delsett reminds us, specimens can be removed from collections or may only be partially included in collections. Moreover, Shephard shows in her contribution that established knowledge of Arctic geography, and the cartographic representations of this geography had to be rectified when explorers discovered that the island of Crocker Land did not exist. Northern knowledge – like any other knowledge – can disappear, reappear or be amended. Furthermore, objects are often mobile in the sense that they circulate between cultural contexts and languages. In her chapter on early nineteenth-century Danish and Norwegian children's magazines, Kaasa demonstrates that articles and images on the North were often retrieved from British or German printed materials. The children's magazines, then, are best approached as a transnational space that re-circulates, re-contextualizes and re-collects northern imagery. Likewise, the interest in Norwegian stave churches as architectural masterworks can be traced back to a German-language publication (albeit published by a Norwegian), as Bjordal points out in her contribution. Sometimes artists or writers may obscure previous knowledge on their object of interest in order to adjust it to a specific national context, as Hansson shows in her discussion of Paul du Chaillu's search for British Viking identity. At other times, objects change cultural affiliation during their itineraries. Lindbach presents the complex history of a Sámi archive originating in the eighteenth century, of which some materials were kept in Denmark, while others were transferred to Norway. Recently, parts of the archive were moved to the north of Norway, a result of efforts to return Sámi historical evidence and objects to Sápmi.

Objects also provide us with information about the past that goes beyond the object's history and opens up a variety of northern narratives. Databases about everyday objects in museums and archives provide information about social living conditions in nineteenth-century Iceland (Baldursdóttir); illustrations and texts in children's magazines tell us about images of the North in Denmark and Norway in the first half of the nineteenth century (Kaasa); manuscripts of Icelandic farmers in the nineteenth century can provide insights into their writing and living conditions (Ólafsson and Magnússon); the photograph of an Austrian actor informs us about tourism discourses in late nineteenth-century Europe (Spring); drawings and diaries of the North are a path to Nordic languages (Carbone & Hansen); and a diary by a Central European aristocrat and maps of an imagined island help us to trace polar knowledge production (Mattes; Shephard). And what can a page itself tell us about the North, and more specifically about Icelandic manuscript culture (Magnússon)?

3 Collections and Collecting

How do these various contributions help us trace ideas of the North and these objects' 'northernness'? Is 'northernness' determined by their place of origin or the context in which they are collected, published or displayed? Or do the places and spaces in which objects are kept, determine whether they are considered northern or not? In other words, how does the perspective of collections and collecting play into our notion of northern objects as well as of the North?

Collections and collecting can, like objects, be considered nodes in a complex network of relations. We draw here on insights from actor-network theory (ANT) and assemblage thinking, in which the ordering of entities, here the collecting of objects, is understood as multiple and relational, without one inherent and fixed hierarchizing principle (see e.g. Deleuze & Guattari 1980; Latour 2005; see also Müller 2015). Recent collection research has emphasized the constantly changing nature and identities of collections, focusing on their mobility, their endings or ruptures and relevance for the present and future rather than on their role as containers of the past and an accumulation of objects (Driver, Nesbitt & Cornish 2021; Jardine, Kowal & Bangham 2019; Thomas 2018). Thus, our approach to collecting and collections is based on the notion of such assemblages as shifting, never given, and with agency. Collecting practices, then, form a central part of understanding concepts of the North: they involve ordering and thus selection, which makes manifest the relational and the power dynamics of collecting and collections as well as of how – and why – certain objects are made meaningful to our notion of the North, whereas others are not. The objects that make up a collection are already part of an assemblage before they enter the collection, and therefore always already collected. By being inserted into collections, their meaning is often temporarily stabilised and fixed by collection inventories, exhibition contexts or book covers. Collections are also ideological constructions (Vergo 1989). They are repositories of a carefully chosen and selected past and are used as such in current geopolitical discussions about their authentic belonging. In recent times, much research has discussed attempts to make museums open participatory places, as well as efforts by Western cultural institutions to decolonize their collections and to critically reflect on them from ethical perspectives (e.g. Hetland, Pierroux & Esborg 2020; Pearce 1993; Troelenberg, Oțoiu & Bodenstein 2022).

In this volume, we have opted for an extended notion of the term 'collection' to allow for the many variations of how the North can be collected and to extend the repertoire of existing narratives of northern objects and the North. Our contributors examine museum collections and archives, and they analyse books and periodicals as collections in their own right. Thus, not only three-dimensional objects but also illustrations and texts in magazines or in books form collections (Hansson; Kaasa). Some collections were created intentionally, others more haphazardly, whereas some changed their ordering and dynamics with objects entering or leaving or being evaluated anew. As Conrad shows in her contribution on so-called 'ethnographic photographs' of Sámi, collecting impacts how histories are told.

Following this line of thought, we emphasize 'collecting' rather than 'collection' in the title of our volume, acknowledging the processual and open-ended character of gathering objects and of knowledge. One meaning of the verb 'to collect' is 'the bringing together of things', moving them from one place or space to another and in the process taking along various forms of knowledge. In that sense, collecting is always a 'glocal' process, exceeding the mere process of acquiring objects at/from a specific place.

The objects discussed in this volume were collected in the Nordic countries and the European Arctic during the nineteenth and early twentieth centuries, when museums and archives were established across Europe in interplay with nation-building and imperialist expansion. Collecting in the nineteenth century, then, did often occur in the context of colonialism and nation-building, as many chapters in this volume demonstrate. Likewise, the purposes of collecting varied greatly. Collecting may be scientifically and financially motivated, or it may also be driven by personal needs and desires. Collecting may be intentional or random, or it may also set out to collect the unique and particular or the ordinary and general. Some of the objects explored in this volume were collected with the purpose of increasing knowledge of the North, whereas others were not collected with such an aim. Some of the objects were collected during touristic tours and scientific expeditions in the North, others were collected as part of industrial activities such as whale hunting or by individual travellers and residents. Often, the processes of collecting include some or all of the overlapping and conflicting purposes outlined above. Collecting is never neutral (Vergo 1989). It implies power dynamics and biases in every stage, with a capacity of inducing suffering and inflicting pain. Whether intentional or not, and whatever the intentions may be, collecting always entails selection, inclusion, and exclusion.

Collecting thus may be connected to human and animal suffering, be it the archival gathering of information on the oppressed Sámi population (Lindbach), the hunting and displacement of polar bears (Brenna) or the collecting of whales contributing to extinction risk (Delsett). Collecting could, however, also have healing powers, as Weber shows in his chapter on composer Grainger's personal collection. Sometimes collected material may turn suffering into healing, such as when Sámi today use early ethnographic photographs of Sámi, mostly taken within an oppressive context, as a means to learn about their family histories, to remember and even repair the past (Conrad), or when archives on the Sámi population are returned to Sápmi.

The latter example reminds us of the changes collection practices as well as the interpretation of collections have undergone in the past hundred years. As Carbone and Hansen point out in their chapter on the collections of University College London, the initial aim of collecting may be different from later uses of the collected objects. Attitudes, interpretations and evaluations change: while many saw the hunting of polar bears and whales as a heroic endeavour in the nineteenth century and considered them valuable souvenirs or museum specimens, museums today need to critically examine the provenance of their collections. While taking pictures of oneself in Sámi clothes was a common pastime among tourists in the North during the nineteenth century, today many consider this a disrespectful act (Spring). Moreover, the terminology used to describe Greenlanders to Scandinavian child readers in the 1830s is in our time acknowledged as patronizing and unacceptable (Kaasa).

Collecting, at least that which is intentional, ascribes some kind of value to the object being collected. At the very least, the object collected is made valuable through the collecting and its inclusion in a collection. However, the understanding of what is worth collecting is culturally and historically dependent, as the chapters in this volume remind us. During the past centuries, the perspective on what objects were worth collecting or not, changed drastically: from the Wunderkammer tradition with its focus on valuable and aesthetically pleasing objects, to natural specimens and ethnographic objects as evidence for scientific research, as well as everyday objects as part of nation-building in the second half of the nineteenth century (Miller 2017). Certainly, most of the objects discussed in this volume are part of what Manfred Sommer (1999) has called "economic collecting", that is, the objects had value in the sense of being of use, not being primarily sources for aesthetic pleasure. All the selected objects have – in one way or another – shaped European knowledge and contributed to the understanding and interpretation of the North, as their itineraries demonstrate: they became part of or themselves constituted collections in the extended sense we apply in this volume, being assigned relevance as keepsakes.

Our emphasis on collecting and collections implies an understanding of the North as a geography of collecting, in and from which objects have been collected and have entered new contexts. Such re-contextualization entails re-collecting and re-discovering objects and their relevance to narratives and imaginings of the North. These perspectives are amplified when collections and collecting practices are digitized and mediated, altering our access to and experience of objects, providing new biographies and itineraries, and expanding their relationality and inter-connectedness. To write about the origins of collections and collecting practices, then, necessitates reflection on one's own position as a researcher and an acknowledgment that we are all part of a continuous process of rewriting and reinterpreting objects and collections.

•••

In concurrence with our theoretical and conceptual discussions above, we see the present volume as both an object and a collection. Complying with the volume's focus on objects, each chapter of *Collecting the North* is structured under an object heading, signalling the type of object that is studied. We have opted for a structure that acknowledges the 'glocal' belonging of each object, and have roughly curated the sixteen chapters in keeping with their differing focus on the written – the book, the manuscript – and the visual – the 'artefact'. This organization is of course somewhat arbitrary: a book, for example, is a three-dimensional artefact that features two-dimensional text and possibly illustrations, themselves objects within the object. However, it helps us to emphasize the relevance of textual objects to the field of Northern material studies, to acknowledge their specific features as well as to showcase the many possible approaches to collected objects. The structure thus reinforces the volume's aim to pay close attention to the selected objects, their specific features, and their agency and relevance to the narratives, imaginings and interpretations of the

North rather than to write a comprehensive history of collections, collection practices and northern objects.

The structure and the objects illustrate the radical cross-disciplinarity that underpins the volume. The contributing authors have their background in a wide range of research fields, including Cultural studies, Nordic literature, English literature, History, Cultural history, History of science, Scandinavian studies, Museology, Palaeontology, Musicology, Anthropology and Geology. While we asked the authors to focus on one singular object or a collection of same-category objects, we encouraged them to apply their own theoretical and methodological approaches. The chosen objects reflect this multifacetedness and include human made objects (periodical, encyclopaedia, expedition diary, book illustrations, stave church, archive, fragment, writing tools, photographs, treasures), natural objects (polar bear, whale), the sensory/immaterial that is materialized through objects (knowledge, song and pain, the vernacular, everyday life), and lost objects (island). The objects form part of various collections (small, large, private, public, etc.), testifying to the different purposes and circumstances of collecting as well as to the variety in the composition of collections.

Acknowledgements

This chapter is a result of the UiO:Nordic research group Collecting Norden (2021–2023, University of Oslo). Special thanks go to Sandra McElroy for proofreading and to Hans Kristian Strandstuen Rustad for his constructive comments on an earlier draft of this chapter.

Bibliography

Braidotti, Rosi. 2013. *The Posthuman*. Cambridge: Polity Press.

Deleuze, Gilles & Félix Guattari. 1980. *Mille Plateaux*. Paris: Les Editions de Minuit.

Driver, Felix, Mark Nesbitt & Caroline Cornish (eds). 2021. *Mobile Museums: Collections in Circulation*. London: UCL Press. https://library.oapen.org/handle/20.500.12657/51827.

Finger, Matthias & Gunnar Rekvig. 2022. *Global Arctic: An Introduction to the Multifaceted Dynamics of the Arctic*. Cham: Springer.

Fjågesund, Peter. 2014. *The Dream of the North: A Cultural History to 1920*. Amsterdam: Rodopi.

Fontijn, David R. 2013. "Epilogue: Cultural Biographies and Itineraries of Things: Second Thoughts." In *Mobility, Meaning and the Transformations of Things*, edited by

Hans Peter Hahn & Hadas Weiss, 183–96. Oxford: Oxbow. https://doi.org/10.2307/j.ctvh1dn08.16.

Hamelin, Louis-Edmond. (1975) 1979. *Canadian Nordicity: It's Your North, Too*. Translated by William Barr. Montreal: Harvest House.

Harman, Graham. 2017. *Object Oriented Ontology: A New Theory of Everything*. London: Pelican Books.

Hetland, Per, Palmyre Pierroux & Line Esborg (eds). 2020. *A History of Participation in Museums and Archives: Traversing Citizen Science and Citizen Humanities*. London: Routledge.

Jardine, Boris, Emma Kowal & Jenny Bangham. 2019. "How Collections End: Objects, Meaning and Loss in Laboratories and Museums." *BJHS Themes* 4: 1–27. https://doi.org/10.1017/bjt.2019.8.

Kopytoff, Igor. 1986. "The Cultural Biography of Things: Commoditization as Process." In *The Social Life of Things: Commodities in Cultural Perspective*, edited by Arjun Appadurai, 64–91. Cambridge: Cambridge University Press.

Latour, Bruno. 2005. *Reassembling the Social*. Oxford: Oxford University Press.

Law, John. 1994. *Organizing Modernity*. Oxford: Blackwell.

Law, John. 2009. "Actor-Network Theory and Material Semiotics." In *The New Blackwell Companion to Social Theory*, edited by Bryan Turner, 141–58. Oxford: Blackwell.

McDonnell, Terence E. 2023. "Cultural Objects, Material Culture, and Materiality." *Annual Review of Sociology* 49 (1): 195–220. https://doi.org/10.1146/annurev-soc-031021-041439.

Miller, Peter N. 2017. *History and Its Objects: Antiquarianism and Material Culture since 1500*. Ithaca, New York, London: Cornell University Press.

Morton, Timothy. 2010. *The Ecological Thought*. Cambridge, MA: Harvard University Press.

Morton, Timothy. 2013. *Hyperobjects: Philosophy and Ecology after the End of the World*. Minneapolis: University of Minnesota Press.

Müller, Martin. 2015. "Assemblages and Actor-networks: Rethinking Socio-material Power, Politics and Space." *Geography Compass* 9 (1): 27–41. https://doi.org/10.1111/gec3.12192.

Olsen, Bjørnar, Mats Burström, Caitlin DeSilvey & Þóra Pétursdóttir (eds). 2020. *After Discourse: Things, Affects, Ethics*. London: Routledge. https://doi.org/10.4324/9780429200014.

Pearce, Susan M. 1993. *Museums, Objects, and Collections: A Cultural Study*. Washington, DC: Smithsonian Institution Press.

Plumb, Oisín, Alexandra Sanmark & Donna Heddle (eds). 2020. *What is North? Imagining and Representing the North from Ancient Times to the Present Day*. Turnhout: Brepols Publishers.

Roudometof, Victor. 2016. "Theorizing Glocalization: Three Interpretations." *European Journal of Social Theory* 19 (3): 391–408. https://doi.org/10.1177/1368431015605443.

Schimanski, Johan & Ulrike Spring. 2020. "The Arctic, Germany and Austria-Hungary: Narrative Borders in Discourses around the Austro-Hungarian Polar Expedition (1872–1874)." In *German Representations of the Arctic (17th–19th Centuries): Writing the Arctic,* edited by Jan Borm & Joanna Kodzik, 246–67. Newcastle upon Tyne: Cambridge Scholars Publishers.

Simon-Ekeland, Alexandre & Lene Liebe Delsett. 2024. "Whales as a Nordic speciality: Collecting Cetaceans for Natural History Museums (1860s–1910s)." *Nordisk Museologi/The Journal Nordic Museology. Special Theme: Objects of the North* 36 (1): 28–43. https://doi.org/10.5617/nm.11591.

Sommer, Manfred. 1999. *Sammeln: Ein philosophischer Versuch.* Frankfurt am Main: Suhrkamp.

Spring, Ulrike & Janicke S. Kaasa (eds). 2024. "Foreword". *Nordisk Museologi/The Journal Nordic Museology. Special Theme: Objects of the North* 36 (1): 6–8. https://doi.org/10.5617/nm.11777.

Spufford, Francis. 1996. *I May Be Some Time: Ice and the English Imagination.* London: Faber and Faber.

Sundqvist, Anneli. 2021. "Things That Work – Meditations on Materiality in Archival Discourse." *Journal of Contemporary Archival Studies* 8 (7). https://elischolar.library.yale.edu/jcas/vol8/iss1/7.

Thomas, Nicholas. 2018. "Introduction: Presence and Absence." In *Pacific Presences, Volume 1: Oceanic Art and European Museums*, edited by Lucy Carreau, Alison Clark, Alana Jelinek, Erna Lilje & Nicholas Thomas, 13–35. Leiden: Sidestone Press.

Troelenberg, Eva-Maria, Damiana Oţoiu & Felicity Bodenstein (2022). *Contested Holdings: Museum Collections in Political, Epistemic and Artistic Processes of Return.* New York, NY: Berghahn Books.

Vergo, Peter (ed.). 1989. *The New Museology.* London: Reaktion Books.

OBJECT: Periodical

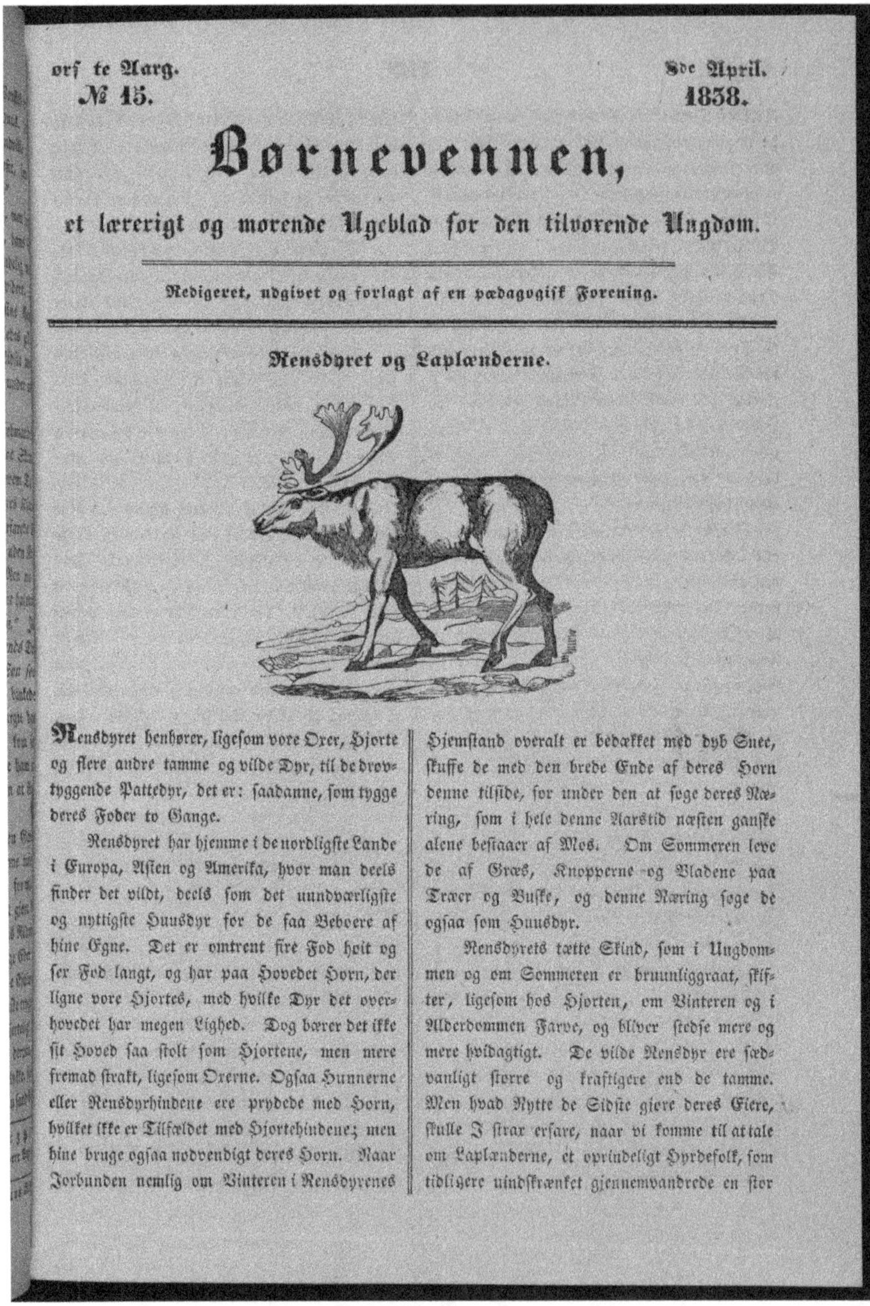

ørste Aarg.
№ 15.

8de April.
1838.

Børnevennen,

et lærerigt og morende Ugeblad for den tilvorende Ungdom.

Redigeret, udgivet og forlagt af en pædagogisk Forening.

Rensdyret og Laplænderne.

Rensdyret henhører, ligesom vore Oxer, Hjorte og flere andre tamme og vilde Dyr, til de drøvtyggende Pattedyr, det er: saadanne, som tygge deres Foder to Gange.

Rensdyret har hjemme i de nordligste Lande i Europa, Asien og Amerika, hvor man deels finder det vildt, deels som det uundværligste og nyttigste Huusdyr for de faa Beboere af hine Egne. Det er omtrent fire Fod høit og sex Fod langt, og har paa Hovedet Horn, der ligne vore Hjortes, med hvilke Dyr det overhovedet har megen Lighed. Dog bærer det ikke sit Hoved saa stolt som Hjortene, men mere fremad strakt, ligesom Oxerne. Ogsaa Hunnerne eller Rensdyrhindene ere prydede med Horn, hvilket ikke er Tilfældet med Hjortehindene; men hine bruge ogsaa nødvendigt deres Horn. Naar Jordbunden nemlig om Vinteren i Rensdyrenes Hjemstand overalt er bedækket med dyb Snee, skuffe de med den brede Ende af deres Horn denne tilside, for under den at søge deres Næring, som i hele denne Aarstid næsten ganske alene bestaaer af Mos. Om Sommeren leve de af Græs, Knopperne og Bladene paa Træer og Buske, og denne Næring søge de ogsaa som Huusdyr.

Rensdyrets tætte Skind, som i Ungdommen og om Sommeren er bruunliggraat, skifter, ligesom hos Hjorten, om Vinteren og i Alderdommen Farve, og bliver stedse mere og mere hvidagtigt. De vilde Rensdyr ere sædvanligt større og kraftigere end de tamme. Men hvad Nytte de Sidste giøre deres Eiere, skulle I strax erfare, naar vi komme til at tale om Laplænderne, et oprindeligt Hyrdefolk, som tidligere uindskrænket gjennemvandrede en stor

FIGURE 2.1 Front page of the Danish children's periodical *Børnevennen*, 8 April 1838
PHOTO: ROYAL DANISH LIBRARY

CHAPTER 2

Collecting, Producing and Circulating the North: the Inuit and the Sámi in Danish and Norwegian Children's Magazines

Janicke S. Kaasa

Abstract

Børnevennen (1838–1839) and *Billed-Magazin for Børn* (1838–1839) were early, illustrated magazines aimed at child readers in Denmark and Norway respectively, modelled on similar publications in France, Germany and Great Britain. The two children's magazines provide fruitful vantage points from where we can explore periodicals as collections, and their role in collecting, producing and circulating cultural imagery. As ephemeral and affordable forms of print, the magazines disseminated verbal and visual materials in various genres and covered a wide range of topics, mostly reproducing imported materials. Although they were published in national contexts, then, the magazines were highly transnational in both form and content. This chapter investigates the collection, production and circulation of northern knowledge and imagery in the two periodicals, with a focus on ethnographic texts and images portraying the Inuit and the Sámi. In particular, the chapter investigates if and how these ethnographic materials form part of the notion of Denmark and Norway as distinctly northern nations.

Keywords

print culture – children's magazines – periodical press – ethnography – Inuit – Sámi

In her much-cited discussion of the periodical as a genre, Margaret Beetham (1990, 19) asserts one of its most characteristic traits: "Periodicals are among the most ephemeral of printed forms. Read today and rubbish tomorrow, each number of a periodical becomes obsolete as soon as the next comes out". The

 | DOI:10.1163/9789004749665_003

ephemerality that Beetham emphasizes ties into both contents and materiality. Published at frequent intervals, the periodical's subjects are often that of current topics, and its binding and paper of low quality. Furthermore, due to their ephemerality, periodicals have often not been considered worth collecting, and given Beetham's description it seems somewhat counter-intuitive to include such publications in a volume on collecting and collections. After all, why would we collect something that will be tomorrow's trash?

Nonetheless, the periodical makes an interesting prism through which we may approach the themes of collecting and collections – and, as I will return to, of the North. First, the periodical can itself be regarded as a collection – the term 'magazine' originally comes from French, referring to 'warehouse' or 'storehouse' – that assembles and prints written and visual materials. Second, the boundaries between the periodical and the book were blurred during the eighteenth and early nineteenth centuries. The individual issues were intended to be collected by the reader who would bring them to a professional book binder to group them into a bound volume featuring an index.

Often, the magazines would themselves call attention to this change in format. We find this, for instance, in the first Swedish children's magazine, *Weckoblad för swenska ungdomen* (Weekly for the Swedish Youth), published in 1785–1787. In his address to the readers, the publisher refers to how the weekly magazine transforms into a book every quarter of the year ("så snart et Quartal är flutadt, forwandlar det sig til en Bok", Gjörwell, 1786–1787, n. p.), suggesting that both publishers and readers were used to thinking about the single issues as part of a larger collection. The change in format also entailed a change in materiality, from ephemeral leaflets to bound books, which increased both their durability and value as an object. Thus, the very collection of periodicals, turning them into volumes that were worth keeping and downplaying their periodicity, had an impact on their status in the literary field and as material objects. Third, such bound volumes, themselves collections, evidently had a greater chance of entering larger collections than the less robust single issues, many of which have not been collected and thus have been lost (Beetham 1989).

The two magazines that I investigate here have entered the collections precisely as bound books. It is in this format that they are available as objects of study – I have not been able to locate any single issues – which evidently impacts my understanding of and approach to these periodicals as collections. The Danish *Børnevennen* (The Children's Friend, 1838–1839) is located at the Royal Danish Library in Copenhagen and the Norwegian *Billed-Magazin for Børn* (Picture-Magazine for Children, 1838–1839) is located at the National Library of Norway in Oslo; both have been digitized and can be accessed online.[1] Both address a child audience ('børn' means 'children') and were indeed ephemeral forms of print – *Børnevennen was* published weekly and *Billed-Magazin for Børn* published monthly – but were not merely intended to be "read today and rubbish tomorrow". As argued above, the change in format from single issues into bound volumes allows for an understanding of these publications to be permanent rather than transitory reading.

In her study of the Victorian illustrated press, Toni Weller argues against the notion of periodicals as just momentary publications. Examining the two popular weeklies the *Penny Magazine* and *Illustrated London News* in the period 1842–1843, she finds that these "were not just ephemeral products of the moment, but rather, they were conscious and deliberate attempts to preserve themselves as part of the popular historical record" (Weller 2008, 206). Weller brings in another perspective than the obsoleteness that Beetham underlines, namely the periodicals as a publishing form with an ambition to preserve knowledge and themselves become "objects of reference" (Weller 2008, 200). Such a perspective certainly echoes the motivations for collecting and is relevant to *Børnevennen* and *Billed-Magazin for Børn*. First, and linked to the issue of materiality, the very fact that the single issues of these periodicals were intended to be bound together into volumes testifies to how they were attempts to be something other than "just" ephemeral publications. Second, and linked to the issue of contents, it was not current news that the two magazines disseminated first and foremost, but informative texts on topics such as history, geography, and zoology, as well as moral tales, poems, and fables.

Importantly, *Børnevennen* and *Billed-Magazin for Børn* are examples of early illustrated magazines aimed at child readers in the Danish and Norwegian contexts. They were significant efforts to reach and circulate texts and images to a larger audience, and they provide insight into the written and visual culture for children at the time. During the first half of the nineteenth century, illustrations were still scarce in Danish and Norwegian publications. The production of woodcut clichés was underdeveloped, especially in Norway (Økland 1981, 96), which lacked the necessary technology. As a result, the clichés had to be imported from abroad, which meant both economic and practical challenges. Indeed, the editors of *Billed-Magazin for Børn* mention specifically the costs and difficulties of importing the clichés. This was the case especially during winter as the editors explain in their note on why they must limit the number of illustrations, from six to eight images each month in the first volume,

1 See Billed-Magazin for Børn (nb.no) and Børnevennen: et lærerigt og morende Ugeblad for den tilvoxende Ungdom. – Royal Danish Library (kb.dk).

to four to five images each month in the second.[2] Thus, for the child reader (indeed, for any reader) in Denmark and Norway, the images in the periodicals would be a big draw and a reason to save and collect the issues. Furthermore, the illustrations are central if we are to see these publications as attempts to enter the popular historical record in the way that Weller argues in her study of the British illustrated magazines. Given the fact that images were far from abundant at the time, the illustrations in *Børnevennen* and *Billed-Magazin for Børn* would most likely play a key role in circulating and consolidating certain visual tropes.

The present chapter approaches the two children's periodicals as collections and, in Weller's phrasing, as objects of reference, focusing on one specific theme, namely the North. As several scholars have argued, 'the North' may be approached not merely as a geographical term, but as something far less tangible. For instance, in her discussion of Canada and the North, Sherrill E. Grace rejects the notion of the North as a national given, and as a stable and unitary idea. The North, she writes, "is multiple, shifting and elastic; it is a process, not an eternal fixed goal or condition" (Grace 2001, 16). Similarly, in his cultural history of the North, Peter Fjågesund (2014, 15) refers to the North as "an idea, a concept, or a spiritual or even mystical landscape", and understands it as "a mental horizon", "a dream", and, importantly, as "an essential part of western cultural history, and hence of our identity".

I draw here on the understandings of the flexible North and acknowledge that the knowledge and imagery I discuss are only one of a multitude of possible 'Norths' and interpretations. In my "locating" of the North in the two magazines, I searched for terms in both Danish and Norwegian that I considered relevant, such as 'North*', 'Winter', 'Polar', 'Arctic', 'Eskimo', and 'Lapp'. Although this approach risks being tautological, merely confirming existing assumptions, it gave me an overview of the material that was related to the North as a geographical place. In the periodicals, this mainly refers to the polar region, Greenland, and the northern areas of North America, Russia, Sweden, and Norway. Examining textual and visual material in *Børnevennen* and *Billed-Magazin for Børn*, the chapter explores their cultural imaginary of the North. What knowledge, which ideas and images were disseminated to Danish and Norwegian child readers during the first half of the nineteenth century? In particular, I have focused on ethnographical texts on northern Indigenous peoples, exploring the extent to which they are made meaningful to and representative of an understanding of Denmark and Norway as northern.

2 "Afbildningenes Kostbarhed, i Forbindelse med Vanskeligheden, især om Vinteren til rette Tid at reholde dem" (*Billed-Magazin for Børn* December 1838, 288).

Typically for the period, the periodicals collected, recycled, and translated material from other, often foreign, publications intended at both children and a general readership. As Iver Tangen Stensrud (2023, 240) has pointed out, periodicals were "transnational products", and so was *Børnevennen* and *Billed-Magazin for Børn*. They exemplify how the cultural imaginary of the North circulated transnationally, and the chapter contextualizes some of the textual and visual material by tracing its origins. This perspective entails an approach to the periodicals as what Johan Östling, David Larsson Heidenblad, and Anna Nilsson Hammar (2023, 10, 12) have termed "knowledge actors", referring to contributors "to the production and/or circulation of knowledge". Because the two magazines mainly reproduced materials from other publications, circulation is key here. Even so, I will argue that the periodicals did not merely circulate and translate existing imagery. By collecting textual and visual materials into these new publication contexts, they also produced knowledge and cultural imagery through their circulation.

As children's periodicals, *Børnevennen* and *Billed-Magazin for Børn* are representative of two rather underexplored fields: both (seemingly) ephemeral forms of writing, and publications aimed at children, have often been deemed of little literary value and thus of minor academic interest. This means that periodical publications, and those for children especially, have largely been neglected in literary studies. Moreover, as "rubbish tomorrow", these objects do not seem to lend themselves well to studies on collecting practices or collections. However, the significance of such publications for the understanding of popular reading, of the circulation of texts and images, as well as of their part in material culture, should not be underestimated. Thus, the two children's magazines provide fruitful vantage points from where we can explore periodicals as collections, and their role in disseminating cultural imagery – here of the North specifically – entering the popular historical record.

1 *Børnevennen* and *Billed-Magazin for Børn*

The two publications share several features. Published in 1838–1839, they form part of the rise in children's magazines that occurred during the nineteenth century also in Denmark and Norway (Weinreich 2006, 294; Økland 1981, 96). The ambition was to enlighten and entertain, with features such as informative texts on history, natural history and ethnography, and fables, fairy tales, parables, poems, and moral tales, mirroring the current ideas of children's texts and their dual purpose to instruct and amuse. Likewise, and as I have argued elsewhere, they mirror the educational magazines aimed at a general audience from the same period, such as the British *Penny Magazine* (1832–1845) and the Norwegian *Skilling-Magazin* (1835–1891). *Børnevennen* and *Billed-Magazin for Børn* explicitly addressed a young readership, were almost identical in their visual expression and layout, and often circulated and translated (versions of) the same texts and images.

Variances of the title *Børnevennen* (The Children's Friend) appear on numerous children's periodicals in Denmark and Norway throughout the nineteenth century, testifying to the influence from the German (*Der Kinderfreund*, 1775–1784) and the French (*L'ami des enfants*, 1782–1783) from the late eighteenth century onwards (Weinreich 2006, 296).[3] The specific "children's friend" of interest here is the Danish children's magazine that was published weekly in Copenhagen in two volumes in 1838–1839. The final issue was published 30 June 1839, providing the readers with a small note announcing a halt due to changes in the editorial staff. The halt turned out to be permanent and the second volume of *Børnevennen* was terminated halfway. It is not clear who was responsible for the publication (Kristensen 1913, 276). Below the title on the front page, we read that it is edited and published by a pedagogical association ("Redigeret, udgivet og forlagt af en pædagogisk Forening"), but no names are provided except for the printing house (Fabritius de Tengnagel, January–March 1838; S.U. Rissens Enke, March 1838–June 1839). The periodical was published each Sunday in 8-page leaflets with two columns on each page, and one illustration covering almost the whole first page, and its subtitle – "an educational and amusing Weekly for the upgrowing youth" ("et læreright og morende Ugeblad for den tilvoxende Ungdom") – certainly points to the magazine's ambition to educate and entertain.

Billed-Magazin for Børn (Picture-Magazine for Children) is considered Norway's first children's magazine, being printed, and published in Christiania (today's Oslo) in 1838–1839 (Økland 1981, 94). The literary field in Norway was largely oriented towards Denmark, even after 1814 when the Dano-Norwegian union ended and Norway as a constitutional monarchy entered a union with Sweden that would last until 1905. Copenhagen continued to be the literary centre, and the written language was Danish until the mid-nineteenth century. *Billed-Magazin for Børn*, then, was part of the Danish-Norwegian literary field, which is further underlined by its many similarities with *Børnevennen*. Moreover, advertisements in Norwegian newspapers suggests it is likely that *Børnevennen* reached the educated bourgeois children in the larger cities of

3 Examples include the Danish *Børnevennen* (1861–1862), *Børnevennen* (1867–1919) and *Den nye Børneven* (1868–1869), and the Norwegian *Børnevennen* (1843–1850) and *Illustreret Børneven* (1856), cf. Shine 1971, 97–98; Weinreich 2006, 296; Økland 1981, 97–99.

Norway.[4] However, because *Billed-Magazin for Børn* was published in Norway and edited by Norwegians, it represented something completely new in the Norwegian context and became an important precursor for later children's magazines in Norway (Økland 1981). The magazine was published by Carl August Guldberg and Adam Dzwonkowski (who also published the *Skilling-Magazin*) and was edited by the writers and educators Maurits C. Hansen and Knud O. Knudzsen, the latter superseded by Peter Chr. Asbjørnsen, who with Jørgen Moe today is best known for his collections of Norwegian folktales. Published monthly in two volumes with 12 24-page issues each in 1838 and 1839, *Billed-Magazin for Børn* provided readers with a rich content of text and images. And, as the title of the publication emphasizes, the *illustrations* were the big draw. In total, the publishers promised 100 (later 50–60) woodcuts for each volume, an overwhelming number if we compare it with other illustrated publications for children in Denmark and Norway during the same period, including *Børnevennen*, which featured far fewer illustrations.

As stated, both periodicals circulated materials from other publications. But, to whom? In their discussion of Danish children's periodicals, Charlotte Appel and Nina Christensen (2023, 248) suggest that the readers of *Børnevennen* were most likely the privileged children of the Copenhagen bourgeois. Similarly, in his overview of Norwegian children's magazines, Einar Økland (1981, 97) mentions privileged children as the plausible readers of *Billed-Magazin for Børn*. Unfortunately, there are no subscription lists that can give us information on who bought the magazines, but the fact that both were short-lived endeavours suggests that subscribers were not numerous enough, and that there was not yet a big enough market for this type of publication. Indeed, the editors of *Billed-Magazin for Børn* (1839, 288) bring this up in their note to the readers in the final issue, in which they express their disappointment at the lack of attention and support from their readers (Kaasa 2019, 310). Yet, their relatively low prize – *Børnevennen* cost 3 speciedaler and 8 skilling each quarter, *Billed-Magazin for Børn* was cheaper and cost 1 speciedaler for one volume or 10 skilling per issue – highlights their similarity to the educational magazines, which relied on a low retail price and large distribution, and at the very least testifies to the *ambition* of reaching a larger audience. The low retail

4 These advertisements are, however, not numerous. Searches in the digital newspaper archives of the National Library of Norway provide three results: one advertisement in the Bergen-based newspaper *Bergens Adressecontoirs Efterretninger* on 4 July 1838 (including also *Billed-Magazin for Børn*) and two advertisements in the Christiania-based newspaper *Den Constitutionelle* on 14 and 17 August 1840. Searches for *Billed-Magazin for Børn* in the newspaper archives of the Royal Danish Library yielded zero results, implying that the Norwegian magazine did not circulate in Denmark.

price is one important reason why so much of the material was circulated from other sources since it was less expensive and less time-consuming than providing the magazines with original texts and illustrations. And although the revised copyright regulation of 1828 was extended to reprints of foreign publications, these regulations seem not to have been complied with in practice (Tveterås 1950, 194; 1979, 101–02). Evidently, such practical circumstances impacted which texts and images were included, and which knowledge and cultural imagery of the North was circulated to and produced for the readers of *Børnevennen* and *Billed-Magazin for Børn*. Yet, the selection of materials should also be seen in relation to the actors behind the publications: in their choices of what to include and to exclude, the editors indeed edited their own collections and were central to the shaping of these magazines as objects of reference and in their role as knowledge actors.

2 Neighbours of the North Pole

So, which North is presented to the readers of these two magazines? In what way, if any, is the North and northernness made relevant to and representative of an understanding of Denmark and Norway and the magazine readers as northern? What we could consider typically northern topics do not make up a large portion of the material in either periodical, and the texts we do find are mainly informative pieces on natural phenomena such as the northern lights, the northern climate, reindeer, polar bears, whaling, and seal hunting. Indeed, the emphasis on natural science tells us something about what was seen as representative of the North at the time, and points to the persistent idea of the North as a space empty of culture and human activity (see for example Davidson 2005). However, we also find contributions on Viking kings and northern mythology and, important in this chapter, ethnographic texts on "the Neighbours of the North Pole ("Nordpolens Naboer", *Børnevennen* 13 January 1839, 11).[5] These ethnographic examples are of particular interest because they acknowledge the North as an inhabited space. Moreover, they may tell us something about the extent to which the child readers were encouraged to think

5 Among the texts on northern topics in *Børnevennen* are "Iisbjørnens Ømhed for sine Unger" (25 March 1838), "Rensdyret og Laplænderne" (8 and 15 April 1838), "Eskimoeren og hans Hund" (1 July 1838), "Vinteren høit oppe I Norden" (16 December 1838), "Eskimoerne" (13 January 1839), "Et Tilbageblik paa Lapland" (10 February 1839). And, in *Billed-Magazin for Børn*: "Eskimoerne" (July, August, and September 1838), "Reensdyret" (November 1838), "Eskimoerne og deres Snehytter" (July 1839).

about themselves as northern and whether the North was positioned in relation to national identity, be it Danish or Norwegian. In other words, the texts offer an insight into what and who is, and is not, associated with the North.

One of the texts that centres on the peoples inhabiting the region, "Eskimoerne" (The Eskimos), is published in three instalments in *Billed-Magazin for Børn*, covering approximately 13 pages in the issues of July, August, and September 1838. A text with the same title appears in *Børnevennen* on 13 January 1839 but is much shorter covering only three pages. Both texts provide ethnographic knowledge on the Indigenous peoples of the North, more specifically the "Eskimos"[6] of North America, describing the cold and harsh climate in which they live and hunt, their physiognomy, clothing, housing, customs, and beliefs.

Despite their informative purpose, the texts are not merely descriptive but feature moralizing, patronizing and idealizing characterizations of the Inuit, drawing on current tropes. For example, we find here the common stereotype of the "Eskimos" as ignorant and good natured ("Eskimoerne ere saa uvidende", *Billed-Magazin for Børn* August 1838, 177; "de godmodige Eskimoere", *Børnevennen* 13 January 1839, 9). In both texts, the most negative characteristics concern hygiene, or the seeming lack thereof, amongst the inhabitants of the North, echoing a contemporary preconception of northern Indigenous peoples perhaps most famously articulated by D. H. Lawrence nearly one hundred years later in his characterization of the "greasy little Eskimo" (Lawrence 1964, 158). In *Billed-Magazin for Børn*, for instance, we learn that all are to a high degree unclean ("Alle ere i høi Grad ureenlige", *Billed-Magazin for Børn* July 1838, 167), and that this uncleanliness seems to correspond to the degrees of latitude in the sense that the closer they live to the North Pole, the filthier they get.[7] The text in *Børnevennen*, too, highlights the peoples' uncleanliness, referring to their filthy brown hue, their filthy way of life and the repulsive smell

6 In my discussion, I make use of the term 'Eskimo' when quoting from the magazines, otherwise I employ the term 'Inuit'. Although the latter term today is more widely accepted in Alaska, I acknowledge that it is imprecise and, for some communities, problematic. As Lawrence Kaplan at the Alaska Native Language Center notes: "Alaska Natives increasingly prefer to be known by the names they use in their own languages, such as Inupiaq or Yupik. [...] The Inuit Circumpolar Council prefers the term 'Inuit' but some other organizations use 'Eskimo'." (Kaplan n.d.). The use of 'Eskimo' in the magazines, however, does not indicate which specific groups are referred to, and I therefore find 'Inuit' most accurate in this context.

7 "Ikke alle Eskimoere leve imidlertid nedsjunke i ligestor Ureenlighed, men, jo nærmere man kommer Polen desto større og mere frastødende Ureenlighed hersker der hos dem" (*Billed-Magazin for Børn* July 1838, 167).

in their dwellings ("smudsigbrun Ansiktsfarve"; "smudsige Levemaade og den modbydelige Lugt i deres Boliger" (*Børnevennen* 13 January 1839, 10; 11).

The cultural imagery of the Inuit in both periodicals, then, is indeed a negative one. Yet, we find praise of their hunting skills, tenacity, and ability to adjust to the harsh living conditions of the North.[8] Often though, such praise shifts into dehumanizing characteristics such as in *Billed-Magazin for Børn*'s passage on female Inuit that, if the filth on their faces was removed, they would have been deemed beautiful even in Europe.[9] When the women turn old, however, they become the most "repulsive thing that the human eye can ever behold" and "one hardly recognizes in them human beings" (July 1838, 168, my translation).[10] This passage is typical in its portrayal of the Indigenous peoples, here the women "Eskimos", as almost, but not quite like "us". And, given the reference to European beauty standards, the readers here seem to be identified with Europe rather than Norway and/or with the North. We find a similar distinction in *Børnevennen* when the text on "Eskimoerne" concludes with an account of their "spiritual education", which, the text states, lags far behind. However, during the last 60 years, "our missionaries" have converted several families to Christianity. Some tribes, we can read, have also learnt to sing several of "our psalms" in "their language" (*Børnevennen* 13 January 1839, 11).[11] Again, the distinction is made between "us and them", though this passage from *Børnevennen* seems to suggest at least the possibility of them becoming like us when it comes to spiritual education. It is also noteworthy – and possibly due to the import of material – that the text here does not mention Greenlanders or Greenland, a Danish colony until 1953 and the site of Danish missionary activity throughout the nineteenth century, which would further underline the work of "our missionaries" and the northernness of the Danish.

8 For example: "Disse paa Menneskeslægtens laveste Trin"; "Ikkedestomindre forraade de I mange Ting en Duelighed og Færdighed, som man maa beundre" (*Billed-Magazin for Børn* August 1838, 167; 177).

9 "Selv kvindernes Ydre er, omendskjøndt de ikke kunne gjøre Fordring paa nogen regelmæssig Skjønhed, ofte behageligt, og forenet med et Udtryk af Aabenhed og Godmodighed, saaat det, dersom det var befriet fra den tykke Fedt- og Smudsbark, saaat den egentlige Hudfarve, som ligner en mørk Brunettes, kunde komme tilsyne, selv i Europa vilde blive anseet for smukt" (*Billed-Magazin for Børn* July 1838, 168).

10 "Men naar Kvinderne blive gamle, saa ere de de modbydeligste Gjenstande, som det menneskelige Øie nogensinde kan see. Antændte Æine, skrumpet Hud, sorte Tænder og vanskabte Ansigtstræk gjøre, at man neppe i dem erkjender menneskelige Væsener".

11 "Med Eskimoernes aandige Dannelse staaer det naturligvis endnu langt tilbage; dog tilbede de Fleste et et godt og et ondt høieste Væsen, men i de sidtste 60 Aar have vore Missionairer undervisst adskillige Familier i Christendommen og omvendt dem hertil. Hos nogle Stamme have de fundet en særdeles Modtagelighed for Sang, og lært dem at synge flere af vore Psalmer i deres Sprog".

The intertextual references that we find in both texts point to a European rather than a national and/or northern cultural imagery, and several of the texts on the Inuit make explicit references to previous writings on the polar region: *Børnevennen* and *Billed-Magazin for Børn* refer to accounts by explorers such as John Ross and William Edward Parry. The latter's text "Eskimoerne og deres Snehytter" (The Eskimos and their Snow Huts, July 1839) renders an excerpt covering nearly three pages from John Ross's account "Reise i Polaregnene" (Journey in the Polar regions), probably taken either from a British magazine, possibly via *Skilling-Magazin*, or/and from *John Ross's anden Reise til Egnene ved Nordpolen. Bearbeidet for Ungdommen*, which was published in Copenhagen (like *Børnevennen* printed by S.U. Rissens Enke) in 1838, translated from the German and adapted to young readers and features nine copper plates.[12]

Such references point to an important feature of these periodicals' dissemination of knowledge on the North. First and foremost, they circulate knowledge and cultural imagery produced by British explorers. This is not surprising. As Fjågesund has pointed out, from the 1820s onwards there was "a frantic level of polar activity, especially from the British side" (Fjågesund 2014, 268), in which explorers such as Ross, Parry, and John Franklin were prominent figures not merely in Britain but in all of Europe. Seen in the context of Norwegian polar history in particular, it is interesting how the cultural imagery of the North in these texts is not yet linked to national identity and Nordic heroism, as it would be in the heyday of Norwegian polar exploration from the 1880s. As such, the way the texts draw on the accounts of British explorers is a reminder of how our knowledge and cultural imaginary of the North are continuously in flux, and how the North is a shifting concept. The fact that the child readers of 1838 and 1839 meet a North that is first and foremost the site of British exploration and has not yet been invested with national identity, testifies to how the North as a marker of the Danish and Norwegian, largely taken for granted today, is not a natural given. Furthermore, it attests to how the child readers here seem to be identified with the European rather than with the national.

Introducing the magazines, I mentioned the importance of the illustrations. In both periodicals, the texts on the Inuit are accompanied by images. In *Billed-Magazin for Børn*, the text is supplemented by a total of three illustrations whereas the two first images appear with the first instalment (July 1838) and the third image with the third instalment (September 1838). In *Børnevennen*, the text is preceded by an image covering almost the entire cover page (see Figure 2.2). Like the texts, the images are uncredited. However, the image in

12 John Ross's anden Reise til Egnene ved Nordpolen (nb.no).

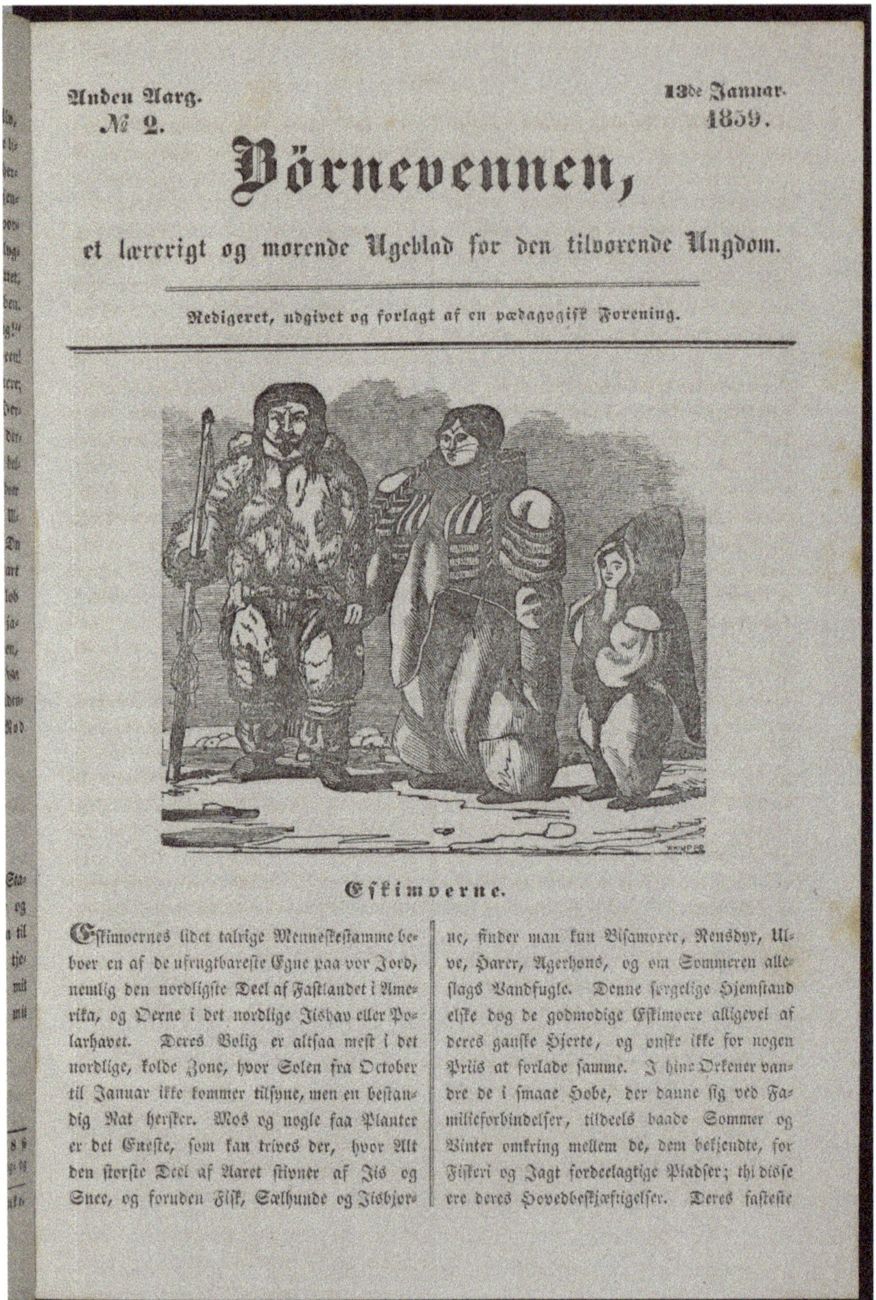

Anden Aarg. № 2.

13de Januar. 1839.

Börnevennen,

et lærerigt og morende Ugeblad for den tilvoxende Ungdom.

Redigeret, udgivet og forlagt af en pædagogisk Forening.

Eskimoerne.

Eskimoernes lidet talrige Menneskestamme beboer en af de ufrugtbareste Egne paa vor Jord, nemlig den nordligste Deel af Fastlandet i Amerika, og Øerne i det nordlige Iishav eller Polarhavet. Deres Bolig er altsaa meest i det nordlige, kolde Zone, hvor Solen fra October til Januar ikke kommer tilsyne, men en bestandig Nat hersker. Mos og nogle faa Planter er det Eneste, som kan trives der, hvor Alt den største Deel af Aaret stivner af Iis og Snee, og foruden Fisk, Sælhunde og Iisbjørne, finder man kun Bisamoxer, Rensdyr, Ulve, Harer, Agerhøns, og om Sommeren allehaande Vandfugle. Denne sørgelige Hjemstand elske dog de godmodige Eskimoer alligevel af deres ganske Hjerte, og ønske ikke for nogen Priis at forlade samme. I hine Ørkener vandre de i smaae Hobe, der danne sig ved Familieforbindelser, tildeels baade Sommer og Vinter omkring mellem de, dem bekjendte, for Fiskeri og Jagt fordeelagtige Pladser; thi disse ere deres Hovedbeskjæftigelser. Deres fasteste

FIGURE 2.2 Front page of *Børnevennen*, 13 January 1839
PHOTO: ROYAL DANISH LIBRARY

Børnevennen seems to be a cropped and slightly altered version of one of the plates from Parry's account, *Journal of a second voyage for the discovery of a north-west passage from the Atlantic to the Pacific: performed in the years 1821–22–23, in His Majesty's ships Fury and Hecla, under the orders of Captain William Edward Parry*, published in 1824.[13] Evidently, this is in line with the texts' circulation of British knowledge and imaginary of the North, as argued above.

The circulation of images points to the material and more practical conditions for periodicals as knowledge actors. Indeed, the fact that so many of the texts in these two periodicals were not original, but collected and translated from other sources, most likely had to do with the editors' lack of time and money to prioritize them. However, the circulation of images from foreign publications testifies even more clearly to the technological and economic constraints of such production and circulation. As previous studies on early Norwegian children's literature show, the international orientation and use of foreign materials was partly due to the limited access to original illustrations (Hagemann 1965; Kaasa 2019; Økland 1981). For *Billed-Magazin for Børn* specifically, we know that the publisher Guldberg had bought woodcut clichés from German and British penny magazines before starting up *Skilling-Magazin* in 1835 (Just 1969, 26; Tveterås 1950, 234), and many of these clichés were used in *Billed-Magazin for Børn* (Kaasa 2019). *Skilling-Magazin* was a pioneering endeavour in the Norwegian context, printing text and image jointly, which was far more costly than the usual illustrations on separate pages (Tveterås 1950, 235–36). However, since these illustrations had to be imported, the visual imagery, including that of the North, circulated to Norwegian readers, children, and adults alike, was necessarily transnational.

3 Thrifty Northerners

Another northern Indigenous group portrayed in the two periodicals, are the Sámi, offering a slightly different vantage point from where we may reflect upon the notion of what and who were included in the collection and understanding of the North. *Børnevennen* publishes the text "Rensdyret og Laplænderen" (The Reindeer and the Lapp, Figure 2.1) in two instalments in the first

13 Details – Journal of a second voyage for the discovery of a north-west passage from the Atlantic to the Pacific: performed in the years 1821-22-23, in His Majesty's ships Fury and Hecla, under the orders of Captain William Edward Parry: illustrated by numerous plates – Biodiversity Heritage Library (biodiversitylibrary.org).

volume (8 and 15 April 1838) and the very short piece "Et Tilbageblik paa Lapland" (A Lookback at Lappland) in the second volume (10 February 1839), whereas *Billed-Magazin for Børn* publishes the text "Reensdyret" in the November issue of 1838. With some variations, the texts emphasize the importance of the reindeer to the thrifty Northerners ("nøisomme Nordboere", *Børnevennen* 10 February 1839, 42), elaborating on their usefulness when it comes to food and clothing. In addition, we recognize the same stereotypes as in the texts on the "Eskimos", for example when *Børnevennen* refers to the uncleanliness of "Laplænderne", resulting in their dark, blackish-brown hue.[14]

As the title suggests, the text in *Billed-Magazin for Børn* accentuates the reindeer and gives a thorough account of its appearance and whereabouts. To the extent that the text comments upon the Sámi, it is mainly about their use and husbandry of the reindeer. Yet, in its introduction to both the reindeer and the Sámi, the text's first association is that of the Bedouin on his camel riding through the desert before it continues its comparison with other peoples and their animal husbandry:

> Within the Polar Circle live the Lapps,[15] a tribe that in ancient times was widespread throughout Scandinavia. In these regions, where they now live, a nine-month-long Winter reigns; the summer's Midnight sun exhorts only every third year to produce a scanty Grain crop, and they would not be able to here maintain Life, if they lacked the Reindeer. This is for the Lapps and other inhabitants of the Polar Circle of even greater Importance than the camel is to the Arabs, than the Lama is to the inhabitants of the highlands of the southern America; for it is them One and All. (*Billed-Magazin for Børn* November 1838, 241, my translation)[16]

14 "Lapplænderne ere smaae af Vækst, ligesom alle i Nærheden af Nordpolen boende Folkeslag, dog temmelig undersætsige, især Rensdyrlapperne, om hvem Talen fornemmelig her er. Deres mærke, sortcbrunc Hudfarve er kun en Fælge af Ureenlighed, hyppigt Ophold i fri Luft og Rægen i deres Telte" (*Børnevennen* 8 April 1838, 114).

15 'Lapp' and 'Lapper' were common terms for the Sámi, but today considered derogatory. In this chapter, I make use of these terms when quoting the magazines, otherwise I use 'Sámi'. For more information about the Sámi and naming, see for example Kulonen, Seurujärvi-Kari & Pulkkinen 2005.

16 "Indenfor Polarkredsen boe Lapperne af en Folkestamme, der i Oldtiden var udbredt over hele Skandinavien. I hine Egne, hvor de nu boe, hersker en ni Maaneder lang Vinter; den sommer lige Midnatssol formaner kun hvert tredie Aar at frembringe en sparsom Korn afgrode, og de skulde her ei kunne opholde Livet, hvis de manglede Reensdyret. Dette er for Lapperne og Polarkredsens ovrige Beboere af endnu storre Vigtighed end Kamelen er for Araberne, end Lamaen for Beboerne af det sydlige Amerikas Hoiegne; thi det er dem Eet og Alt".

On the one hand the reindeer and the Sámi are here located in Scandinavia, but from a distance, emphasising the long and harsh winter and the midnight sun. On the other hand, they are placed in a global setting with the comparison to the Arabs and the camel, and the Southern Americans and their lama. As such, the Sámi do not seem to be representative of the North first and foremost in this piece, but of a nomadic way of life that we may find throughout the world. Later in the same text, the Sámi is situated within yet another context, more specifically a European one:

> Now, however, they [the reindeer] are only found in the northernmost Part of the World, hardly further south than 61 Degrees Latitude, but everywhere in the Polar countries. In addition to the *Lapps in Europe*, Koræk-erne[17] [*sic*] and the Samoyeds on the shores of the Polar Sea also keep reindeer as Domestic animals in Herds, sometimes numbering several thousand; in Greenland, on the other hand, and on Spitsbergen, just like on the Shores of the Arctic Ocean in America, it exists only in the wild. [...] *With us*, there are wild Reindeer in Dovre, Filefjeld and the highest from these mountain ranges projecting Main branches, sometimes in Herds of several hundred. (*Billed-Magazin for Børn* November 1838, 245, my translation and italics)[18]

Whereas the Sámi are referred to as "Lapperne in Europa", distinct from "Korækerne" and the Samoyeds, the text situates the reindeer in a different way. We find the reindeer also here, "hos os", which can best be translated as "in our place" or "with us". The fact that the reindeer, but not the Sámi, are included here in the possessive "us" underlines the contrast to the more

17 "Korækerne" probably refers to the Koryaks, an Indigenous group in Kamchatka Krai, in the far east of Russia. The term does not seem to have been widely used, and searches in the digital archives of the National Library of Norway provide only three results: in H.B. Melchior, *Den danske Stats og Norges Pattedyr* (Copenhagen: Gyldendal, 1834), *Norsk Penning-Magazin: et Skrift til Oplysning, Underholdning og nyttige Kundskabers Udbredelse* (Christiania, no. 4 1838), and H. Mortensen, *Naturhistorie for Ungdommen: Nærmest efter F. Martin's* "Naturgeschichte für die Jugend" *Med 262 illuminerede Afbildninger* (Copenhagen: Philipsens forlag, 1859). For a thorough discussion of the term's etymology and use, see Rink 1906.

18 "Nu derimod treffes de blot i den nordligste Deel af Verden, neppe sydligere end paa 61 Graders Bredde, men overalt i Polarlandene. Foruden *Lapperne i Europa* holde ogsaa Korækerne og Samojederne paa lishavets Kyster Renen som Huusdyr i Flokke, under tiden paa flere tusinde; i Gronland deri mod og paa Spidsbergen, ligesom og paa lishavets Kyster i Amerika, findes det blot vildt. [...] *Hos os* findes vilde Reens dyr paa Dovre, Filefjeld og de hvieste fra disse Fjeldkjeder udstydende Hovedgrene, undertiden i Flokke paa flere hundrede".

general reference to the Sámi as Scandinavian, European and a nomadic tribe, comparable to peoples from other parts of the globe.

The contrast becomes more conspicuous when we compare this locating of the Sámi to a passage from the text "Rensdyret og Laplænderne" in *Børnevennen*, where they are specifically situated in the northernmost regions of Sweden and Norway:

> The Laplanders, an original Shepherd people, who formerly roamed unrestrained through a large Part of Sweden and Norway, but now only inhabit the northernmost regions of these countries and the neighboring, European Russia, and have not yet fully converted to Christianity. (*Børnevennen* 8 April 1838, 114, my translation)[19]

Whereas the Sámi are European in *Billed-Magazin for Børn*, they are here situated more specifically in the northern parts of Sweden and Norway, as well as Russia. Thus, the Sámi are directly associated with and included in the identified nations. In the Danish periodical, then, the Sámi are regarded as part of Norway, significant also to the association of Norway (and Sweden) as northern. The Norwegian periodical, however, situates the Sámi not within any national context, but within a global and transnational community of nomads that, one could argue, implies exotification and foreignization.

Like the texts on the Inuit, the texts on the Sámi are accompanied by illustrations. In *Børnevennen* we find a total of three images that accompany the texts, two illustrating the text "Rensdyret og Laplænderne", showing a reindeer and a Sámi in a sleigh being pulled by a reindeer. In *Billed-Magazin for Børn*, two images are featured in the text "Reensdyret", one of them showing a reindeer, similar but not identical to the image in *Børnevennen*. Importantly, one and the same image is circulated in the two magazines, namely an illustration showing a Sámi camp, with two Sámi milking a reindeer in the foreground (Figure 2.3).

In both magazines, the image covers a whole page; in *Børnevennen* it is placed on the issue's front page. The image is not credited in any of the periodicals, but is an engraving taken from *The Penny Magazine* of 20 April 1833 (148), with the caption "Milking of the Rein-Deer". In *Børnevennen*, the caption reads "A Lookback at Lapland" ("Et Tilbageblik paa Lapland"), probably referring to a lookback on the previous volume's text on the Sámi; in *Billed-Magazin*

19 "Laplænderne, et oprindeligt Hyrdefolk, som tidligere uindskærnket gjennem vandrede en stor Deel af Sverige og Norge, med nu kun beboer de nordligste Egne af disse Lande og det tilgrændsende, europæiske Rusland, og endnu ikke fuldkommen er omvendt til Christendommen".

Anden Aarg. 10de Februar.
№ 6. 1839.

Börnevennen,

et lærerigt og morende Ugeblad for den tilvorende Ungdom.

Redigeret, udgivet og forlagt af en pædagogisk Forening.

Et Tilbageblik paa Lapland.

244 Billed-Magazin.

En Lappefamilie, mellende sine Reensdyr.

FIGURE 2.3
Billed-Magazin for Børn, November 1839 and front page of *Børnevennen* 10 February 1839
PHOTOS: NATIONAL LIBRARY OF NORWAY; ROYAL DANISH LIBRARY

for Børn, corresponding more directly to the caption in the *Penny Magazine*, it reads "A Lapp Family milking their reindeer" ("En Lappefamilie, melkende sine Reendsyr"). Whereas the captions in *Penny Magazine* and *Billed-Magazin for Børn* are mostly descriptive, highlighting the activity foregrounded in the image, the caption in *Børnevennen* serves a different function. It does not highlight the milking but rather contextualizes the scene as representative of Lapland as such, and, perhaps, as representative of the Lapland of the past through its use of the word "lookback", supplementing the reading of the caption as a reference to the periodical's former piece on the Sámi. Nevertheless, the differing captions for one and the same image are of interest in the discussion on how new knowledge and cultural imagery are produced in the circulation of material in new collections and contexts.

None of the periodicals circulate the text from *Penny Magazine*, which begins with a reference to the failed attempts to introduce reindeer in Scotland and explicitly draws on an account entitled *Travels in Lapland*, by M. De Broke.[20] In *Billed-Magazin for Børn*, the introduction highlights the Polar region as one of the few places on Earth that are either so hot or so cold that they are almost uninhabitable, before it situates, as we have seen, the Sámi within a global, nomadic community. In *Børnevennen*, in contrast, the introduction is devoted to an elaborate account of the milking scene depicted in the illustration:

> It is towards Evening; The Reindeer, and especially those that were to be milked, have already been driven to their Owner's Tents, and in the Foreground You see a female Laplander engaged in milking, with a square-built Laplander helping her, as he has thrown a Rope around the impatient Animal's Horn, and thus is holding it. (*Børnevennen* 10 February 1839, 42, my translation)[21]

In their rendering of this image, in combination with the new texts, the periodicals not only circulate, but produce knowledge and cultural imagery of the Sámi. As such, the periodicals become knowledge actors in the way that Östling, Heidenblad, and Hammar have suggested, providing new (con)texts

20 I have not succeeded in identifying this author or title.

21 "Det er henimod Aften; Rensdyrene, og fornemmeligt de, som skulle melkes, ere allerede drevne hen til deres Eiers Telte, og i Forgrunden see I en Laplænderinde ifærd med at melke, hvorved en fiirskaaren Laplænder er hende behjelpelig, idet han har slynget et Reeb om det utaalmodige Dyrs Horn, og saaledes holder det".

for the circulated image that generate new interpretations of the image and representations of the Sámi.

4 Collecting and Circulating the North

The two periodicals do not circulate a translation of the text in *Penny Magazine*, but provide the readers with other texts, which may, or may not, have been translated from other sources. One could claim that the very image is translated through its juxtaposition with these texts, appearing in new contexts and providing new interpretations of the visual imagery. Moreover, bringing in perspectives from the cultural turn in translation studies and what Gideon Toury (1980) has referred to as the target-orientedness of translation, one could assume that these texts would be more adapted to their target audiences, namely Danish and Norwegian child readers respectively. In the texts on the Inuit and the Sámi, such a target-orientedness could for example be made manifest in an identification with the Northern region, with the neighbours of the North Pole and the thrifty northerners.

Perspectives from translation studies are relevant in the discussion about circulation and collection, which we may also approach as target-oriented. The actors that were behind these publications often had several roles, as editors, publishers, authors, and translators, tailoring the collected material to the taste of their readers and to the specific linguistic and cultural context in which they operated. Circulation and collection may also be shaped by the specific audiences, purposes, and contexts to which they are oriented, and if we approach the actors as collectors and curators of the periodicals, we may see them as facilitators of transnational material to national readerships. However, such a target-oriented translation strategy is *not* dominant in the texts on the Inuit and Sámi specifically, or in the periodicals' contents more generally. For instance, Greenland as a Danish colony is not mentioned in *Børnevennen*, and the Sámi are not situated in the north of Norway in *Billed-Magazin for Børn*. Rather, the collected and circulated materials on the North in the two magazines must be considered part of a transnational, predominantly British, knowledge circulation and cultural imagery.

The periodicals' collection and circulation of texts and images on the North are partly due to practical circumstances and based on what was already available, as outlined in this chapter. Nevertheless, the inclusion and exclusion of materials is also the result of selections and decisions made by those responsible for the magazines. Even though little is known about how the editors went about collecting and editing texts and images, the magazines – themselves knowledge actors – are made by human actors choosing which North to include and which to exclude in their printed "storehouses". Consequently, the North that was collected, produced, and circulated in *Børnevennen* and *Billed-Magazin for Børn*, whether by accident or intention, is not target-oriented and not made meaningful to the idea of Denmark and Norway as northern nations. The readers are not invited to associate themselves with the North and northernness, either as individuals or citizens. They, the Inuit and the Sámi are northern; we, the editors and readers, are not.

Bibliography

Appel, Charlotte & Nina Christensen. 2023. *På sporet af børn og bøger*. Copenhagen: Gads forlag.

Beetham, Margaret. 1989. "Open and Closed: The Periodical as a Publishing Genre." *Victorian Periodicals Review* 22 (3): 96–100.

Beetham, Margaret. 1990. "Towards a Theory of the Periodical as a Publishing Genre." In *Investigating Victorian Journalism*, edited by Laurel Brake, Aled Jones & Lionel Madden, 19–32. Palgrave Macmillan: London. https://doi.org/10.1007/978-1-349-20790-9_2.

Billed-Magazin for Børn. 1838–1839. Christiania: Guldberg & Dzwonkowski.

Børnevennen. Et lærerigt og morende Ugeblad for den tilvoxende Ungdom. 1838–1839. Copenhagen.

Davidson, Peter. 2005. *The Idea of North*. London: Reaktion Books.

Fjågesund, Peter. 2014. *The Dream of the North: A Cultural History to 1920*. Amsterdam & New York: Rodopi.

Grace, Sherrill E. 2002. *Canada and the Idea of North*. Montreal & Kingston, London & Ithaca: McGill-Queen's University Press.

Gjörwell, Carl Christoffer. 1785–1787. *Weckoblad för swenska ungdomen [Bibliothek för swenska ungdomen: innehållande: dels hela wetenskaper, dels delar af någon särskild wetenskap, samt åtskilliga, i synnerhet historiska och sedolärande, stycken. Med hans kongl. maj:ts allernådigste privilegium. Stockholm], 1785–1787.*

Hagemann, Sonja. 1965. *Barnelitteratur i Norge inntil 1850*. Oslo: Aschehoug.

Just, Carl. 1969. *Fabritius & sønner. Institusjonen Scheibler: I Gutenbergs teneste gjennom 125 år*, edited by Carl Just & Øivind Seip Berggrav. Oslo: Fabritius & sønner.

Kaasa, Janicke S. 2019. "'Saavel fra fjerne Lande som fra vort eget Hjem.' Importert materiale i *Billed-Magazin for Børn*." In *Litterære verdensborgere: Transnasjonale perspektiver på norsk bokhistorie 1519–1850*, edited by Aasta Marie Bjorvand Bjørkøy, Ruth S. Hemstad, Aina Nøding & Anne Birgitte Rønning, 310–29. Oslo: Notabene.

Kaplan, Lawrence. Undated. "Inuit or Eskimo: Which Name to Use?" The Alaska Native Language Center, University of Alaska Fairbanks. Inuit or Eskimo: Which name to use? | https://www.uaf.edu/anlc/research-and-resources/resources/archives/inuit_or_eskimo.php. Accessed 16 May 2024.

Kristensen, Niels K. 1913. "Lærerne og Børnelitteraturen." In *Lærerne og Samfundet. Folkeskolens kendte Mænd og Kvinder. Jubilæumsskrift 1814–1914*, vol. 1, edited by Carl Poulsen & W. Th. Benthin, 267–344. Copenhagen: København, Fr. Bagges.

Kulonen, Ulla-Maija, Irja Seurujärvi-Kari & Risto Pulkkinen. 2005. *The Saami: a cultural encyclopaedia*. Vammala: Vammalan Kirjapaino OY.

Lawrence, D. H. (1924) 1964. "Whitman." In *Studies in Classic American Literature*, 154–68. London: Heinemann.

Penny Magazine. 20 April 1833. "The Rein-deer," 148–49. London: The Society for the Diffusion of Useful Knowledge.

Rink, Signe. 1906. "Kalâlek=Karâlek at udlede fra Koræk og ikke fra Skrælling." *Geografisk Tidsskrift* 18: 149–52, 224–27. https://tidsskrift.dk/geografisktidsskrift/article/view/49277/63006.

Shine, Norman. 1971. "Børneblade i Danmark fra 1770–1900." *Børn og Bøger* 4: 91–98.

Stensrud, Iver Tangen. 2023. "Diffusing Useful Knowledge: *Skilling-Magazin*, Transnational Images and Local Communities." In *Literary Citizenship in Scandinavia in the Long Eighteenth Century*, edited by Ruth Hemstad, Janicke S. Kaasa, Ellen Krefting & Aina Nøding, 238–58. Suffolk: Boydell & Brewer.

Toury, Gideon. 1980. *In Search of a Theory of Translation*. Tel Aviv: Porter Institute for Poetics and Semiotics, Tel Aviv University.

Tveterås, Harald L. 1950. *Den norske bokhandels historie 1: Forlag og bokhandel inntil 1850*. Oslo: Cappelen.

Tveterås, Harald L. 1979. *I pakt med tiden: Cappelen gjennom 150 år: 1825–1879*. Oslo: Cappelen.

Weinreich, Torben. 2006. *Historien om børnelitteratur: Dansk børnelitteratur gennem 400 år*. København: Branner og Korch.

Weller, Toni. 2008. "Preserving Knowledge Through Popular Victorian Periodicals: An Examination of *The Penny Magazine* and the *Illustrated London News*, 1842–1843." *Library History*, 24 (3): 200–7. https://doi.org/10.1179/174581608X348087.

Økland, Einar. 1981. "Norske barneblad." In *Den norske barnelitteraturen gjennom 200 år. Lesebøker, barneblad, bøker, tegneserier*, edited by Tordis Ørjasæter, Halldis Leirpoll, Jo Lie, Gunvor Risa & Einar Økland, 86–134. Oslo: Cappelen.

Östling, Johan, David Larsson Heidenblad & Anna Nilsson Hammar. 2023. "Revisiting Agency in the History of Knowledge." In *Knowledge Actors: Revisiting Agency in the History of Knowledge*, edited by Johan Östling, David Larsson Heidenblad & Anna Nilsson Hammar, 9–23. Lund: Nordic Academic Press.

FIGURE 3.1 The title page of the encyclopedia written by Jón Bjarnason from Þórormstunga in Vatnsdalur, Iceland. National and University Library of Iceland (handrit.is)

CHAPTER 3

Monsters on a Page and Microhistory: Textual Embodiment and an Encyclopedia

Sigurður Gylfi Magnússon

Abstract

Pages offer a particular form of 'thingness'. The page is structured by a certain person and affects the mind, the thoughts, and the self. The letters or the art on the page create a specific form of 'textual embodiment'. The process of writing or drawing up an art piece gives us an opportunity to reflect on the way we create our ideas – how the art is placed on the page; where the artist was when he or she created the art piece; how they make excuses for their primitive motives, etc. The artwork may provide important evidence of the impact that such an 'object' made on people. This chapter presents the research model termed 'textual environment' and is related to the history of emotions, outlining how the page, as a phenomenon, affects our lived experience in everyday life, drawing on examples from Icelandic text/manuscript culture. In short, parts of real-life embodied logic interact with abstract mental faculties in ways that have been overlooked; how does a page with a 'text' affect our mental being and bodily function?

Keywords

egodocuments – microhistory – embodiment – barefoot historians – manuscript culture – creative space – counter-archive

1 One Small Question!

One fundamental question is on the table in this chapter: "What takes place, when nothing happens?" This viewpoint was expressed by Italian microhistorian Giovanni Levi in a workshop in which I took part in Budapest at the end of

 | DOI:10.1163/9789004749665_004

September 2015. That question was put forward, as it was seen as summing up admirably the focus which microhistorians should have in mind in their work (see Magnússon & Szijártó 2013, 121–23; Simon 2015, 237–48).

The plan in this article is to discuss the importance of focusing on one specific instance or event in people's life – that will be introduced later – in order to be able to fully grasp or gain insight into the complexity of human society in northern Europe. I will discuss the importance of taking the methods of microhistory seriously, and try to fulfill the promises this methodology has given historians all over the world in the last few decades. Microhistorians argue that their specific focus sharpens our understanding of past times, but in the end their arguments have just been a footnote in macro-analysis by traditional historians. The reason is that most historians see the cases of microhistory as insignificant at best, and sometimes even obscure. I will explore in this chapter how we can approach subject matters that are difficult to analyze; unusual cases that relate to the lives of ordinary people. I am convinced that many of these cases are likely to draw out the importance of the daily affairs of people in the periphery, and their dealings with the formal structure of the society. For that reason, I find it important to pay special attention to concepts like 'textual embodiment' and 'counter-archives' – how the methods of microhistory can bring meaning to these concepts when we are dealing with the gaps in our understanding of past times. I have referred to this scholarly focus as "potential history" (Magnússon 2022, 117–43).

In order to pursue these essential principles mentioned earlier, the ideology must consist in investigating with great precision each and every fragment connected with the matter in hand for which there are sources, and in bringing up for consideration all possible means of interpretation that bear directly upon the material. The focus here is both on the idea of studying as thoroughly as possible the material directly relating to the subject – examining every detail exhaustively – and also striving to bring into the study as much material as possible that relates to the subject, from its immediate environment. This effort is likely to give historians a chance to understand the narration which the sources contain. Such efforts often open doors to unexpected connections and "voices" which may offer competing explanations of specific aspects of the study (Magnússon 2003, 701–35). The paths taken by microhistorians in order to attain that objective are in many ways unusual in the context of the conventional historian's approach. I am referring to those who do not make a clear distinction between the past and history, those who deal with history as a constant (with the right method you can bring back the past). Instead, microhistorians emphasise that each individual finds their own voice or method in

FIGURE 3.2 The encyclopedia written by Farmer Jón 1845–1852 about the creation of the world and the development of species. A most impressive work of scholarship by a man with no formal education. His passion was, however, astronomy, and he wrote to professors in Copenhagen to have his calculations, for example about the distance from his home to some specific stars, confirmed. National and University Library of Iceland (handrit.is)

historical scholarship to deal with often difficult cases, as will be explained later.[1]

2 Far-Reaching Farmer Jón

The focus in this chapter is on a text that was written in 1845–1852 by farmer Jón Bjarnason of Þórormstunga in Vatnsdalur, north Iceland. The manuscript is a multi-volume work (ÍBR 67–73 4to), totaling nine thick books, penned by Jón about natural science, geography and various other useful knowledge: quite an impressive encyclopedia on the creation of the world and the development of species. With my co-author for that research, fellow historian Árni H. Kristjánsson (a former student of mine), I have sought to explore how, in the mid-nineteenth century, an uneducated member of the Icelandic peasantry was able to make use of available information to produce works of this nature, in the

1 See such classic microhistorical works as: Ginzburg 1980; Zemon Davis 1983; Ladurie 1979; Kates 1995; Ozment 1989; Reay 2002.

conditions of the time – how he managed to collect in the North knowledge that was mostly foreign to his fellow countryman.[2]

Jón Bjarnason is an ideal subject for a study such as the one in this chapter; his life story and position in society need to be explored at the outset and his work gives us an opportunity to understand the meaning of his work for other people. Farmer Jón grew out of a dual cultural heritage which combined orthodox Christian doctrine with a mystical belief in nature that dated back to time immemorial among the Icelanders (Oslund 2011; Magnússon 2010). Despite living in a remote valley, he was in touch with far-reaching ideological influences from abroad, which had a deep effect upon him. He believed in scientific rigor, based on rational thinking and the realist principles of the Enlightenment. The influence of the Enlightenment is clearly seen in Jón's book, which he intended to be read by the common people. Due to Enlightenment influence, Jón was guided by the principle of *the idea of progress*. By the same token, the influence of the dual Icelandic cultural heritage is also clear: Jón took religious dogma literally, was superstitious, and believed in age-old shibboleths and folk traditions. In this, he was a typical member of the Icelandic peasantry (Simpson 2004).

Jón Bjarnason was receptive to the Enlightenment only up to a certain point, due to conflicts with his religious and cultural heritage. This may well have been due to the fixed ways of the peasantry, and the stagnation of society in many fields. Social structure and cultural heritage thus indisputably hindered the spread of Enlightenment ideas in Iceland. All of this is dealt with in the writing of Jón Bjarnason, and with his work opens up, for example, an opportunity to discuss how the writer's illustrations – a total of more than 500 in volume three of the manuscript – may be used to shed light on nineteenth-century Icelandic society. Also, the way in which the 'barefoot historians' – poor peasants and farmhands who sat and copied manuscripts during their limited leisure hours, day in and day out – approached that peasant society and how their writing affected the general outlook of the public (Magnússon & Ólafsson 2002, 175–209; Magnússon & Ólafsson 2012, 495–524).

Farmer Jón sat pondering all kinds of questions about the development of species – his own take on evolutionary theory. He appears to have had an

2 Part of this article is based on the research done by me and my co-author Árni Kristjánsson in 2014, published that year as Árni H. Kristjánsson and Sigurður Gylfi Magnússon 2014. Árni was a MA student at the time at the University of Iceland and his thesis dealt with Farmer Jón. The book was published after he finished his thesis and I have dealt with the history of Farmer Jón independently in many publications after our book was published, mostly though with the illustrations in his work. See for example: Magnússon & Ólafsson 2017.

FIGURE 3.3 Monsters with many wings and heads, something that the general public in Iceland wrote about and heard stories of, but had never seen represented pictorially. National and University Library of Iceland (handrit.is)

insatiable desire to learn, to collect material about the North, and in fact about the whole world. He looked at his neighbors and wanted them to have the opportunity to share his experience, through his answers to such questions. We know this thanks to Jón's encyclopedic writings about the wonders of the world, which he based upon books in Icelandic and other languages – Danish, English, and German – in which he was self-taught, as well as oral sources, and all the illustrations, which he drew himself. The combination of text and images makes the manuscript probably unique among nineteenth-century Icelandic manuscripts.

It is safe to say that the outside world – foreign affairs – touched the lives of people in Jón's home territory through his scholarship. His manuscript opened up a perspective which was not easily found, even for readers who were keen to learn. That fact is an interesting subject for consideration, which changes our ideas about the horizons of the peasantry at that time, before reading matter was widely disseminated in Iceland. Publication of books at this point in time was mostly in the hands of the church, but other material was distributed by the so-called 'barefoot historians' (like Farmer Jón) who copied non-religious material from past centuries, created new works, and sent their writings from farm to farm in their local community. This network of popular scribes was in fact responsible for full literacy in nineteenth-century Iceland, a country with minimum infrastructure and no formal education system (Magnússon 2023, 31–43).

It is tempting to focus primarily on the book's illustrations, and exploring their meaning: what the drawings depicted, and how they brought information about exotic parts of the world into Icelandic rural homes in the mid-nineteenth century. The material is abundant and complex – and highly unusual in the context of its time, and that is precisely what makes it rich material for microhistorians to deal with.

The temptation is to look specifically at the images that did not belong to the known animal kingdom as it was laid out by the champions of the Enlightenment in the learned circles of the society. Farmer Jón was keen to make drawings of monsters and odd-looking people, but only if he had reliable oral sources for their existence. These illustrations struck a nerve with the public that had been raised up with tales told of the monsters of the sea or those who lived in the mountains. They were as real as any other animals that lived in Iceland, that people had in front of their eyes every day. The people who had the opportunity to read the encyclopedia were able for the first time to actually look at them with their own eyes and comprehend their physical size and features. We can only imagine what kind of impact it had on their mental state, but this was a real-life experience for all of them. It is not known how

FIGURE 3.4 Monsters in all their glory! Farmer Jón only wrote about them if he had a reliable source for their existence. National and University Library of Iceland (handrit.is)

people reacted to his work, but all the 'barefoot historians' thrived on networking with their fellow countryman – sharing their scholarship with others. For that reason we can assert that people were exposed to his writing, and that its importance was discussed in the farmhouses of his community.

3 How to Deal with Farmer Jón?

I encourage historians of any persuasion to work inward, looking into every corner – leaving no stone unturned – to fill out the 'space' that the material at hand takes up. I am afraid that historians tend to be occupied with the larger context, sacrificing the gain that a detailed analysis might bring to their research. As will be stated later in this chapter, I argue for a more intense analysis of the historical sources than historians have grown accustomed to – even microhistorians – to deconstruct their context and inner being in as much detail as possible (Munslow 1997; Merwick 1999; Munslow 2010).

Even if the scale is reduced in the way envisioned here when the case of Farmer Jón is under scrutiny, one must still expect some structural orientation within the frame of reference. But this structure must always be subject to laws other than those imposed by the traditional metanarratives and, because of their scope, must be more malleable – i.e. the frames must be more limited and more easily controlled. What is proposed here is that the essential idea behind microhistory be taken literally: that scholars place the main emphasis on the small research unit, and confine themselves largely to that. I have named this approach "the singularization of history," and in this sense it provides the researcher with a means of bringing out the contradictions that exist between the different 'discourses' of individual groups; and that is a precondition for our being able to approach ideas and points of view that in the general run of events do not come to the fore (Mímisson & Magnússon 2014, 131–56).

Care is needed here because this idea almost reads like a call for total decontextualization. In fact I am arguing for the need to separate the desire to contextualize from the temptation to expand that contextualization into realms of the abstract – where contextualization morphs into generalization. The argument here is about generalization – and even then, this raises all kinds of questions about what kinds of generalization, etc. The point is that there is a tangle of issues which I have too easily glossed over in my previous writing – contextualization, generalization, abstraction. In fact, one needs to be more specific in what lines are drawn, and where, around what counts as the *singular*.

American archaeologist Charles E. Orser Jr. wrote an introductory article for a forum on the "Singularization of History," published in the *International Journal of Historical Archaeology*. Orser here addressed the potential of the *singular* in the broad context. Orser and another archaeologist, Kristján Mímisson, author of a paper in the same publication, have both taken up the idea of the "Singularization of History" and developed it further, including the idea that "the concept of singularization can be expanded to the point where its intensive form of examination can be employed to larger social networks and still not violate [Sigurður Gylfi] Magnússon's concept of the 'manageable unit'", as Orser (2016) argues.[3] He simply points out that it is not possible to ignore what he calls "metaprocesses of modernity" when working in the context of modern scholarship. We shall return to this point below.

The key concern – for history and other disciplines, for example archaeology – is to make the leap from talking about things and their concrete properties and relations to dealing with society in terms of abstract processes and categories.

The strength of the microhistorical approach is precisely that each subject addressed is discussed both with respect to known facts, and to what is not

3 This subject is discussed at length in an issue of the *International Journal of Historical Archaeology*, published early in 2016, as mentioned above. In my view Orser and Mímisson have developed this research model in a very productive way. See Orser, Jr. 2016; Mímisson 2016; Magnússon 2016. A few doctoral dissertations have been written where the historian uses the Singularization of History model in their research. See for example: Walsh 2023.

known. The latter is, in other words, no less important to the microhistorian's research than the former. But discussion of the unknown naturally requires research methods and approaches which are quite different from those used by historians whose narrative is broad-brush – those who work with the "grand narrative". The above-mentioned approach is thus a great leap away from the conventional routes taken by historians in their research and to the approach that I have referred to as "potential history".

In my view it is this concept of singularization that needs greater attention. Its power seems to reside in the fact that it is both an epistemological and an ontological position – that is, it argues for a research program which works from the bottom up, or even just stays at the bottom, looking for lateral or horizontal connections, not vertical generalizations and abstractions. This approach calls for an "integral history", as Brad S. Gregory argues when he discusses the importance of microhistory: "an endeavor that seeks to identify and integrate everything – all relevant material, social, political, and cultural data – that permits the fullest possible reconstruction of ordinary life experience in all their varied complexity, as they are formed and transformed" (Gregory 1999, 102).

FIGURE 3.5
It is evident that each page is structured in the spirit of the Enlightenment, where the animals are divided into categories with reference to other volumes in the whole work. It needs to be stated that the Enlightenment filtered down to the general public in Iceland in the late nineteenth century and was manifested in books like Farmer Jón's, long after it was founded in France in the eighteenth century. National and University Library of Iceland (handrit.is)

Like Ginzburg's evidential paradigm, ontologically, it is equally concerned to focus on the small and ordinary – events, individuals, etc. – as opposed to more abstract subjects such as society or class or ideology. If one accepts this argument, how on earth can the global perspective fit into the research model of microhistory?

4 Local and World Space and Scale

None of this, mentioned earlier, appears to cause any concern to Jo Guldi and David Armitage, authors of the book *The History Manifesto* (2014), which has been greeted with considerable attention and discussion. At the heart of their argument is the assertion that historians must now turn away from the "bean-counting" of recent years – what they call "short-termism" – in order to grapple with the large-scale trends and developments in history, if they want to be taken seriously[4] (see exchange in Guldi & Armitage 2014; Cohen & Mandler 2015; Armitage & Guldi 2015). They also maintain that signs of the problems of history are to be seen everywhere – for instance in students showing less interest in pursuing the discipline. Their view is that the big digital data that embrace the whole world have revolutionized the potential for historians to get to grips with much larger-scale issues, and hence to make the discipline more interesting and influential.

The problem remains – in my view – that it can be difficult to deal with big history without ignoring what historians do best, i.e. working with specific resources in a meaningful manner: addressing their origins and form; working with the context of those who appear in the narrative; analyzing the many "voices" to be heard. For that reason, those who work with long-term developments have increasingly looked to microhistorians, by no means avoiding the small-scale as proposed by the authors of *The History Manifesto*.

In a special issue (2011) of *International History Review* edited by Bernhard Struck, Kate Ferris and Jacques Revel, on the advantage of using the methods of microhistory for analysis of transnational history (global or world history), three important factors are mentioned: a) opening the opportunity to bring actors and agency back into the analysis;[5] (Sewell 2005), b) opening the

4 See also a critical discussion about the book – *The History Manifesto* – in *American Historical Review*: Cohen & Mandler 2015. A reply from the authors of the book appeared in the same issue: Armitage & Guldi 2015.

5 Agency is here defined as the capacity "of exerting some degree of control over the social relations in which one is enmeshed, which in turn implies the ability to transform those social relations to some degree" (Sewell 2005).

opportunity to examine individuals who lived "global lives" in olden times, moving from one continent to another and recounting their experiences;[6] (Davis 2011, 118–228), c) opening the opportunity to zoom in and out "from grand and large-scale questions to microanalysis, case studies of individuals or small groups and vice versa" with the intention to "fulfill [the historian's] craft and the ethic of the discipline by working close to primary sources" (Stuck, Ferris & Revel 2011, 577). One can only agree with every point here: this all goes to make a "grand study" fuller and more appealing.

In the final analysis, I do not want to reject the big context. I feel, quite simply, that history is not well-equipped to explore it critically, on its own terms, with the tools it has at its disposal. But it is hard to reject that viewpoint, to pretend that it does not exist; and for that reason I feel that it is useful in many ways as a focus for an inquiry in the present day in an intellectual context, as an influence on the practice of microhistory. By this I mean that, while global history can scarcely become the subject of microhistory in the manner generally accepted, the microhistorical approach is, in the end, a methodology rooted in a contemporary culture which is definitely informed by the global perspective. In other words, the world/global vision inevitably influences those who apply the methods of microhistory, and makes its mark on their research. So, the question is: how can we work systematically with the global reality in applying microhistorical methods in historical analysis when we, for example, try to focus on collecting the North?

In short: I have come to the conclusion that my attention has to be on the micro perspective, but at the same time I need to bring to the micro analysis the assurance that the global vision has had an impact on every individual in his/her own local environment throughout the history of mankind. In other words, I do not attempt to fit the microhistorical approach to the conceptual framework of world or global history, as Francesca Trivellato (2011) proposes in her paper on the question of whether there is room for microhistory in the age of global history. (See also Stuck, Ferris & Revel 2011, 577).

One might ask: on what foundation is my argument built? It is important to give oneself time to consider the qualities of people who have walked a rocky road in their lives, and not the smooth, straight path. Every person's manner and body language have their own qualities which give meaning to life; but these can only be discerned by scrutinizing all the details of each subject. In fact, the idea behind normative behavior tends to force us to concentrate on the common way of doing things. Once we manage to escape the heavy hand

6 See a special forum on the importance of this approach by Professor Davis in *History and Theory* (2011, 188–228). Participants: Natalie Zemon Davis, Joan W. Scott, Bonnie G. Smith, David Abulafia.

of the "normative" we get a different meaning of life, that is when we take our analysis to the individual and their way of living.

This ideology opens up an opportunity for microhistorians to examine their subjects minutely, and to discuss them in an enlightened manner. This ideology often leads to such detailed examination of objects and materials on the scholar's desk that the result may be *neo-materialism* – an ideology which requires minute scrutiny of the material under consideration, an attempt to answer the question: *What takes place, when nothing happens?*[7] (Orser 2010).

5 Creative Space and Peasant Society

It is interesting to observe that a man like Farmer Jón – a poor farmer in Iceland – was clearly in touch with the outside world. Despite his isolation in the North, he must have been diligent in acquiring or collecting new information – seizing all opportunities to learn from people he knew or met, and those who passed through his district. His mind was clearly always on learning more and adding to his store of knowledge. Jón had no formal education, other than his preparation for confirmation as a youngster, which comprised a few weeks of home schooling.

Farmer Jón did not seek to conceal his sources, and often described them in some detail. Some of these works were large in scale, published in many volumes – which led to the question of how he had access to them. He had only a few books in his own home, and no-one nearby had a book collection. Jón was the son of poor farming folk, but he managed to make quite a good living for himself. After the death of his parents, he appears to have devoted much of his time and resources to his passion: grappling with the mysteries of the world.

We know that Jón had a good "nose" for Icelandic writings that might be useful to him, and these are conscientiously listed in his encyclopedia. When he knew of people who had books, he would write and ask to borrow them. He adapts the content of the image found in these works to his own interest in unusual phenomena of all kinds: giants, dwarves, or anyone who was "deformed" in some way (Henderson 2013, 17–33).

At that period, the Icelandic peasantry had almost no experience of visual representation of their surroundings. Pictures and visual expression were practically unknown, with the exception of biblical images in churches. There are cases, however, of manuscripts from earlier centuries which include pictures of

7 Archaeology has much to contribute here, as it has many features in common with what has been described. See Orser (2010) on the development of historical archaeology.

people and animals. It is clear that he made his illustrations of fantastical creatures from an Icelandic setting – again, in which he had a keen interest – based on accounts from people who believed they had personally observed them. Hence he was following certain models, while also adding his own touches.

In this way Jón carried out his task, either using specific models in books to which he gained access, or based on personal accounts, or drawing on his own imagination. Thus he introduced a new dimension of perception to his friends and relatives – a dimension with which people were unfamiliar: illustrations of phenomena of which they had little or no knowledge. I tend towards the view that a work of this nature is not only highly unusual in the context of Icelandic country folk, but also in a European context. Works are known, made by bourgeois writers in the early modern period, which have similarities with Farmer Jón's efforts. One such example is the early-seventeenth-century Thomas Trevelyon Miscellany, a beautiful facsimile edition of which is to be found on the Folger website.[8]

The question is, how best to deal in a scholarly manner with a subject such as Farmer Jón and his life. How can we explore specific aspects of his life, and at the same time fathom its meaning in the larger context – for the society in which he lives, and the world he described in his works? I have urged microhistorians to stick to the principle of departing as little as possible from the research unit in question – to keep to the point, so to speak (Magnússon 2015, 77–94). That means that a document like the one Jón created is full of interesting details that traditional history writing has overlooked. Why would a historian dealing with political or intellectual history take Farmer Jón as a subject matter, or even just social historians? They have historically been more eager to deal with synthesis of larger scale developments and big structures.

The story of Farmer Jón is consistent with the ideology – his world view – which has been described above. His writings are an example of material which has received little attention – or, frankly, none at all – due to the scholarly attitudes discussed earlier. But if we bear in mind that Farmer Jón was a member of a small and relatively isolated community in the mid-nineteenth century, his encyclopedia opens up an entirely new view of his relationship, and that of his neighbors, with the outside world. The task he set himself was large and bold: to gather information about the creation of the world, and connect that story with the epistemology with which he was familiar in an isolated community in remote Iceland. And it is truly a remarkably wide-ranging and complex undertaking, which clearly had a major impact on the way he thought about himself and the world outside.

8 For information on Trevelyon/Trevelian, see http://collation.folger.edu/2012/12/a-third-manuscript-by-thomas-trevelyontrevelian/.

The pictorial aspect of the work, which was created in an environment where visual representations of the world were almost unknown – a society which did not even possess mirrors – is huge, much greater than we today can imagine. To be able to see pictures of exotic creatures such as monsters must have transformed the ideas of people like Jón about the world – thrusting the outside world in all its glory into the homes of Icelandic peasants.

It must be borne in mind that the encyclopedia was written and drawn before printed matter was readily available to the Icelandic peasantry. Reading societies, which were to become influential in the last decades of the nineteenth century, did not yet exist, and publication of books and periodicals was very limited (Magnússon 1997, 69–164). Again, book-publishing was dominated by religious subjects at this time. And we may ask: what was the significance of this opportunity to learn about the world and the heavenly bodies, for those who made the effort?

And that brings us to the hypothesis that, firstly, personal expression in writing – as practiced by Farmer Jón – was more widespread in Iceland than elsewhere in Europe, and secondly, that the same is true of creative thinking among the peasantry. A certain kind of "creative space" came into existence in Iceland, within which people flourished and had the opportunity to place their ideas and thoughts in the public eye. If we consider the two different forms of cultural memory – collective memory and individual memory – there are good reasons to believe that collective memory was exceptionally weak in Icelandic peasant society. As a result of the geographical conditions in the country, with almost no urban centers and the population widely scattered on farms with considerable distances between them, it was difficult for groups to coalesce within firm enough bonds for them to be able to share their memories and hold them in common. In consequence of this, people had much greater scope to create and shape their own individual memories than in most other parts of Europe. The strength of the individual memory manifests itself in a wide variety of ways in Icelandic society, but perhaps most notably in the massive outpouring of written material by ordinary working-class peasants in the eighteenth and nineteenth centuries (Magnússon 2010, 239–43). Those conditions provided fertile soil for creative global connections within the local community, as exemplified in the life of Farmer Jón. Had the "creative space" not developed in the way envisioned here, the impact of the global connections would have been much less readily seen.

In Icelandic peasant homes, manuscripts – handwritten copies of a wide variety of material – were generally read aloud and discussed at the *kvöldvaka* or "winter-eve gathering": the members of the household sat together in semi-darkness at various tasks, while a single lamp provided light for the reader. This custom is described in innumerable egodocuments extant from the eighteenth, nineteenth and twentieth centuries.[9] Jón reveals in his foreword that he had a readership in mind as he compiled his encyclopedia. He wanted the books to be passed from farm to farm, and read. What we do not know, on the other hand, is what his intended readers thought of his work. But we know that at this period of Icelandic history an unprecedented ferment was taking place in Icelandic society, within the aforementioned "creative space" – and that led to near-destitute individuals embarking on creative projects in a manner that is hard to understand or explain (Magnússon 2020). Jón's encyclopedia is the product of that environment or creative space – a work that was grounded both in the cosmos of the Icelandic peasantry, and also in faraway territories, so that the wonders of the world became part of the daily life of Jón's rural neighbors. The hypothesis is proposed here that works such as that made by Jón – and they were of an astonishing diversity – together formed, in essence, what I have called "creative space". In that way world history became part of the life of people living in remote rural communities in faraway Iceland.

6 Embodiment and Counter-Archive

When we look back and consider Farmer Jón's achievements, and all that he created and made public in one way or another, the result is astonishing. He creates, or simply lives his life, and succeeds in leaving behind a paper trail which provides us with the opportunity to examine his ideas about himself and others. With the methods of microhistory and the emphasis on ordinary poor people we manage to draw out the importance such individuals had in their local community with their endless quest for knowledge and education. Their passion was based on the urge to discover unknown ideas and address them in their own particular way. In terms of traditional scholarship this local knowledge was just "minor" knowledge!

It is tempting to try to link the concept of *counter-archive*, and *embodiment* in a textual sense, with Jón's work when we try to understand the idea behind his deeds. The author – Jón himself – plays with the presentation of his own character through his text: a certain form of discourse that he uses to give himself a voice. At the same time, he embodies himself in the reader's and listener's mind, manifested in direct speech, heteroglossia, personification, in his text. This is a kind of "embodied connectedness", whereby the thing takes form as an "embodied" phenomenon by the signifiers in his text merging with the recipient (Harvey 2019, 455–69; 2020, 136–39). This entails that Jón applies the idea

9 See a book written by Davíð Ólafsson and me, on the importance of manuscripts in modern times: Magnússon & Ólafsson 2017. See also: Magnússon 2023.

of *embodiment* in his collecting of various objects – like the encyclopedia – and puts it forward in opposition to the idea of the "learned" culture. He contrives to make his materials into an organized counter-archive, and the explanatory model is his "emotional community", if we refer to the concept developed by American historian Barbara Rosenwein (Rosenwein 2002, 826–36).

It is at least tempting to think about his collection of manuscripts, a total of 400 different manuscripts on religion, natural sciences, chronicles on weather and agriculture, dictionaries on Danish/Icelandic and Latin terms for natural phenomenon, as a counter-archive – something that gives a totally different image of the life of an ordinary person, an image which is distinct from what we learn from the educated culture.

Embodiment is, of course, a difficult concept to deal with, as it has so many possible meanings. In the encyclopedia it applies to cases where the voice is embodied – whether in literal embodiment, e.g. in performance of a work, in which the performer gives the work a voice as Jón does in his writings. He does this e.g. in a textual sense, where the author plays with presentation of characters through text, a certain form of discourse and other factors in order to give them a voice, and at the same time he succeeds in embodying them in the mind of the reader/listener especially when he gives his readers a peek into the world of the unknown – the monsters. He also succeeds in doing so, for instance, most effectively in his illustrations of different races, using language that is, of course, not acceptable in modern communication. Let us take an example from the encyclopedia:

> Negroes have in general black, smooth, soft skin, and short hair which is as soft as down. The beard is like the hair, yet it turns grey at an early age, while the hair very rarely does so. Thus there are many who are white-bearded and black-haired at the same time. Their teeth are as white as ivory, and the white color contrasts sharply with the black skin. But it is not only in their facial features that most of them are ugly. The breasts of women who have had children hang down to the navel, and they invariably suckle the children they carry on their backs by flinging their breasts back on their shoulders. Their bodies too are to our eyes unhandsome, and the souls of most of them have been endowed with little intelligence, and the heat appears to have impaired them mentally even more than physically. The majority of them are thus feeble-minded, lazy and envious.

And he continues:

> From this one may deduce about the religion of Africa, or of the people of the continent, that most of them are heathens, worshippers of pagan

> deities, Mahometans. Some are said to be Christians, especially in Abyssinia, and Jews are also found on this continent. But against this there are, in the inner lands, even cannibals. But regardless of how senseless, unintelligent and uncivilized the people there may be, it appears that those Christian people of the North have little reason to regard themselves as better than them, who every year buy and sell many thousand negroes into lifelong servitude, subject to the most godless treatment, as if they were beasts to the slaughter or inanimate objects. Alas! At such deeds that bring shame on all mankind, any humanitarian must shudder! (Kristjánsson & Magnússon 2014, 151–53)

We can tell that Farmer Jón is sincere when he expresses himself about people from distant places. This is what we may term "embodied connectedness", i.e. that the thing takes form as an "embodied" phenomenon in this dialogue between the signifier and the recipient and bring into the local community a totally new knowledge.

We are in effect working with an unusual scenario and the life experience of one individual, and we have the opportunity to show how these are "embodied" in Jón's presentation in so many fields. It is this transformation that takes place from the unclear – Jón's life – into an embodied thing or phenomenon relating to him, that reveals both his own attitude to life and the grand narrative of the nineteenth century. In this way we have the opportunity to examine Jón's life and work and his perspective as a poor farmer, and measure him against the science of his time, nineteenth-century society and the people who were part of his life. All that Jón did may be viewed as a "counter-archive", an attempt to counteract the idea of "normal" behavior of a farmer in an Icelandic peasant society with the aid of the methods of microhistory. What we have in our hands is a potential history!

Bibliography

Cohen, Deborah & Peter Mandler. 2015. "*AHR Exchange*: The History Manifesto: A Critique." *American Historical Review* 120 (2): 530–42.

Davis, Natalie Zemon. 1983. *The Return of Martin Guerre*. Cambridge, Mass.: Harvard University Press.

Ginzburg, Carlo. 1980. *The Cheese and the Worms: The Cosmos of a Sixteenth-Century Miller*, translated by J. & A. Tedeschi. Baltimore: Johns Hopkins University Press.

Gregory, Brad S. 1999 "Is Small Beautiful? Microhistory and the Writing of Everyday Life." Review essay in *History and Theory* 38 (1): 100–10.

Guldi, Jo & David Armitage. 2015. *American Historical Review* 120 (2): 543–54.

Guldi, Jo & David Armitage. 2014. *The History Manifesto*. London: Cambridge University Press.

Harvey, Karen. 2019. "Epochs of Embodiment: Men, Women and Material Body." *Journal for Eighteenth-Century Studies* 42 (4): 455–69.

Harvey, Karen. 2020. "One British Thing: A History of Embodiment: Ann Purvis, *ca.*1793–1849." *Journal of British Studies* 59: 136–39.

Henderson, Gretchen E. 2013. "The Ugly Face Club: A Case Study in the Tangled Politics and Aesthetics of Deformity." In *Ugliness: The Non-Beautiful in Art and Theory,* edited by Andrei Pop & Mechtild Widrich, 17–33. London: I. B. Tauris.

Kates, Gary. 1995. *Monsieur d'Eon Is a Woman: A Tale of Political Intrigue and Sexual Masquerade*. Baltimore: John Hopkins University Press.

Kristjánsson, Árni H. & Sigurður Gylfi Magnússon. 2014. *Bóndinn, spendýrin og fleiri undur alheimsins. Alfræðiverk fyrir alþýðu sem Jón Bjarnason frá Þórormstungu í Vatnsdal vann á árunum 1845–1852*. Sýnisbók íslenskrar alþýðumenningar 17. Reykjavík: Háskólaútgáfan.

Ladurie, E. Le Roy. 1979. *Montaillou: the Promised Land of Error*, translated by B. Bray. New York: Penguin Books.

Magnússon, Sigurður Gylfi & Davíð Ólafsson. 2002. "Barefoot Historians: Education in Iceland in the Modern Period." In *Writing Peasants: Studies on Peasant Literacy in Early Modern Northern Europe,* edited by Klaus-Joachim Lorenzen-Schmidt & Bjørn Poulsen, 175–209. Århus: Landbohistorisk selskab.

Magnússon, Sigurður Gylfi & Davíð Ólafsson. 2012. "Minor Knowledge: Microhistory and the Importance of Institutional Structures." *Quaderni Storici* 47 (2): 495–524.

Magnússon, Sigurður Gylfi & Davíð Ólafsson. 2017. *Minor Knowledge and Microhistory: Manuscript Culture in the Nineteenth Century*. London: Routledge.

Magnússon, Sigurður Gylfi & István M. Szijártó. 2013. *What is Microhistory? Theory and Practice,* 121–23. London: Routledge.

Magnússon, Sigurður Gylfi. 2015. "Tales of the Unexpected: The 'Textual Environment,' Ego-documents and a Nineteenth-century Icelandic Love Story – An Approach in Microhistory." *Cultural and Social History* 12 (1): 77–94.

Magnússon, Sigurður Gylfi. 2003. "The Singularization of History: Social History and Microhistory Within the Postmodern State of Knowledge." *Journal of Social History* 36: 701–35.

Magnússon, Sigurður Gylfi. 2016. "Views into the Fragments: An Approach from a Microhistorical Perspective." *International Journal of Historical Archaeology* 20: 182–206.

Magnússon, Sigurður Gylfi. 2022. *Archive, Slow Ideology and Egodocuments as Microhistorical Autobiography: Potential History*. London: Routledge.

Magnússon, Sigurður Gylfi. 2023. *Autobiographical Traditions in Egodocuments: Icelandic literacy Practices*. London: Bloomsbury.

Magnússon, Sigurður Gylfi. 2020. *Emotional Experience and Microhistory: A Life Story of a Destitute Pauper Poet in the 19th Century.* Microhistories. London: Routledge.

Magnússon, Sigurður Gylfi. 1997. *Menntun, ást og sorg. Einsögurannsókn á íslensku sveitasamfélagi 19. og 20. aldar*. Sagnfræðirannsóknir 13. Reykjavík: Háskólaútgáfan.

Magnússon, Sigurður Gylfi. 2010. *Wasteland with Words: A Social History of Iceland.* London: Reaktion Books.

Merwick, Donna. 1999. *The Death of a Notary: Conquest and Change in Colonial New York*. Ithaca: Cornell University Press.

Mímisson, Kristján & Sigurður Gylfi Magnússon. 2014. "Singularizing the Past. The History and Archaeology of the Small and Ordinary." *Journal of Social Archaeology* 14 (2): 131–56.

Mímisson, Kristján. 2016. "Building Identities: The Architecture of the Persona." *International Journal of Historical Archaeology* 20: 207–27.

Munslow, Alun. 1997. *Deconstructing History*. London: Routledge.

Munslow, Alun. 2010. *The Future of History*. London: Palgrave Macmillan.

Orser, Charles E., Jr. 2010. "Twenty-First-Century Historical Archaeology." *Journal of Archaeological Research* 18: 111–50.

Orser, Charles E., Jr. 2016. "Introduction: Singularization of History and Archaeological Farming." *International Journal of Historical Archaeology* 20: 175–81.

Oslund, Karen. 2011. *Iceland Imagined: Nature, Culture, and Storytelling in the North Atlantic*. Seattle: University of Washington Press.

Ozment, Steven. 1989. *Magdalena & Balthasar: An Intimate Portrait of Life in 16th-Century Europe Revealed in the Letters of a Nuremberg Husband and Wife*. New Haven: Yale University Press.

Reay, Burry. 2002. *Watching Hannah. Sexuality, Horror and Bodily De-Formation in Victorian England*. London: Reaktion Books.

Rosenwein, Barbara. 2002. "Worrying About Emotions in History." *American Historical Review* 107, 826–36.

Sewell, William H., Jr. 2005. *Logics of History: Social Theory and Social Transformation*. Chicago: University of Chicago Press.

Simon, Zoltán Boldizsár. 2015. "Microhistory: In General." *Journal of Social History* 49 (1): 237–48.

Simpson, Jacqueline. 2004. *Icelandic Folktales and Legends*. Foreword by Magnus Magnusson. London: Tempus Publishing.

Stuck, Bernhard, Kate Ferris & Jacques Revel. 2011. "Introduction: Space and Scale in Transnational History." *The International History Review* 33 (4): 573–84.

Trivellato, Francesca. 2011. "Is There a Future for Italian Microhistory in the Age of Global History?" *California Italian Studies* 2 (1): http://escholarship.org/uc/item/0z94n9hq.

Walsh, Ian Thomas. 2023. "The Dublin City Coroner's Court, 1891–1900: A Microhistory." Ph.D. dissertation, University of Limerick.

OBJECT: Expedition diary

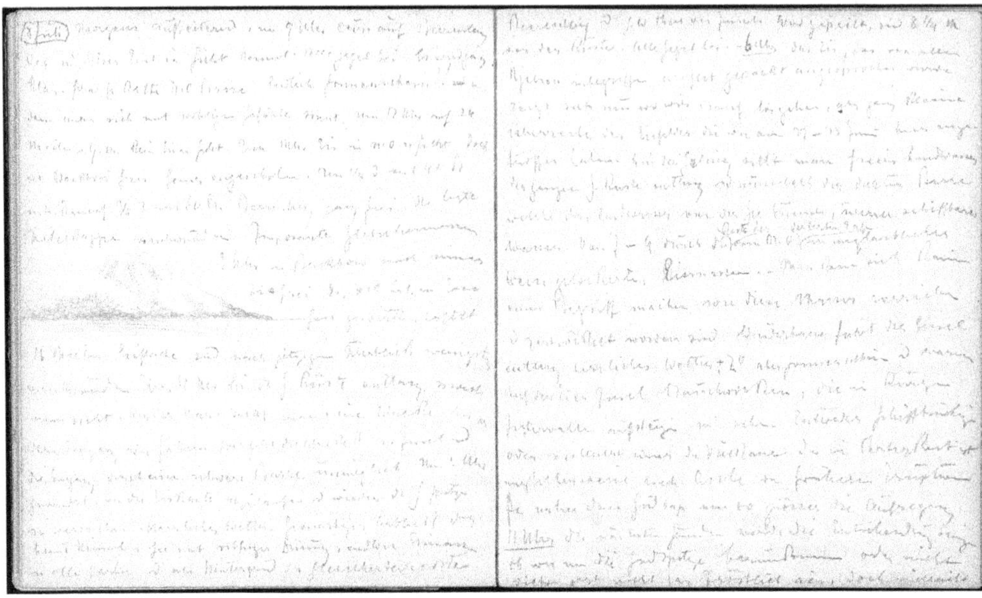

FIGURE 4.1 Extract from the diary of József Pálffy (1882, 43–44). Private collection of Dr. Martin Pálffy (Vienna). Digitisation by the Woldan Collection of the Austrian Academy of Sciences under the Creative Commons PDM 1.0 (Public Domain) licence

CHAPTER 4

Sensing, Observing, Measuring: József Pálffy's Diary and the Austro-Hungarian Jan Mayen Expedition of 1882–1883

Johannes Mattes

Abstract

This article examines an untapped diary kept by the aristocratic traveller József (Jozef) Pálffy of Erdöd during his involvement in a scientific mission of the Austro-Hungarian Navy to Jan Mayen, a 55-km-long volcanic island in the Arctic Ocean, now under Norwegian sovereignty. Taking place within the framework of the First International Polar Year of 1882–1883, the expedition was one of 14 ventures to gather and compare geoscientific data at (temporary) observatories in the Arctic and sub-Antarctic. Using this rare field diary as a lens for analysis, my chapter explores Pálffy's encounters with Nordic environments, revealing new insights into the epistemic, political, and social dimensions of knowledge production in early polar research from a microhistorical perspective.

Keywords

International Polar Year – Jan Mayen – expedition – diary – Habsburg monarchy – Arctic – polar research – knowledge – scientific observation – geographical surveying

On 25 January 1883, naval officer Emil von Wohlgemuth (1886, 49), head of the temporary Austro-Hungarian research station on the isolated, storm-battered island of Jan Mayen in the Greenland Sea, noted a significant event: "After more than 2½ months, we had lunch again at noon in daylight, without the need for lamps". At around the same time, 3,000 kilometres south in Vienna, the bustling capital of the Habsburg monarchy, Count Hans Wilczek (1883, 2), a prominent science patron, was experiencing very different sensations and perceptions. In his patriotic speech, delivered in honour of the naval crew wintering on this island in 1882–1883 in service of what would become known as the First International Polar Year (IPY), he eloquently emphasized the significance of the observations made there as part of "the greatest [...] international scientific enterprise ever launched". Unlike previous expeditions, which often resulted in difficult-to-verify reports of the areas explored, this mission – along with 13 simultaneous ventures by 11 countries – aimed to collect "objective" measurement data every hour for a year using stationary instruments at predetermined locations in the Arctic and sub-Antarctic (Barr 1985; Lüdecke 2004; Barr et al. 2010). Wilczek substantiated his speech by presenting photographs to the Vienna Geographical Society. This evidence, along with the expedition diary maintained by his nephew, Count József Pálffy, as examined in this chapter, provided a narrative of the observation crew's journey to the 55-kilometre-long, uninhabited island, now under Norwegian sovereignty (Figure 4.1).

The German-Austrian naval officer Carl Weyprecht (1838–1881) initiated this coordinated and cooperative study of the polar regions, propelled by early scientific internationalism, imperial ambitions, and competition among participating countries (Berger, Besser & Krause 2008; Mazzoli 2016). Drawing from his experience as the co-commander of the Austro-Hungarian North Pole Expedition (1872–1874), he campaigned for a novel, "scientific" vision of *Polarforschung* (polar research), distinct from geographical explorations.[1] As with oceanography or alpine research, combining accurately observed and thus reliable quantitative mass data from geophysics, meteorology, and hydrography, along with collecting natural history specimens, should open avenues for a more integrative understanding of polar environments. Moreover, by using these areas as testing grounds for global processes, the inductive study of phenomena best observed under the extreme conditions prevailing there aimed to provide solutions to universal questions involving the Earth's magnetic field, solar activity, and ocean currents (Weyprecht 1875, 38). Conducted after Weyprecht's untimely death under the auspices of the International Polar Commission, the IPY marked a lasting shift in the forms, practices, and credibility of knowledge through which scientists conceptualized the High Arctic, transforming it from a mere obstacle into an "object of legitimate inquiry and appropriation" (Walsh 2014, 249; Tammiksaar, Sukhova & Lüdecke 2010). This elite "culture" of polar research, reflected not only in the ventures' instructions

1 In German, the term *Forschung* (research) encompasses both scientific inquiry and geographical exploration. Consequently, in German-speaking countries like Prussia and Austria-Hungary, the distinction between these two fields of inquiry was less distinct in polar research compared with that of Britain or the United States (Krause 1992, 309).

and outcomes but also in the observers' monitoring routines, discipline, and ethos, was particularly pronounced in the Jan Mayen Expedition, serving as a tribute to Weyprecht's legacy (Wohlgemuth 1882). The endeavour mirrored earlier Habsburg Arctic ventures in its ties to the imperial house, private sponsorship, and implementation as a naval affair; however, it stood out with its use of laboratory-based instrumentation and claim to precision "surpassing amateur capabilities" (Weyprecht 1876, 38). The expedition's primary focus, as outlined in the official documentation, centred on the concerted sensing, observation, and measurement of the natural world around the station, departing from romanticized notions of polar exploration.

The history of Arctic science has traditionally been researched in the context of anniversaries of founding figures, institutions, or undertakings, with a focus on International Polar Years. Of particular interest are studies by Barr and Lüdecke (2010) and Launius (2010) on research programs and cross-border collaboration, Doel et al. (2015) and Shadian and Tennberg (2009) on science-policy interactions and long-term changes in research, and Bravo and Sörlin (2012) and Elzinga (2009) on fieldwork practices and infrastructures. Additionally, Fleming (2016), Friedman (1989, 1994), and Wood and Overland (2006), among others, explored international cooperation in Arctic meteorology and oceanography. Research on the Habsburg monarchy, in turn, has primarily focused on the Austro-Hungarian North Pole Expedition, standing out for its public impact and cross-class support. Among them, notable studies include those by Rack (2012) on its social history, Schimanski and Spring (2015) on the public discourse around the venture, and Walsh (2015) on the intersection with liberal and imperial agendas. Conversely, contributions dedicated primarily on the Jan Mayen Expedition are scarce, with the exceptions being Walsh's (2016) discussion of the topographic surveys and a master's thesis by Justnik (2007) on landscape photography. More broadly, Klemun and Mattes (2022) compared this venture with those supported by the Imperial Academy of Sciences in Vienna.

This chapter shifts focus to an untapped diary penned by the aforementioned 29-year-old Hungarian aristocrat József Pálffy (1853–1920),[2] who joined the venture as a representative of his uncle, Hans Wilczek (1837–1922), the

2 Born into an esteemed Hungarian aristocratic family, József Pálffy of Erdöd developed a keen interest in economics and natural sciences, attending lectures at the Technical College (*Technische Hochschule*) and the School of Commerce (*Handelsakademie*) in Vienna. Following the footsteps of his uncle and father-in-law Hans Wilczek, Pálffy established a chemical industry on his estates in the Little Carpathians (Slovakia), oversaw the reconstruction of Smolenice Castle in a neo-Romantic style, and cultivated a passion for art collecting. See Reiszig (1898, 727), Hupko and Orosová (2023), Pálffy (2025), and the short biography compiled by me in the exhibition booklet by the Österreichische Akademie der Wissenschaften (2023, 12–13).

doyen of the Habsburg polar endeavours. Made available to me by Pálffy's descendants, this 97-page document,[3] dating from 16 May to 4 August 1882, chronicles his two-month voyage aboard the *SMS Pola* via Bergen and Tromsø, culminating in a multiweek stay on Jan Mayen (Pálffy 1882a). As with other field diaries (Punch 2012), this source served as an intermediary between private and public spheres, fulfilling two primary roles: first, as a personal record for reflecting on (scientific) encounters and interactions; and second, as the basis for three anonymous travelogues written by Pálffy (1882b–d) and printed in the German-language liberal newspaper *Pester Lloyd* (Budapest) on his outward and return cruise from different seaports. Additionally, in the official four-volume expedition report, Wohlgemuth (1886) quoted a section of Pálffy's diary detailing his attempt to climb the glacier-covered Beerenberg volcano (2,277 m), which explains the absence of records for his final ten days on the island. This blending of personal experiences with measurement data and field observations provides rare insights into the conceptual claims and realities of Weyprecht's vision of *Polarforschung*, particularly when comparing the diary with the expedition's printed findings.

This chapter argues for a reconsideration of this novel "culture" of researching high-latitude regions as it was implemented during the IPY. I seek to shift the understanding of these early studies from being a "purely" scientific and esoteric endeavour to a hybrid practice involving "non-professional" approaches and geographical exploration, such as mountaineering and surveying. Thus, this chapter contributes to a deeper understanding of the origins of scientific polar research as a distinct field grounded in shared aims, methodologies, and instruments. Based on a microhistorical analysis of this diary and its socio-political contexts, I will ask: What specific knowledge, cultural assets, and imaginative frameworks shaped Pálffy's perception? Furthermore, what methods of collection and appropriation did he employ, and how did they manifest in his private and public writings? To what extent did

3 This collection, made available to me by Dr. Martin Pálffy (Vienna) in 2022 and presented in the exhibition "Land, Land, Land at Last!" at the Austrian Academy of Sciences the following year, comprises a 97-page diary of József Pálffy, enriched with letters and photographs. Alongside the diary, there is a brief note by Pálffy's wife, Lucia Wilczek-Pálffy (1862–1958); a printed transcript of a speech delivered by Hans Wilczek (1883) to the Vienna Geographical Society; and an undated letter from the expedition leader, Emil Wohlgemuth, likely penned in February 1885, requesting Pálffy to forward his notes on the Beerenberg ascent (i.e., the absent second part of the diary, covering the period from August 5). Additionally, the collection includes three landscape photographs of Jan Mayen by naval officer Richard Basso (reproduced in Anonymous 1884); a photograph of the transport ship *SMS Pola*; a popular illustration of the station building; and portrait photographs of Gustav Beer (1844–1921), an officer on the *Pola*, and Wollert Danckert Krohn (1819–1908), an Austrian consul in Bergen. See Mattes (2023).

Pálffy's approach align with the expedition's objectives, and what lessons can be drawn for (Habsburg) polar research around the IPY? Therefore, I will focus less on the participants and logistics of the venture, and instead use the diary as a tool to analyse how Pálffy and the crew gathered insights about the "Arctic" across three interrelated dimensions: socially, on deck and at the station; geographically, at the ice edge and on the island; and knowledge-based, through scientific work and exploration.

1 Social Spaces: Ship and Station

As Pálffy (1882a, 1) departed from Vienna, traveling through Hamburg en route to Bergen to embark on the *Pola*, he sketched a sociogram of the 68-man venture on the first page of his diary. Organized by status and area of responsibility, he prioritized listing the ship's "command staff" and "machinists", followed by the officers of the "expedition" (research station), and finally the "passengers", including his uncle and himself.[4] Notably, he also listed Georg Neumayer (1826–1909), head of the German Maritime Observatory and co-organizer of the IPY. During the stopover in Hamburg, Wilczek and Pálffy seized the opportunity to confer with Neumayer and other esteemed German polar explorers, from whom they received additional guidance on conducting observations (Kinsky-Wilczek 1933). Class distinctions prevailed as Pálffy (1882a, 3–4, 16) strolled through Bergen accompanied by aristocratic naval officers, met the Norwegian-born Austrian consul Wollert Danckert Krohn (1819–1908) – whom he described as a "terribly boring" yet "dear fellow" – and, as a *bon vivant*, praised the local delicacies of "lobsters, shrimps" as well as the allure of "pretty girls".

The well-defined structure of his sociogram began to shift as Pálffy struggled to adapt to the new daily routines and discipline on board the *Pola* during the first two-week attempt to reach Jan Mayen. The 50-meter transport steamer of the Austro-Hungarian Navy (Figure 4.2), so far limited to Adriatic operations and unequipped for sea ice, lacked luxury of any kind (Wohlgemuth 1882). Additionally, sharing his cabin with two ship officers, cramped like "anchovies", interacting with the "ordinary" seamen on deck, and adapting to naval shift work and the extended daylight of the polar summer all formed part of Pálffy's (1882b, 5) encounter with the "Arctic". Moreover, as a passenger, Pálffy (1882a, 34) perceived himself as a mere external observer, feeling "not officially part" of the venture. Nevertheless, he endeavoured to "swindle himself in" and

4 Pálffy's diary and newspaper articles (1882a–d) quoted in this chapter were originally written in German; all translations into English are my own.

FIGURE 4.2 Transport steamer SMS *Pola*, *ca.*1882. Author: B. Ćirković (Pula). Private collection of Dr. Martin Pálffy (Vienna). Digitisation by the Woldan Collection of the Austrian Academy of Sciences under the Creative Commons PDM 1.0 (Public Domain) licence

integrate into the ship's isolated, hierarchical community, caught between the bulwarks, sea, and sky. Drawing on his background as an auditor at the Vienna Technical College, Pálffy (1882c, 5) volunteered for watch duty and to learn the use of nautical instruments, as he emphasized in his report to the *Pester Lloyd*:

> After my four-hour watch, I often find myself completely drenched, necessitating a full change of clothing from head to toe. Drying my soaked garments and linen proves to be exceedingly challenging in these unfavourable conditions, a difficulty that extends to astronomical observations as well. Frequently, I am compelled to spend precious time waiting for brief glimpses of the sun through the clouds, and I must be cautious of frostbite as I hold the instrument with my exposed hands.

On board the *Pola*, Pálffy's social interactions were limited to the class of officers. He engaged with them on the bridge, below deck, and at the dinner table, exchanging information on weather conditions and scientific measurements while also participating in manoeuvres and positioning. A passionate hunter, he was also the driving force behind replenishing the ship's food supplies with sea birds and, occasionally, larger marine mammals. When the *Pola* crossed the Arctic Circle, the captain, Franz Müller, dispensed with a formal ceremony but raised a toast to the Austro-Hungarian North Pole Expedition, which had traversed the same latitude a decade earlier (Pálffy 1882a, 10–11). Despite the absence of its former crew members, the legacy of this expedition as a patriotic, empire-wide endeavour permeated the atmosphere of the diary. Conversely, potential nationalist-inspired tensions aboard the *Pola* or the international scientific context of the current mission went unmentioned. Although Pálffy initially hoped to prove himself as an explorer by disembarking on the Greenland coast after a stopover at Jan Mayen, his dreams were short-lived when the crew encountered drift ice 120 nautical miles southeast of the volcanic island, navigating through it for days. Faced with the danger of the ship being trapped or damaged, Müller decided to return to Tromsø.

The two-week stay in this town of 6,000 inhabitants, initially portrayed in Pálffy's diary (1882a, 19) as a "miserable nest, inhospitable, unfriendly", proved to be an unwelcome interruption to his life on board. Nevertheless, it offered an opportunity for exploration beyond the ship's confines, enabling Pálffy to immerse himself in the local culture and establish connections with fellow travellers and residents alike. During his stay in Tromsø, besides attending a Catholic mass together with the crew and participating in festive events ashore, some of which Pálffy (1882a, 21) found "incredibly stiff and formal", the *Pola* and the nearly "Grand Hotel" were the main hubs of social interaction for him. There, Wilczek and the officers met the local Austrian consul, acquaintances from previous ventures, and foreign explorers currently in port, including members of the "Eira Rescue Expedition", with whom they exchanged information on ice conditions. Although telegraphed back to Vienna, Wilczek proceeded to hire the Tromsø skipper Johan Kjeldsen (1840–1909), who had previously worked with him on the *Isbjørn* Expedition of 1872,[5] as ice-master to ensure the *Pola*'s safe voyage to its destination.

Back on the open sea, with Jan Mayen's Beerenberg looming in the distance as the ship sailed along the ice edge, a new social dynamic emerged: the division between the ship's command and expedition. This was fuelled by the absence of Wilczek, the undisputed authority on board, combined with adverse weather conditions and a belt of drift ice that forced the *Pola* to navigate off the island for weeks before reaching its clear west coast. Wohlgemuth's firm resolve to fulfil the scientific mission, coupled with his insistence on the drop-off of the expedition members on the boat *Weyprecht*,[6] heightened

5 In 1872, Hans Wilczek led the *Isbjørn* Expedition to Novaya Zemlya to set up a depot for the Austro-Hungarian North Pole Expedition. For this purpose, Wilczek chartered the schooner *Isbjørn* and appointed Johan Kjeldsen as the skipper (Kinsky-Wilczek 1933, 199).

6 The 27-foot-long *Weyprecht* was one of three boats available to the expedition crew. It had previously served in the Austro-Hungarian North Pole Expedition during the retreat across the pack ice to the island of Novaya Zemlya and held great symbolic value (Wohlgemuth 1886, 79).

FIGURE 4.3 Station building in the winter of 1882–1883. From left to right: astronomical observatory, anemometer house (each standing alone), two magnetic observatories, thermometer house (in front), connecting corridor with main entrance, residential building. Author: Richard Basso. In Anonymous (1884, 5). Digitisation by the Woldan Collection of the Austrian Academy of Sciences under the Creative Commons PDM 1.0 (Public Domain) licence

tensions with Müller. According to Pálffy (1882a, 48), the latter was "tirelessly concerned about the ship's safety but does not want to risk anything". In addition, Kjeldsen, who had to act as "half a diplomat", found himself navigating the sea ice and "warlike conditions" on board. The situations eased in mid-July as weather conditions changed, enabling the safe landing of the expedition crew and cargo. The erection of the rigorously planned station (Figure 4.3), housing astronomical, meteorological, and geophysical observatories and boasting an inventory of 1,800 items alone, created a separate social hub alongside the ship. The unstable ice conditions further compounded this distinction, forcing the *Pola* to cruise offshore.

Although both the ship and station housed naval communities, with the expedition's prefabricated buildings resembling "upturned ships", their social fabric relied on different objectives and levels of commitment (Wohlgemuth 1882, 22). During the *Pola*'s month-long stay on the island, the crew primarily engaged in offloading tasks as part of routine operations, with eight hours of rest per day, while officers undertook exploratory trips. Conversely, the 14 expedition members, regardless of rank, began their first routine observations while constructing the station, receiving only half the rest time of the ship's crew and being restricted to the work site. Moreover, their temporary accommodation in tents provided little respite, as the focus was on quickly setting up instruments. Due to the "challenging climate", "repetitive nature of observation tasks", and need for "subordination" to the scientific agenda, only naval personnel were considered for the expedition (Wohlgemuth 1882, 17). Officers were selected on the basis of their expertise in relevant fields of observation and their place of birth to ensure representation throughout the multinational monarchy. Additionally, the predominantly Dalmatian seamen working at the station were recruited from volunteers in the naval reserve, drawing on their skills as craftsmen from their civilian occupations. While Pálffy (1882a, 56–58) kept an eye on the so-called "colonists" and lent a hand on the construction site, he soon preferred the ship's comforts over the ascetic station life. Instead, he engaged in the *Pola*'s survey work, such as sounding the bays, and accompanied Kjeldsen on (hunting) trips, thereby turning his encounter with the "Arctic" into a social leisure activity.

This leads to two preliminary conclusions: first, in the case of the Jan Mayen Expedition, the implementation of Weyprecht's concept of stationary observations was based on the transfer of maritime techniques and nautical expertise to laboratory-like work on land. The ship and station functioned as hierarchical and isolated spaces, united by discipline, rituals, and shared values. The introduction of systematic methods in polar research built upon these factors but surpassed them in the observers' dedication, supplementing naval virtues like duty and teamwork with scientific internationalism and accuracy. Second, Pálffy's lack of naval affiliation and aristocratic background granted him a special status within the venture, allowing him to navigate between the respective communities of the ship and station. Drawing on his technical experience and kinship with the sponsor, he actively engaged in routine tasks and undertook his own projects. His diary thus illuminates the enduring influence of individual agency and social dynamics within the IPY's "scientific" vision of research. Building on this, in the following section, I will argue that Pálffy's engagement with polar environments as geographical frontiers was influenced by cultural preconceptions and practices of approbation.

2 Geographical Spaces: Ice Edge and Island

In Pálffy's diary, the "North" did not merely represent a cardinal point on a map; it embodied an array of (sub-)Arctic landscapes, ranging from rugged mountains and nestled settlements to icy fjords and drifting sea ice. These scenes captivated Pálffy (1882b, 5), offering an exotic contrast to his life in Hungary and his trip to East Africa, undertaken with peers from the Esterházy and Liechtenstein families in 1880–1881 (Hudec 2023). Notably, he found "the difference between the temperate zone and the equatorial zone [...] less pronounced than that between the former and the far north", yet the "low temperatures" reminded him of "harsh Transylvanian winters". Moreover, Bergen, a settlement "entirely in the German style", set into "this gloomy rocky land in the far north", evoked for Pálffy (1882b, 5) memories of Kronstadt (today's Brașov in Romania), a town founded by Saxon settlers and located near the Habsburg-Ottoman border. Through these comparisons, he overlaid elements of Orientalism onto his encounter with the "Arctic" (Ryall, Schimanski & Wærp 2010), perceiving it as both a natural and cultural frontier, inhabited by what he deemed "the most politically liberal people in the world". In Pálffy's descriptions, which are more critical in the diary than in his newspaper reports, he drew on clichés of the Norwegian population, portraying them as provincial, stuffy, and law-abiding. Despite these stereotypes, he alternated between admiration and astonishment, noting that the middle class in Norway had achieved a significant level of social development.

Furthermore, Pálffy (1882a, 5) explored the coastal towns visited by the ship, considering factors that typically appealed to tourists, such as the local infrastructure and recreational activities. In Bergen, following his companions' advice, he delved into the "splendid collections" of the Natural History Museum and browsed antiquarian bookstores and, in a fur shop, was tempted to purchase a padded polar bear skin despite being aware of the challenges faced by the *Pola*'s crew in enduring the cold climate without any special provisions. Similarly, as Pálffy (1882a, 22) recounted his excursions from Tromsø to the surrounding fjords, featuring "magnificent glaciers" and "large herds of reindeer", he maintained a tourist-like narrative. It was only on the open sea, aboard an ordinary naval transport ship, with no landmarks on the horizon and exposed to rough weather that his engagement with the natural environment began to deepen. Navigating these geographies required nautical skills, physical endurance, and a willingness to take risks, transforming Pálffy into a keen observer of the operations on board. While the Norwegian coast was undoubtedly within Pálffy's conception of a polar environment, the "Arctic" became a tangible frontier for him when the *Pola* encountered drift ice for the first time, preventing the ship's passage to Jan Mayen:

> At half past seven, my uncle wakes me. The engine turned on and off, and shocks went through the ship. *We are in the ice!* Through the cabin door, we see fantastic pieces of ice passing by. Everybody hurries on deck! […] In the fog, visibility is barely a ship's length. Water −0.6 °C. We change course at 10 o'clock and head N.W. (Greenland) to avoid the ice. […] The captain shouts himself hoarse on the forecastle deck. At half past twelve, we receive the report that we are surrounded by ice. (Pálffy 1882a, 12–13)

Pálffy perceived the ice edge, the boundary between the pack ice and open sea where the *Pola* manoeuvred for weeks, as a liminal space. Fog and dense clouds made it challenging to determine the latitude with a sextant while changing weather conditions and ocean currents shifted the position and extent of the pack ice. His daily records of meteorological phenomena, ice conditions, and the ship's position relative to Jan Mayen were not meant to be scientific. Instead, they served as a means to navigate this liminal space and tackle its challenges by supplementing personal observations with measurement data. Furthermore, according to Pálffy (1882a, 36), human senses were prone to being misled in that environment. For example, the crew members, "wearing simple loden clothes" appeared "strangely acclimatized", despite "all of them being affected by frostbite". Similarly, Kjeldsen, although dressed warmly but suffering from a cold, conveyed to Pálffy (1882a, 36) that the climate around Jan Mayen seemed harsher to him than that of the more northerly Spitsbergen. This prompted Pálffy to generalize that "southerners" like the *Pola*'s seamen were "more resilient" than "northerners" and thus better suited for polar exploration. In doing so, he echoed the public discourse on the Austro-Hungarian North Pole Expedition, particularly Weyprecht's personal assessment, bolstering the monarchy's ability to participate in polar exploration alongside other "civilized" countries (Schimanski & Spring 2015, 313).

The symbolic act of land seizure staged by the expedition members upon arrival at Jan Mayen demonstrates the enduring presence of geographical inquiry within the endeavours of the First IPY. Intended to be performed by Wilczek, a representative of the Austrian House of Lords and confidant of the Emperor, the ceremony ultimately witnessed Pálffy stepping in, as indicated by differing accounts between his newspaper report and diary:

> The expedition leader with Count Pálffy […] and some members of the expedition immediately reached the shore. Count Pálffy was the first to step on the flat, driftwood-covered lava soil of the island and to plant the flag of honour on it. (Pálffy 1882d, 9)

> Wohlgemuth, Fischer, Basso, Sobieczky, and me on shore. […] As the uncle's representative, I was the first and unfurled the flag with great ceremony. […] With hurrahs and all kinds of hocus-pocus, as if we had discovered a new part of the world. We toasted to the uncle with a field bottle brought with us for this occasion. (Pálffy 1882a, 52–53)

As the second quote from his diary illustrates, Pálffy was keenly aware of the performative nature of his actions and felt discomfort regarding the patriotism and colonial attitudes involved, reminiscent of earlier European overseas ventures. Even in his uncritical newspaper report, Pálffy (1882d, 9) contrasted the heroic narrative of the land seizure with the stark reality of the "barren island, devoid of all vegetation, making the worst impression on visitors". Nevertheless, in the weeks that followed, he actively immersed himself in touring, collecting, and writing about this "desert place", captivated more by its untapped topography and wildlife than by the measurements taken there. Although the expedition refrained from making any territorial claims to the apparently only scientifically exploitable island, its use of natural resources and Pálffy's personal approach reflected colonial attitudes.

Pálffy's engagement with Jan Mayen was informed by his experiences in the Egyptian Sudan, where he traversed the Nubian Desert with Indigenous guides and brought back numerous trophies (Hudec 2023). In addition to acquiring topographical knowledge, his eagerness for hunting and proving his skills in polar climates prompted Pálffy (1882a, 61), guided by Kjeldsen and armed with a rifle, a collection box, and navigational tools, to embark on "great exploratory marches toward Beerenberg". Fascinated by the glacier-covered volcano, he used the Alps as a template for his romantic descriptions. Such comparisons with alpine peaks, common among German-speaking travellers, placed their encounters with the "North" in a familiar context and highlighted their aptitude for Arctic exploration (Walsh 2012, 74). The whale bones, minerals, and lava sand that Pálffy collected served as both souvenirs and natural history objects. His records of the animals he shot, including now-extinct blue foxes, bird species, and seals, reveal no concern for the potential impact of his actions on their populations. This contrasts with the humanized portrayal of Arctic wildlife in the Viennese humoristic press, which caricatured polar expeditions as attempts to colonize wild animals (Figure 4.4). In this context, a drastic account of the seal hunt on Newfoundland, as quoted in the official expedition report, in which animals killed with clubs were compared to children, can be interpreted as a justification for the Habsburg monarchy's colonial ambitions. Accordingly, Wohlgemuth (1886, 33) emphasized that under Austrian rule on Jan Mayen, "only a few old seals were shot", and that they were immediately harpooned.

In summary, Pálffy perceived the diversity of geographies he associated with the "Arctic" as culturally preformed spaces. Orientalism, travellers' imaginaries, and explorative-colonial attitudes, all rooted in (Central) European stereotypes about "civilization" and "wasteland", shaped his encounter with polar environments. Depending on the setting – coastal town, ice edge, or island – he adopted different practices, shifting between tourist and explorer roles. As his confidence in sensory perception waned, measurements and observations, crucial for navigating these remote areas, became prominent in his diary. Even

FIGURE 4.4 Caricature of Hans Wilczek playing tarock with polar animals on Jan Mayen, published in a Viennese newspaper. In Wilczek (1882, 3)

the expedition crew, which incorporated aspects of colonial endeavour into its identity-forming rituals, did not view Jan Mayen exclusively as a "laboratory" setting for scientific work. Rather, as an exotic field site, the island provided opportunities for geographical inquiry and self-probation that transcended the divisions of class, nation, and profession. In the following section, I will argue that the expedition employed a knowledge-based approach that combined rigorous scientific practices with exploratory endeavours.

3 Knowledge Spaces: Science and Exploration

When the International Polar Commission decided in mid-1881 to establish the Habsburg station on Jan Mayen, the island was far from a blank slate. Originally, Wilczek had proposed to set up an observatory on the Russian archipelago Novaya Zemlya, the site of earlier Austro-Hungarian polar efforts. However, owing to tensions with the St. Petersburg Geographical Society, he eventually opted for the unclaimed island of Jan Mayen, far from the shipping routes and Arctic territories of other IPY participating countries (k. Akademie der Wissenschaften

1881; Wild 1882). Consequently, the Habsburg venture relied heavily on the knowledge and experiences of the few prior visitors to the island, including Carl Vogt (1863) and Henrik Mohn (1878). The latter, head of the Meteorological Observatory at Christiana University (today the University of Oslo), had spent two busy days on Jan Mayen during the Norwegian North-Sea Expedition (1876–1878), collecting natural history specimens and sketching the most data-rich map hitherto rendered. This map, largely based on offshore observations, became instrumental in planning the Austrian venture, with 200 copies added to the station's inventory.

Furthermore, the mission benefited from the expertise gained from previous Habsburg naval ventures, particularly in terms of armaments, boats, and provisions. This involved the renowned Austro-Hungarian North Pole Expedition, whose publications filled the rich library of the Jan Mayen station. Moreover, Wohlgemuth sought advice on fur clothing and sledges from the Austro-American traveller Heinrich Klutschak (1881), who had spent several years among the Canadian Inuit in search of the British Franklin Expedition. By testing this and other equipment in Arctic conditions, the Habsburg venture aimed to establish a model for the future practice of polar research, extending far beyond its scientific agenda (Walsh 2016). The publication of the meticulously prepared planning documents, which included work schedules, menus, or hygiene measures, further aimed to serve as "valuable pointers for [...] forthcoming expeditions" (Wohlgemuth 1882, 16).

Pálffy made little notes of these ambitions in his diary. However, he was impressed by the instruments (over 100) that had been unloaded on the island as well as by the venture's vivid naval spirit, reflected in the expertise of the station's officers. Even though the officers had to strictly adhere to internationally agreed-upon instructions and time regimes for hourly data collection (i.e., two days a month, even at five-minute intervals), they did not perceive themselves as mere "observers", limited to reading and maintaining instruments (Chavanne 1884) (Figure 4.5). Instead, their integrity as naval officers guaranteed the credibility of the findings and, despite the increasing professionalization of research, justified their role as (co-)authors in the subsequent academic publications (Sondhaus 1993). Here, the scientific agency of the Austro-Hungarian Navy, and in particular that of Bernhard von Wüllerstorf-Urbair (1816–1883), Vice-Admiral and former head of the Vienna Meteorological Society, was pivotal. From the 1860s, he had promoted the transformation of the monarchy's Naval Academy in Rijeka into a research institution specializing in geophysics and oceanography and the establishment of a network of coastal (and later high alpine) observatories (Salcher 1902; Matis 2018). These intra-imperial cooperative endeavours (Mattes 2024), in turn, provided a springboard for Weyprecht's vision of international polar research, with the

FIGURE 4.5 Austro-Hungarian Navy officers participating in the expedition. 1: Emil von Wohlgemuth; 2: Richard Basso; 3: Adolf von Bóbrik; 4: August Gratzel; 5: Ferdinand Fischer (physician); 6: Adolf Sobieczky. In Chavanne (1884, 24). Digitisation by the Woldan Collection of the Austrian Academy of Sciences under the Creative Commons PDM 1.0 (Public Domain) licence

erected observatories within the monarchy serving as a model for the novel one in the Arctic.

While the station's routines reflected maritime habits, with officers on watch duty following a rotational schedule for monitoring the instruments, the work transcended mere scientific endeavours, taking on the essence of a way of life. As Pálffy (1882c, 5) noted, non-expedition members were welcome to participate in measurement tasks, yet the station's crew formed a family-like community, bound by loyalty, Weyprecht's legacy, and the fatherly presence of "Uncle" Wilczek. This group operated as a "thought collective" (Fleck 1935), prioritizing quantitative data that were gathered collaboratively over individual sensory perceptions. Unlike Pálffy's diary, the station's voluminous logbooks, containing over half a million data entries, excluded personal sensations. Within this framework, the instruments' "mechanical objectivity" emerged as the primary means of collecting information about the "North" (Daston & Galison 2018). This quest for objectivity, however, posed a dilemma: on the one hand, despite measures such as thermal insulation and instrument calibration with thermometers, the cold and ice affected the devices, preventing them from achieving the desired precision in practice. On the other hand, the officers still relied on sensory perceptions to qualitatively assess phenomena, such as the aurora borealis or brightness of polar nights. These assessments yielded varying results, influenced by the observer's physical condition or the artificial light used for documentation (Nielssen & Petterson 1993; Spring 2016). Accordingly, Wohlgemuth (1886, 14) lamented the "overstraining of the eyes [at polar stations] [...] always ready to follow and record the deflections of the air electroscope, the earth current instruments, and the variations of the aurora at the same time". To overcome this epistemic uncertainty, he assigned multiple officers to each watch and introduced eye exercises between observation circles. However, this strategy pushed the crew to their limits, leading Wohlgemuth (1886, 15) to acknowledge the "realization of one's own powerlessness" in recording sensory perceptions in the far north.

Contemporary demands on the objectivity of photography appeared to address these challenges effectively. As early as the 1850s, polar explorers used this new medium for documentation; however, owing to its realism and (still) limited quality, photography could not compete with the often-exaggerated artistic depictions (Wamsley & Barr 2009). Even in the American, British, and German IPY expeditions researched by Fjellestad (2022, 63), photography primarily served "illustrative and promotional purposes." Conversely, in the Austro-Hungarian case, scientific-military considerations about the potential of this technology for surveying and astronomy led to its testing on the Arctic island (Justnik 2007). Wilhelm Burger (1882), known for his contributions to expedition photography, equipped the station with a camera, drawing on his expertise from previous ventures, including Wilczek's *Isbjørn* Expedition and Pálffy's African journey.[7] Subsequently, Burger also introduced Pálffy, who probably assisted in capturing some of the Jan Mayen photographs, to the Vienna Photographical Society. The 45 preserved landscape photographs from the venture aimed to help the Vienna Military Geographical Institute map the island by capturing the heights and positions of its topographic features (Figure 4.6). Despite their scientific value, the photographs rarely feature people, making them unsuitable for popularization (Anonymous, 1884).

While Weyprecht (1875, 40) had strictly rejected "detailed geographical research", limiting it to a supporting role for scientific purposes, photography and scientific-military interests brought exploration back on the agenda. The honourable mention of Pálffy's attempt to climb the Beerenberg volcano in the expedition's public accounts reflects this shift. Organized at short notice with the *Pola*'s crew days before departure, his party, despite inadequate footwear, managed to climb over 1,500 meters above sea level before being forced to turn

7 Based on the photographs taken on Jan Mayen, the Austro-Hungarian Navy commissioned Burger in 1884 to integrate photography into naval training for remote sensing and espionage (Rosenberg 1984, 40).

FIGURE 4.6 "Southern part of Jan Mayen seen from Danielsen Mountain." Author: Richard Basso. In Anonymous (1884, 30). Digitisation by the Woldan Collection of the Austrian Academy of Sciences under the Creative Commons PDM 1.0 (Public Domain) licence

back halfway. Although Pálffy's lengthy diary report was published in the official expedition volume, it contained limited scientific information, focusing instead on the grandeur of the Arctic landscape:

> At our feet, the glittering glacier sank into a sea of mist that seemed to cover the world for an infinite distance. While the lightly moving clouds, shimmering in the softest white, swirled around the ice-covered flanks of the Beerenberg, invisible forces had compacted it above the sea into a mass whose sharply defined surface [...] undulated rhythmically up and down in immense, perfectly regular waves. (Pálffy in Wohlgemuth 1886, 42)

Undoubtedly, Pálffy's concept of "exploratory journeys", as portrayed in his report, diverged from the geographical exploration envisioned by the expedition crew. Inspired by the mountaineering-like endeavours during the Austro-Hungarian North Pole Expedition on Franz Joseph's Land, resulting in rather vague sketches of discovered territories, Pálffy's approach contrasted with the painstaking military surveying techniques employed by Adolf Bóbrik von Boldva (1886), the officer tasked with mapping Jan Mayen. Following the *Pola*'s departure, Bóbrik's week-long campaigns on foot and by boat covered the entire island, establishing a dense triangulation network. This effort culminated in the creation of a large-scale map, probably one of the earliest professionally surveyed maps from the High Arctic (Walsh 2017). Leveraging this map, Wilczek introduced 23 place names into the island's topography, borrowing them from the Habsburg imperial family for peaks and from aristocrats and scientists for craters and mountains. Instead of altering existing toponyms, the mapmakers meticulously depicted the volcanic terrain to accommodate the new names. Having surveyed the island just five years earlier, the Christiania meteorologist Henrik Mohn (1892, 59) praised the Austrian map in the Norwegian Geographical Yearbook as a testament to the expedition's outstanding research. With diplomatic foresight, Wilczek had immortalized not only his nephew Pálffy but also this Norwegian scientist in Jan Mayen's topography.

Overall, the expedition's scientific agenda and Pálffy's exploratory aims were rooted in different assumptions about the reliability of collecting knowledge about the "North". While the former aimed to minimize errors by excluding human impact, promoting cooperative data collection or introducing new technologies, Pálffy and other participants relied on individual sensory perceptions during their "explorations" while valuing "objective" data. Pálffy's diary served as a hybrid medium for gathering multiple types of knowledge, independent of ownership, and integrating them into prevailing narratives about the "North". Furthermore, the Jan Mayen Expedition highlights the limitations of the late nineteenth-century demand for mechanical objectivity. Neither Weyprecht's successors nor the IPY instructions were able to fully implement his novel vision of *Polarforschung*, taking into account the personal and environmental factors that potentially affected observers and instruments. Consequently, the High Arctic remained a challenging space for scientific and tourist visitors, where assumptions, methods and devices were put to the test and could fail.

4 Conclusion

This chapter employs a diary, maintained during an IPY expedition, to explore how polar travellers appropriated the "Arctic" both individually and as part of a large-scale scientific enterprise. To this end, I have expanded the conventional notion of collecting from the gathering of individual objects to the accumulation of various forms of information, as applied in the data-driven, observing science of the IPY. This knowledge, coupled with the practices of collecting and appropriation, materializes in Pálffy's diary, offering novel insights into the cultural conceptions of encountering the "North" and the aspirations and realities of international polar research.

In exploring these social dimensions, I have shown that individual encounters with Arctic environments were based on shared experiences and imaginaries. Additionally, hierarchies, routines, and cultural values framed how (scientific) travellers like Pálffy documented these environments, whose fascination transcended the idea of class and nation. Although the Austro-Hungarian North Pole Expedition set the template for the public reception of the Jan Mayen venture, their agendas and international contexts differed significantly. The latter's yield – quantitative data, a rather modest flora and fauna collection, and monotonous landscape photographs – could not live up to the popularity of the former expedition.

On the geographical front, I have demonstrated that the travellers discussed here, whether exploratory or scientific, implemented colonial practices in their activities. Unlike the expeditions of other European powers, the Habsburg monarchy's polar ventures refrained from directly claiming Arctic territories or populations. Instead, they employed (scientific) practices such as creating (data) collections, naming topographic features, and exploiting natural resources to assert hegemonic claims over territories. These efforts were propelled not by the state but by an intimate "thought collective" surrounding Wilczek, informed by German networks and aligning with the monarchy's imperial goals in pursuit of personal gain.

Delving into the epistemic dimensions, I have highlighted the challenges of acquiring "objective" knowledge about the "North". Geoscientific (polar) research without heroic explorers or territories in the far north, as envisioned by Weyprecht, addressed these epistemological concerns. Furthermore, it allowed the Habsburg monarchy, established in fields like geophysics and meteorology, to gain status among European superpowers. Apart from official instructions, IPY research still included "non-professional" and exploratory ambitions, eagerly picked up by the press. As Wilczek (1882, 3) humorously remarked about the Jan Mayen Expedition, the "political significance of the polar coldness" lay in easing "friction between nationalities" through a shared scientific pursuit. This approach was intended to promote cohesion within the monarchy and set a precedent for international cooperation.

However, despite joint planning efforts, the First IPY fell short of its lofty goals. Each expedition ultimately operated independently and published their results in isolated volumes, without comparative analysis. In the case of the Jan Mayen venture, the enduring scientific impact came not from the extensive collection of data but from the Habsburg map of the island. This map continued to be used by Norwegian authorities well into the second half of the twentieth century, following Jan Mayen's placement under Norwegian sovereignty in 1929. During the Second IPY (1932–1933), the remote island once again hosted an Austrian expedition, albeit with fewer resources, aiming to continue the Habsburg legacy in polar research.

Acknowledgements

This research was funded by the Marie Skłodowska-Curie actions (MSCA) grant agreement No 101066820 (Poles and People) from the European Commission. I would like to thank Robert Marc Friedman, Gerhard Holzer, Sandra Klos, Tomáš Kowalski, Petra Svatek, and the colleagues of the Woldan Collection of the Austrian Academy of Sciences for commenting on this chapter and/or providing sources. I am also grateful for the constructive and enriching suggestions of the editors of this volume. Last but not least, I especially thank Martin Pálffy for making his grandfather's diary available to me for research purposes and for his unwavering support throughout the writing process.

Bibliography

Anonymous. 1884. *Photographische Aufnahmen von der Insel Jan Mayen.* Woldan Collection, Austrian Academy of Sciences. Accessed 3 March 2024. https://viewer.acdh.oeaw.ac.at/viewer/image/AC03292736/1/.

Barr, Susan & Cornelia Lüdecke (eds.). 2010. *The History of the International Polar Years.* Berlin, Heidelberg: Springer.

Barr, Susan, Louwrens Hacquebord, Erki Tammiksaar & Natal'ya Georgievna Sukhova. 2010. "The Expeditions of the First International Polar Year." In *The History of the International Polar Years*, edited by Susan Barr & Cornelia Lüdecke, 35–107. Berlin, Heidelberg: Springer.

Barr, William. 1985. *The Expeditions of the First International Polar Year.* Calgary: Arctic Institute of North America.

Berger, Frank, Bruno Besser & Reinhard Krause (eds.). 2008. *Carl Weyprecht. Seeheld, Polarforscher, Geophysiker.* Vienna: ÖAW.

Bóbrik von Boldva, Adolf. 1886. "Aufnahme und Beschreibung der Insel Jan Mayen." In *Die internationale Polarforschung. Die österreichische Polarstation,* edited by k. Akademie der Wissenschaften. Vol. 2, 1–50. Vienna: Hof- und Staatsdruckerei.

Bravo, Michael & Sverker Sörlin (eds.). 2002. *Narrating the Arctic: A Cultural History of Nordic Scientific Practices.* Canton, MA: Science History Publications.

Burger, Wilhelm. 1882. "Die Photographie in heissen Ländern auf Reisen zu Pferd, Maulthier oder Kameel." *Photographische Correspondenz* (Vienna) 19 (226): 104–110.

Chavanne, Joseph (ed.). 1884. *Jan Mayen und die österreichische arktische Beobachtungsstation.* Vienna: Hartleben.

Daston, Lorraine & Peter Galison. 2018. *Objektivität.* Frankfurt a.M.: Suhrkamp.

Doel, Ronald E., Robert M. Friedman, Julia Lajus, Sverker Sörlin & Urban Wråkberg. 2014. "Strategic Arctic Science: National Interests in Building Natural Knowledge." *Journal of Historical Geography* 44: 60–80.

Elzinga, Aant. 2009. “Through the Lens of the Polar Years: Changing Characteristics of Polar Research in Historical Perspective.” *Polar Record* 45 (4): 313–36.

Fjellestad, Marthe T. 2022. “Photography in the Arctic Archipelago during the First International Polar Year.” *Scientia Canadensis* 44 (1): 50–67.

Fleck, Ludwik. 1935. *Entstehung und Entwicklung einer wissenschaftlichen Tatsache.* Basel: Schwabe.

Fleming, James R. 2016. *Inventing Atmospheric Science: Bjerknes, Rossby, Wexler, and the Foundations of Modern Meteorology.* Cambridge, MA: MIT Press.

Friedman, Robert M. 1989. *Appropriating the Weather: Vilhelm Bjerknes and the Construction of a Modern Meteorology.* Ithaca: Cornell UP.

Friedman, Robert M. 1994. *The Expeditions of Harald Ulrik Sverdrup: Contexts for Shaping an Ocean Science.* La Jolla, CA: Scripps Institution of Oceanography.

Hudec, Jozef. 2023. “Gróf Jozef Pálffy v severovýchodnej Afrike.” In *Pálffyovci v Smoleniciach*, edited by Daniel Hupko & Martina Orosová, 25–54. Bratislava: Pamiatkový úrad SR.

Hupko, Daniel & Martina Orosová (eds.). 2023. *Pálffyovci v Smoleniciach.* Bratislava: Pamiatkový úrad SR.

Justnik, Herbert. 2007. *Wissenschaftliche Landschaftsfotografie. Die Fotografien der österreichischen Polar-Expedition auf der Insel Jan Mayen.* MA thesis, University of Vienna.

K. Akademie der Wissenschaften in Wien (ed.). 1886. *Die internationale Polarforschung. Die österreichische Polarstation Jan Mayen.* 4 Vol. Vienna: Hof- und Staatsdruckerei.

K. Akademie der Wissenschaften zu St. Petersburg (ed.). 1881. *Bericht über die Verhandlungen und Ergebnisse der 3. Internationalen Polar-Konferenz.* St. Petersburg: Academy printing.

Kinsky-Wilczek, Elisabeth (ed.). 1933. *Hans Wilczek erzählt seinen Enkeln Erinnerungen aus seinem Leben.* Graz: Leykam.

Klemun, Marianne & Johannes Mattes. 2022. “Expeditionen und Forschungsreisen.” In *Die Österreichische Akademie der Wissenschaften 1847–2022*, edited by Johannes Feichtinger & Brigitte Mazohl, Vol. 1, 197–273. Vienna: ÖAW.

Klutschak, Heinrich. 1881. *Als Eskimo unter den Eskimos.* Vienna: Hartleben.

Krause, Reinhard. 1992. “Die Gründungsphase deutscher Polarforschung.” *Berichte zur Polar- und Meeresforschung* 114: 1–450.

Launius, Roger, James Fleming & David DeVorkin, eds. 2010. *Globalizing Polar Science.* New York: Palgrave Macmillan.

Lüdecke, Cornelia. 2004. “The First International Polar Year: A Big Science Experiment with Small Science Equipment.” *History of Meteorology* 1: 55–64.

Matis, Herbert. 2018. “Dual Use Research. Kooperationen der k. k. Kriegsmarine und der Kaiserlichen Akademie der Wissenschaften.” In *Wandlungen und Brüche*, edited by Johannes Feichtinger, Marianne Klemun, Jan Surman & Petra Svatek, 145–154. Vienna: V&R.

Mattes, Johannes. 2023. "Researcher or Tourist? The Hungarian Count Jozef Pálffy on the Shore of Jan Mayen." *Fortid* 20 (2): 58–61.

Mattes, Johannes. 2024. "Collaborative Research in Imperial Vienna: Science Organization, Statehood, and Civil Society." *Austrian History Yearbook* (Cambridge) 55: 174–201.

Mazzoli, Enrico. 2016. *Carl Weyprecht and the International Polar Years*. Trieste: Luglio.

Mohn, Henrik. 1878. "Die Reise der Norwegischen Nordmeer-Expedition nach Jan Mayen." *Petermanns Geographische Mittheilungen* 24: 228–35.

Mohn, Henrik. 1892. "Øen Jan Mayen." *Norsk Geografisk Aarbog* 3 (3): 57–71.

Nielssen, Alf Ragnar & Arvid Petterson. 1993. *Nordlyspionerene: Menneskene og observatoriet på Halddetoppen i Alta*. Oslo: Grøndahl.

Österreichische Akademie der Wissenschaften (ed.). 2023. *Land, Land, endlich Land! 150 Jahre Franz-Josef-Land*. Exhibition booklet. Vienna: ÖAW.

Pálffy, József v. 1882a. *Diary*, 97 p. Woldan Collection, Austrian Academy of Sciences. Accessed 3 March 2024. https://viewer.acdh.oeaw.ac.at/viewer/image/ACxxxxxxxx/1/.

Pálffy, József v. [Anonymous]. 1882b. "Von der Nordpol-Expedition." *Pester Lloyd* 142: 5.

Pálffy, József v. [Anonymous]. 1882c. "Die österreichische Nordpol-Expedition." *Pester Lloyd* 176: 5.

Pálffy, József v. [Anonymous]. 1882d. "Auf der Heimkehr von Jan Mayen." *Pester Lloyd* 247: 9.

Pálffy, Martin. 2025. *Geschichte der Familie Pálffy: 1000 Jahre im Dienst für Ungarn und Österreich*. Vienna: self-published.

Punch, Samantha. 2012. "Hidden Struggles of Fieldwork: Exploring the Role and Use of Field Diaries." *Emotion, Space and Society* 5 (2): 86–93.

Rack, Ursula. 2010. *Sozialhistorische Studie zur Polarforschung anhand von deutschen und österreich-ungarischen Polarexpeditionen*. Bremerhaven: Alfred Wegener Institut.

Reiszig, Ede. 1898. "A Pálffyak." In *Magyarország vármegyei és városai*, edited by Samu Borovszky, Vol. 3, 701–38. Budapest: Apollo.

Rosenberg, Gert. 1984. *Wilhelm Burger. Ein Welt- und Forschungsreisender*. Vienna: Brandstätter.

Ryall, Anka, Johan Schimanski & Henning Howlid Wærp. 2010. "Arctic Discourses. An Introduction." In *Arctic Discourses*, edited by idem, IX–XXII. Newcastle: Cambridge Scholars.

Salcher, Peter. 1902. *Geschichte der k.u.k. Marine-Akademie*. Vienna: Gerold.

Schimanski, Johan & Ulrike Spring. 2015. *Passagiere des Eises. Polarhelden und arktische Diskurse 1874*. Vienna, Cologne, Weimar: Böhlau.

Shadian, Jessica M. & Monica Tennberg (eds.). 2009. *Legacies & Change in Polar Sciences: Historical, Legal & Political Reflections on the International Polar Year*. Burlington, VT: Ashgate.

Sondhaus, Lawrence. 1993. "The Austro-Hungarian Naval Officer Corps." *Austrian History Yearbook* 24: 51–78.

Spring, Ulrike. 2016. "Materializing the Aurora Borealis: Carl Weyprecht and Scientific Documentation of the Arctic." In *Expeditions as Experiments*, edited by Marianne Klemun & Ulrike Spring, 141–162. London: Palgrave Macmillan.

Tammiksaar, Erki, Natal'ya Georgievna Sukhova & Cornelia Lüdecke. 2010. "The International Polar Year 1882–1883." In *The History of the International Polar Years*, edited by Susan Barr & Cornelia Lüdecke, 7–33. Berlin, Heidelberg: Springer.

Vogt, Carl. 1863. *Nord-Fahrt entlang der Norwegischen Küste nach dem Nordkap.* Frankfurt a.M.: Jügel.

Walsh, Stephan. 2014. *Between the Arctic & the Adriatic: Polar Exploration, Science & Empire in the Habsburg Monarchy*. PhD diss., Harvard University.

Walsh, Stephan. 2015. "On Slippery Ice. Discovery, Imperium, and the Austro-Hungarian North Polar Expedition." In *Expedition into Empire*, edited by Martin Thomas, 148–170. New York: Routledge.

Walsh, Stephan. 2016. "Void into Meaning: Geophysics and Imperial Cartography in the High Arctic." In *SpaceTime of the Imperial*, edited by Holt Meyer, Susanne Rau & Katharina Waldner: 394–412. Berlin, Boston: De Gruyter.

Wamsley, Douglas & William Barr. 1996. "Early Photographers of the Arctic." *Polar Record* 32 (183): 295–316.

Weyprecht, Carl. 1875. "Über die österr. Nordpol-Expedition." In *Tageblatt der 48. Versammlung deutscher Naturforscher und Aerzte*, edited by Johannes Frischauf, Vitus Graber & Rudolf Klemensiewicz: 38–42. Graz: Leuschner & Lubensky.

Weyprecht, Carl. 1876. *Die Nordpol-Expeditionen der Zukunft und deren sicheres Ergebniß*. Vienna: Hartleben.

Wilczek, [Hans]. 1883. *Vortrag über die österreichische Polarstation in Jan Mayen*. Separate reprint of a speech at the meeting of the Vienna Geographical Society on 23 January 1883. Woldan Collection, Austrian Academy of Sciences. Accessed 3 March 2024. https://viewer.acdh.oeaw.ac.at/viewer/image/ACxxxxxxxx/1/.

Wilczek, Hans. 1882. "Frostige Briefe des Grafen Wilczek." *Der Floh* (Vienna) 15 (25): 3.

Wild, Heinrich. 1882. "Geschichte des Weyprecht'schen Unternehmens internationaler wissenschaftlicher Polarforschung." *Mittheilungen der Internationalen Polar-Commission* 1: 1–6.

Wohlgemuth, Emil v. 1882. *Die österreichische arktische Beobachtungs-Station auf Jan Mayen*. Vienna: Gerold.

Wohlgemuth, Emil v. 1886. "Vorbericht zur wissenschaftlichen Publication der österreichischen Polarexpedition." In *Die internationale Polarforschung*, edited by Kaiserliche Akademie der Wissenschaften, Vol. 1, 1–118. Vienna: Hof- und Staatsdruckerei.

Wood, Kevin & James Overland. 2006. "Climate Lessons from the First International Polar Year." *Bulletin of the American Meteorological Society* 87: 1685–97.

OBJECT: Book illustrations

Fig. 1.—Earlier runic stone at Tjängvide, Götland, with the eight-footed horse of Odin. —Height about 5 feet; width, 4 feet 4 inches; thickness, 1 foot. Another similar stone with representation (in relief) of an eight-footed horse has been found also in Laivide in Götland.

FIGURE 5.1 Print of runic stone from Paul du Chaillu, 1889, *The Viking Age* Vol. 1, 58

CHAPTER 5

Illustration as Collection: Paul du Chaillu and the Objects of *The Viking Age*

Heidi Hansson

Abstract

Including 1,366 illustrations, Paul du Chaillu's two-volume history *The Viking Age* from 1889 is one of the most richly illustrated works about the early life of the Scandinavian people. The depicted objects are included in chronologically organised chapters and classified according to type, finding-place and function. The current location or publication history of the items is normally omitted, however, which means that their earlier collection history is obscured and the role of the illustrated objects as items in, for example a museum display, disregarded. Recycled as illustrations of du Chaillu's argument that the British were descended from Scandinavian Vikings, the original collections are disordered and the relationship between individual pieces disturbed or even lost. The sheer mass of objects, however, means that they become more than illustrations of the text and can be interpreted as a collection on their own. As a result, du Chaillu's collection of images occasionally embodies a different narrative than the argument he intends to propound.

Keywords

Paul du Chaillu – *The Viking Age* – illustration – collection – historical objects – archaelogical finds – Saga literature – genealogical claims

∴

The first in-text illustration in Paul du Chaillu's two-volume history of the so-called Viking Age is a woodcut of a pre-Christian runic stone discovered on the Gotland farm Tjängvide in 1844 (vol. 1, 58). One of the pictures on the stone shows a man on an eight-legged horse, sometimes identified as the Asa god Odin on his horse Sleipner, and the illustration appears as supporting evidence

 | DOI:10.1163/9789004749665_006

in a chapter about the god. Later interpreters have disputed this identification since the man on the horse is quite puny and far from god-like (Jansson & Wessén 1962, 190–98), but it was commonly accepted in the late nineteenth century, and the idea still circulates, for example in articles based on older encyclopaedias ('Tjängvidestenen'). As a significant find, the runic stone was moved to the Swedish History Museum in 1869, and several sketches of its carvings were published from 1844 onwards. The caption to the illustration in *The Viking Age* however contains no information about the whereabouts of the Tjängvide stone or the scholarship that grew around it. Within the work, the depicted stone appears as an item in a collection, curated and made meaningful by du Chaillu.

As the material memories of the past, collections of historical objects and artefacts are most often associated with museum displays. Sometimes the objects are meticulously selected and organised according to scientific principles, and the exhibition illustrates the state of science at the time of its production. Alongside these more or less official expositions are the assemblages of curiosities built up by enthusiasts which rather illustrate the tastes and interests of private individuals. Whether publicly sanctioned or idiosyncratic, the collections are subject to secondary (or tertiary and more) collecting practices through the selection of objects that seem particularly representative or evocative to exhibit or represent on posters, in museum catalogues, in schoolbooks and scholarly treatises, and today, more and more frequently as digital displays on-line. In this recycling process, the original collection is disordered and the relationship between individual pieces disturbed or even lost. This is especially the case when already-collected items are arranged to illustrate a particular thesis or idea.

While marketed as a history book, Paul du Chaillu's monumental *The Viking Age* from 1889 is dominated by images and extracts from chronicles and sagas to such an extent that it could be regarded as a portable collection. With its more than 1,300 illustrations of archaeological sites and museum objects, the work illuminates the collection process firstly, by the original assembly of objects put on display at the museums and reproduced in the scholarly treatises consulted, and secondly, by du Chaillu's selection of objects to illustrate and enhance his narrative. The bulk of the two-volume work consists of quoted and reproduced source material that is organised to support a specific version of the north, and by extension, the parts of Europe influenced by northern traders and raiders. Together with this ideological bias, the selection of images reveals an aesthetic bias that was probably already present in the exhibitions and literary sources from which du Chaillu collected his material, as well as the preservation bias caused by the natural decay of perishables and an original disinclination to save everyday objects for posterity.

Partly to explain and justify English imperialist ambitions, partly in response to late Victorian gender anxieties, du Chaillu's overall aim is to protect the English-speaking nations from the suspicion that they should be descended from "the original and effete inhabitants" of the British Isles rather than from the Viking invaders (du Chaillu 1889, vol. 1, 3). Thus, he conceives of the Vikings as a genetically coherent group rather than an occupation, as is usually the case today (Ellis 2021, 2). The ethnic view enables him to trace Viking origins from the Stone Age onwards, regardless of the general scholarly agreement of his time that the Viking Age lasted from the end of the seventh to the middle of the eleventh century, or that the Vikings "sprang upon Europe" in "the eighth and three following centuries" (*Edinburgh Review* 1891, 334).

The use of the Tjängvide stone as an unfiltered depiction of Odin and his horse provides a window on du Chaillu's practices. Since his overall purpose is to demonstrate the manliness of Viking culture (du Chaillu 1893, vol. 1, 4), weapons and ornaments are overrepresented in the illustrations whereas farming and household equipment is largely absent. Roman coins validate seafaring skills and superior tradesmanship, and jewellery from various periods is reproduced as proof of refinement and advanced civilisation, but no artefacts illuminate the Celtic culture in the British Isles. The central images on the Tjängvide stone are however surrounded by a knotwork border very similar to Irish decorative art, and the stone has indeed been used as evidence of a Western influence on Nordic runological art, in direct contradiction to du Chaillu's objective (Bugge 1905, 320–25). In addition, the man on the horse seems to be received by a woman holding a drinking vessel, a sign of the public presence and importance of Viking-Age women that goes unremarked in the text.

As the first of 1,366 images (du Chaillu 1889, vol. 1, n.p.), in a book that gives little space to domestic life and consistently sidelines the Celtic heritage in Britain, the Tjängvide picture stone becomes slightly ironic, but more importantly, it shows how text and illustration may support different narratives. Collected and recirculated as illustrations of du Chaillu's account of the Viking Age, that is, subservient to the text, the depicted objects combine into a history of English imperial supremacy based on Norse roots, but individually and separated from the text, they occasionally embody alternative stories. In the light of nineteenth-century gender debates, even the woodcut or xylographic technique becomes meaningful since it provided an opportunity for women to become independent (Johannesson 2018; 2).[1]

1 I am grateful to Dr Ann-Catrine Eriksson, Umeå University, for identifying the illustration technique in Paul du Chaillu, *The Viking Age*, and for information about the xylographic industry in the late nineteenth century and its importance for women's work opportunities.

Objects – or their images – travel, and as Sheenagh Pietrobruno notes about museum objects on social media, "their narratives shift" as they circulate (2021, 1393). Since the use of search engines and social media is based on personal preference, this shift can be seen as a kind of personalisation of the object, or at least its meaning. Pietrobruno's example is the Viking helmet and its occasional association with right-wing nationalist groups, and she cautions that the "circulation of museum objects on personalized SERPS [search engine result pages] may erode or erase meanings integrated within the curatorial aim of museums to nurture critical awareness and endorse social justice through the eradication of rigid identity politics" (Pietrobruno 2021, 1414). The pace of these narrative shifts may be faster and the representations more individualised today, but the phenomenon is by no means new. It is also fundamentally unstable since it emphasises the ability of ostensibly inanimate objects to resist the meanings they are ascribed. The "personalised meaning" of the items illustrating du Chaillu's history of the Viking period may be to prove the Norse roots of British imperial triumphs, but the objects remain open to alternative interpretations that sometimes disprove the author's thesis.

1 Paul Belloni du Chaillu

Paul Belloni du Chaillu was a French-American travel writer and explorer, probably born around 1831 (Edmonds 1998, 110). His father had trading interests in Africa, and after arriving in the Gabon on the Atlantic coast of West Africa in 1848, du Chaillu made several trips inland along the equator before leaving for New York in 1852. In the United States, he wrote articles and lectured about his experiences, and in 1855, he returned to west Africa on a collecting expedition partly sponsored by members of the Academy of Natural Sciences of Philadelphia. The Academy as a whole did not admit responsibility for the trip, however, and would not reimburse his expenses. On his return, du Chaillu therefore sold off his specimens of flora and fauna to other scientists and scholarly associations in Europe and the US (McCook 1996, 182), reserving some of the more spectacular ones, such as stuffed gorillas and gorilla skulls, to use as props on his lecture tours. He also began work on his first book, *Explorations and Adventures in Equatorial Africa* which was published in 1861 on both sides of the Atlantic. This was followed a few years later by *A Journey to Ashango-Land* (1867) and in 1881, the two-volume *The Land of the Midnight Sun* that launched him as an authority on Scandinavia.

du Chaillu's books can be described as ethnographic travel adventures and contain a substantial number of pictures of people, customs, scenery and animals, ranging from seventy-odd illustrations in the first book, to some twenty pictures in the second and 235 images in the last one. Ostensibly, the illustrations were based on the author's own observations, but in reality they were mostly produced by illustrators at the publishing houses or were recycled portraits and drawings from other sources. Despite advances in photography, woodcuts were often preferred in the late nineteenth century, since they were cheaper and worked well with common printing techniques. The need for an original image, like a recent or traceable photograph was also eliminated or at least minimised, which further enabled picture recycling although not always without risk. The disclosure that the frontispiece of a gorilla in du Chaillu first book was plagiarised from a scholarly article published three years earlier damaged his reputation as a field collector and he was heavily criticised for faking part of his books (McCook 1996, 187–88). Even so, he remained a sought-after travel writer and lecturer, using the collected items and illustrations to reinforce the exotic nature of his experiences and substantiate his arguments. In spite of their questionable provenance, the physical specimens enhanced his credibility by creating the illusion of first-hand knowledge. The numerous depictions of Norse objects in *The Viking Age* have a similar effect.

After completing *The Land of the Midnight Sun,* Paul du Chaillu embarked on the Viking Age project. In his Preface, he establishes a relationship between his British and American contemporaries and the ancient Viking warriors and travellers. His stated purpose is to answer some questions about England's successful colonial projects by proving the northern foundation of English character, liberties, imperial ambitions and seafaring skills:

> Why is it that, wherever the English-speaking people have settled, or are at this day found, even in small numbers, they are far more energetic, daring, adventurous, and prosperous, and understand the art of self-government and of ruling alien peoples far better than other colonising nations? Whence do the English-speaking communities derive the remarkable energy they possess; for the people of Britain when invaded by the Romans did not show any such quality? What are the causes which have made the English such a pre-eminently seafaring people? for without such a characteristic they could not have been the founders of so many states and colonies speaking the English tongue! (du Chaillu 1889, vol. 1, VII)

The answers seem to have preceded or at least have been developed in tandem with du Chaillu's thesis that England owed its triumphs to the Viking genealogy of its population:

> To the numerous warlike and ocean-loving tribes of the North, the ancestors of the English-speaking people, we must look for the transformation that took place in Britain. In their descendants we recognise to this day many of the very same traits of character which these old Northmen possessed, as will be seen on the perusal of this work. (du Chaillu 1889, vol. 1, VIII)

The emphasis on Viking prowess in battle continues throughout the work and becomes visually prominent through the weapons of hunt and war selected to illustrate Norse pre-history. Scandinavia is established as the home-place of the ethnic Vikings, which allows du Chaillu to utilise his reputation as a travel writer. As in his earlier works, he suggests first-hand knowledge or a kind of insight through osmosis, thanking his companion George C. Taylor for their "wanderings in the Land of the Midnight Sun, in the home of the old Vikings, while I was engaged on the present work" (du Chaillu 1889, vol. 1, n. p., original text italicised). His methods are rather more rigorous, though, and he states in the preface that his account is based on the dual processes of collecting and selecting evidence from archaeological finds and Icelandic sagas: "After thus collecting my material, which was of the most superabundant character, I went over it and selected what seemed to me to be the best accounts of the various customs with which I deal in these volumes" (du Chaillu 1889, vol. 1, X). Even though his text is agenda-driven, he is at pains to show that his claims are founded on material evidence. That this entails a series of previous collecting and selecting procedures is disregarded and possibly unrecognised by author and readers alike.

The preparation of the manuscript took eight and a half years and according to du Chaillu himself, he spent up to sixteen hours daily doing research for the work, as reported in the *St. James's Gazette* review: "The book shows many traces of the zeal with which Mr. Du Chaillu has gone to work, and for the many years of labour at fourteen or sixteen hours a day (so he tells us) which have enabled him to collect the mass of material which figures in these pages" (1889, 6). His exertions find favour with the reviewer of the *Nonconformist and Independent* who regards the work as "remarkable alike for its thoroughness of investigation, for its admirable method, for its weight of information, and for fine sympathy and tact in dealing with the early Norse literature" (1889, 1057). Others were sceptical of his main thesis, like the London *Daily News* critic: "M. du Chaillu has brought out his promised – as, some might prefer to put it, his threatened – book which is to prove that we in England are not what we seem" (1889, 2). Even critical commentators however acknowledge the value of the illustrations:

> Fortunately, however, M. du Chaillu's book does not depend for its interest upon the truth of an hypothesis. It is essentially a pictorial presentation of the archaeology of the North. The text illustrates the figures, rather than the figures the text. These are good in themselves; they are well chosen, admirably executed, and strew the pages in lavish abundance. (*Edinburgh Review* 1891, 334)

The fact that pictures and saga extracts constitute the greater part of the work makes the reception of *The Viking Age* as itself a store of source materials quite reasonable. Most of the 1,366 illustrations are representations of museum displays, or, as stated in the subtitle, "the antiquities discovered in mounds, cairns and bogs as well as from the ancient sagas and Eddas" (du Chaillu 1889, vol. 1, n. p.). The number of pictures may not be huge in comparison with the available number of finds, but it is certainly huge in relation to the literary medium. The majority of the large production cost of £11,000 can probably be traced to the reproduction of the images (*The East and South Devon Advertiser* 1896, n. p.), while the high printing costs, conversely, add value to the illustrations.

Classified according to type, finding-place, function and chronology, the illustrations order the narrative at the same time as the narrative transforms them from random items into a knowable, scientific collection. Much of du Chaillu's research was undertaken in the Museum of Northern Antiquities in Copenhagen (du Chaillu 1889, vol. 1, XI), and it is hardly a coincidence that the first chapters of *The Viking Age* follow the Danish archaeologist and museologist Christian Jürgensen Thomsen's (1788–1865) organisation of pre-history into three ages based on the respective prevalence of stone, bronze and iron articles (Miller 2017, 98). As the example of the image of the Tjängvide stone shows (du Chaillu 1889, vol. 1, 58), the present location or publication history of the illustrated objects is normally omitted, however, which means that du Chaillu's indebtedness to other scholars is obscured, at least insofar as the interpretation of individual objects is concerned. Such exclusions enable him to ascribe alternative meanings to the finds and integrate them in a British context, regardless of their place of origin or discovery.

2 Historical Evidence

In their focus on the material traces of the past, historians, archaeologists, philologists and ardent collectors can be traced back to the antiquarians of the early modern period. The difference between them primarily lies in how they make artefacts tell their story. In his 1950 re-evaluation of antiquarianism, Arnaldo Momigliano connects the historian to a chronological method and the antiquarian to a systematic method, where all the items connected to a specific subject are collected "whether they help to solve a problem or not" (1950, 286). To interact with objects of the past is an affective bodily experience, which is why antiquarianism, in its preoccupation with things, can be understood as "the study of the past-in-the-present" (Miller 2017, 7). An object belongs both to its time of origin and the time of encounter which, to use Mary Louise Pratt's terminology (1991, 34), gives them an "ability to act as contact zones or zones of contact in their own right" (Poulter 2014, 38). In *The Viking Age*, Paul du Chaillu comes across more as an enthusiastic collector than a historian, with the illustrated artefacts representing embodied contact zones. Any agency they might possess in their own right is however circumscribed by the roles they are made to play in the work.

Writing at the end of the nineteenth century, du Chaillu would have been in the middle of debates about what constituted historical evidence. The term antiquarian was still used alongside the newer job description of archaeologist, but the age of indiscriminate collection was mostly over. Archaeology as well as museum collections had traditionally prioritised ornaments and cult objects before quotidian items, but there was an ongoing shift from an earlier preoccupation with artistic objects to a recognition of the evidentiary value of non-aesthetic items (Miller 2017, 134). Localities also began to matter. Local history associations flourished as part of the nationalist and national romanticist movements in the nineteenth century and established local museums where collections were organised according to finding-place, to give a fairer picture of the life and culture of people who once occupied the area (Miller 2017, 142). Towards the end of the nineteenth century, the centrality of material evidence however began to give way to a document-driven approach that dominated historical research more or less until the advent of the new materialism at the turn of the millennium (Findlen 2020, 270). The importance of objects for dating was nevertheless "generally recognised by archæologists" (du Chaillu 1889, vol. 1, 69), and manifested in the three-age system of prehistory perfected by the Danish archaeologist Thomsen.

Traces of these debates can be found in the way du Chaillu organises his text and in the items he reproduces to illustrate his argument. On the whole, he seems governed by the antiquarian impulse to include as many examples as possible of his main object categories, and his work is primarily a collection of source materials organised into smaller collections based on time period, theme and object type. To transform the 1,366 illustrated objects into sense-making evidence, he uses the Icelandic sagas and the traditions that seem to emanate from them. He is aware of the uncertain truth claims of the sagas (1889, vol. 1, IX), but still treats them as satisfactory sources, disregarding the fact that they describe a period after the heyday of Viking culture and were written down when Christianity had influenced societal structures in Scandinavia for a couple of hundred years (Sigurdsson 2022, 99).

Like the museum objects, the sagas are subjected to a selection process, and the excerpts are particularly prominent in chapters about love, birth, death, burial, justice, religion and social organisation, that is, accounts of human interaction. The material is set off from the narrative text and is excessive enough to appear as a variety of illustrated object as much as written record. The fact that one of the appendices of volume 2 consists of six pages of saga facsimiles emphasises the sagas' double status as documents and artefacts (1889, vol. 2, 544–50), and in a sense, du Chaillu employs two types of antiquities as proof of each other. In a circular process the saga excerpts are used as documentary but occasionally flawed evidence that is then validated by archaeological finds: "as will be seen in the course of this work, the antiquities treasured in the museums of the North bear witness to the truthfulness of the records" (1889, vol. 1, 2). At the same time, archaeological finds are explained with the help of sagas and chronicles: "The few jewels found scattered here and there, either in France or Germany are thoroughly Northern, and show that in these places the people of the North made warfare, as corroborated by the testimony of the Eddas and Sagas" (1889, vol. 1, 159). The testimony of the sagas is in turn validated by the Bayeux tapestry (1889, vol. 2, 362), as another historical object. The circles of authorisation expose du Chaillu's inadequacies as a historiographer, but they underscore the importance he assigns to material sources.

The chapters based on the three-age system of pre-history are among the most richly illustrated, possibly because finds from these periods were overrepresented in museums at the time. While du Chaillu acknowledges overlaps between categories in his text, the material is still organised according to this imposed system of difference. Thus, the Bronze Age chapter is dominated by illustrations of gold and bronze artefacts, and of the 28 pages of the Iron Age chapter, 23 are almost entirely taken up by pictures of objects: weapons, ornaments and a large number of urns and vessels. The numerous illustrations create a familiarisation effect, making the Vikings and Viking culture commonplace simply by virtue of the large number of objects represented side by side. To put it differently: one Bronze Age knife or spear-head draws attention to the advances of a particular culture at a particular time, but page upon page of Bronze Age weapons reduce their exceptionality. This inadvertent emphasis on the quotidian is partly at odds with du Chaillu's stated purpose to provide a nobler and purer heritage for the English-speaking nations, but in line with present-day museological thought where national myths are rejected in favour of "human-centred narratives of everyday life and religion that are grounded in the current state of knowledge regarding gender roles, racial and ethnic identities and mobility during the Viking Age" (Axelsson et al. 2022, 123).

The abundance of clay and metal pots in the Bronze Age chapter seems incongruous in relation to du Chaillu's initial claim that the "vast number of weapons" from the period testify "to the warlike character of the people" (1889, vol. 1, 126), but can be understood as a clash between the desire to represent what is available, following the antiquarian impulse, and the desire to attribute historical meaning to the objects according to an ideological principle. The discrepancy resonates with what Åmund Norum Resløkken identifies as the two main registers of historical conceptualisation when creating a Viking heritage site in Vestfold, Norway: the Bakhtinian chronotope of the road, or adventure time, and the chronotope of the rural village, or folkloric time (2022, 69). Weapons, boats and images of coins reinforce du Chaillu's overall thesis of the Vikings as explorers, invaders and traders, in correspondence with the adventure-time chronotope, and these are the aspects he highlights in his text. At the same time, his depiction of early Norse culture as highly civilised and sophisticated requires the support of folkloric time items and practices, although the associated implication that people must to a great extent have been place-bound is left unstated. As Jón Sigurdsson notes, "[h]ouseholders and their families made up the largest group" in Viking society (2022, 90). Here, as on the few other occasions where household implements are represented, the material evidence sits uneasily with du Chaillu's main argument, which provides an opportunity for reinterpretation. In this respect, the antiquarian impulse offers more scope for alternative versions of the past than a problem-oriented historical approach.

In regard to place, Viking invasion is more important than Viking origin, since it is central to du Chaillu's heredity principle. The organisation of the material according to finding-place or archaeological excavation site allows him to locate similar objects in Scandinavia and England. Defending his argument that Britain was colonised from the north in a Letter to the Editor reproduced from *The Times*, he writes:

> Turning to England, we find copious remains from a very early date which are not British or Celtic nor Roman, which have been variously labelled as Saxon, Anglo-Saxon, Anglo-Roman, and so forth, but which correspond minutely with the remains which I have described in the north. Of every

> such object found in England I claim that I can produce the counterpart in Scandinavia [...]. I ask again, is it or is it not reasonable to infer that these remains found in England indicate that England was invaded from the north? (*Science* 1890, 74)

He returns to the topic in the introduction to his later "romantic history" *Ivar the Viking* (du Chaillu 1893, n. p.), which he describes as a "life-like picture of the period" (du Chaillu 1893, VIII). Unlike *The Viking Age*, and perhaps unexpectedly since the story seems to have been intended for young readers, *Ivar the Viking* is not illustrated, a circumstance that at least some reviewers find worth noting (*Globe* 1893, 3). The novel is to a great extent a retort to his detractors, and in the introduction, he again claims that a comparison of archaeological finds in the Nordic countries and in England proves that Britain was settled from the north (du Chaillu 1893, XVI):

> When I say that the antiquities found in England are the same and of the same type as those found in the Norselands, I call this a fact and not a theory; and when I say also that these are not found in the Saxon lands, I call this a fact and not a theory. When I say that the antiquities found in England are not found in the so-called earlier England of the historian, I call this a fact and not a theory. (du Chaillu 1893, XVIII)

As further proof of his genetic claims du Chaillu appends a letter from W. E. Gladstone in facsimile in which the Prime Minister appears to entertain similar views: "When I have been in Norway, or Denmark, or among Scandinavians, I have felt something like a cry of nature from within, asserting (credibly or otherwise) my nearness to them" (du Chaillu 1893, XX). Ultimately Gladstone is polite, but non-committal, however, referring to his lack of expertise and regretting that he can only offer "feeble generalities" (du Chaillu 1893, XX). Even so, the letter can be related to the evidentiary practices in *The Viking Age* and understood as itself a reproduced artefact as well as a testimony of authority, just like the saga quotations and images of objects validated by museum experts and scholars in this earlier work.

3 The Status of the Image

A few years after the publication of *The Viking Age*, the remains of an Iron Age settlement were discovered in Somerset, near Glastonbury. Excavations were conducted from 1892 until 1910, with a hiatus between 1898 and 1904, and a first description of the findings was published by the original discoverer Arthur Bulleid and the archaeologist Harold St. George Gray in 1911 (Phillips 2005, 74). Bulleid also wrote an article for *The Illustrated London News* which was published together with photographs of some findings and a series of artistic reconstructions by the journal's house artist Amédée Forestier. The imagined scenes were widely reproduced and enormously influential in readjusting popular conceptions of the ancient inhabitants of Britain. To give an illusion of unfiltered access to the past, Forestier inserted pictures of the authentic finds into the scenes, and projected contemporary social and gender roles onto the past (Phillips 2005, 79, 84). Based on the excavation photographs accompanying the article, the depicted objects enhance the sense of reality while the recognisable social organisation provides familiarity. Both these effects interact with the documentary status of a journal specialised in reporting the news. The objects are made meaningful by the place of publication and by their position in a narrative picture that is, in turn, corroborated by photographic evidence.

Although archaeological discoveries figured quite prominently in *The Illustrated London News*, there are no pictorial representations of Viking life and manners in the journal. Instead, Viking artefacts appear as museum items, neither contextualised nor explained (*The Illustrated London News* 1891, 111). The same is true of du Chaillu's history which can be understood as a reorganised archaeological collection where artefacts and saga extracts validate each other: "The testimony of archæology as corroborating the Sagas forms one of the most important links in the chain of my argument" (du Chaillu 1889, vol. 1, x). Since the represented objects fail to engage on a personal level, *The Viking Age* does not stage an encounter with early Norse history in the way that Forestier's reconstructions do. Although the illustrations are orchestrated to support du Chaillu's thesis, the objects are not narrativized or inserted into imagined scenes or stories, which would presumably have deprived them of some of their authority as genuine artefacts. The storied – although unillustrated – object is instead a prominent feature in *Ivar the Viking* which can be seen as a textual correspondence to Forestier's representation of the Iron Age.

The Viking Age is of course different from a regular collection firstly, in that du Chaillu did not collect the objects himself and secondly, because the specimens appear as images and quotations. The museum status and scholarly authentication of most of the items have already fixed them as genuine and significant and this already-established authority partially transfers to du Chaillu as interpreter. The Preface contains information about previous, if not necessarily original collectors, like the runologist George Stephens from whose *The Old Northern Runic Monuments of Scandinavia and England* (1866–1884) du Chaillu reproduced many of the illustrations of rune stones and inscriptions in his text. Other scholars whose published works provided illustrations are the Norwegian archaeologists Oluf Rygh and Anders Lorange, the philologist Sophus Bugge and the historian Nicolay Nicolaysen, the Danish archaeologists Conrad Engelhart and Frederik Sehested and the zoologist Japetus Steenstrup who popularised the term køkken-mødding for the refuse heaps outside prehistoric settlements, the Swedish archaeologists Oscar Montelius and Axel Emanuel Holmberg and the amateur archaeologist and artist Lauritz Baltzer, as well as many others (du Chaillu 1889, vol. 1, XI). The enumeration of respected scholars implies that they have supported or approved of du Chaillu's work and adds to his authority. Whether the illustrations in *The Viking Age* were reproduced with their permission is however unclear. As already mentioned, image recycling was common in the period, and although details about excavation sites or places of display are often given, no information about the original artists is provided in direct proximity to the illustrations. As in du Chaillu's travel books, the impression is that the collection, or at any rate, the choice of items to include, is based on the author's own observations rather than processed through several layers of selection and mediation.

The lack of source information thus obscures how the images are made to participate in the construction of meaning. That they were perceived as independent of du Chaillu's objectives is suggested by the review in *The Illustrated London News*:

> The author can hardly, upon this occasion, be accused of an undue exercise of his own fancy. All the facts stated are expressly authenticated by multitudinous precise citations from ancient documents, or by the drawings and photographs of sculptured figures on monuments, stones with runic inscriptions, and ornamented vessels, weapons, or other implements, abundantly collected in the museums of Northern Europe, where many learned curators and librarians and students of antiquities have aided M. Du Chaillu in his important work (1890, 19).

The illustrations are consequently not criticised, even by reviewers who are vitriolic in their rejection of du Chaillu's ancestry argument. Partly, this is because of a general expectation that archaeological depictions should be neutral, partly because the pictures are received as fundamentally a collection of Norse artefacts for English consumption:

> He undoubtedly makes accessible to English readers a number of unique and valuable illustrations of northern antiquities which have been buried in Danish and Swedish catalogues and periodicals and other inaccessible

> repositories; and an immense number of extracts from the Edda and the Sagas which illustrate the subjects with which he is dealing. (*St. James's Gazette* 1889, 6)

While the relation between image and producer has long been recognised in art history, archaeological representations have frequently been understood as objective, and the role of the artist has been neglected. Today, as Stephanie Moser and Sam Smiles note, we "routinely accept that no pictorial device can be a transparent illustration of the world, but instead deploys technical devices, formal conventions, and ideological assumptions to orchestrate meaning" (2005, 1). The accumulated artefact images in *The Viking Age* are expected to signify in themselves, regardless of du Chaillu's selection bias and any aesthetic bias or technical restraint of the original picture producers. According to *The Illustrated London News* reviewer, the illustrations "incline us to form a high estimate of the moral and intellectual qualities of the whole Scandinavian race" (1890, 19). du Chaillu may flaunt his agenda and openly acknowledge his ideological purposes, but neither he nor his critics regard the depicted artefacts as anything but a transparent window on the Viking world and character.

4 An Ethnic and Gendered Selection

In the past few decades, there has been a great deal of criticism of museum collections, from ethical perspectives as well as from the point of view of representativity. The comparative lack of mundane and domestic objects has been criticised, for instance, for obscuring women's lives and importance in the past, whereas displays of religious objects have been censured for cultural obtuseness. Cultural collection entails an actual encounter with people and habitats, and the memory of the encounter reverberates for generations. The anthropological collections of measurements and remains of non-metropolitan people constitute a particularly disturbing example of how human dignity is violated in the interest of science. The interaction between source communities and collectors is profoundly unequal, as Mary Louise Pratt makes clear when introducing the concept of the contact zone (1991, 34), regardless of whether the physical exchange takes the form of gift, plunder or object for study.

According to James Clifford, cultural collection forms a response to "particular histories of dominance, hierarchy, resistance, and mobilization" (1997, 213). Cultural objects become contact zones in material form, especially when they form part of museum displays. Some interpretations of the contact zone metaphor – including Clifford's own – have been questioned for over-emphasising the dialogic elements while suppressing the asymmetrical relations entailed (Boast 2011, 56), but the notion has revitalised museum practices and led to efforts in recent decades to unearth alternative histories and trouble received narratives. Co-creation and consultations with present-day stakeholders have been introduced and understood as a means of honouring the silenced Other in the original encounter. The issue becomes slightly more complex when ancient, archaeological remains are concerned, however. Not only is no actual, physical contact involved, but the opportunity to confer or collaborate with the people who once handled the objects, ruins and burial remains is forever lost.

In a collection that takes place at several removes from the original encounter, the boundaries of the contact zone become even fuzzier since the exchange includes all previous collecting enterprises both before and after archaeological rediscovery. The question arises of who may be the silenced Other in the series of interactions that precedes the collection, and whether there can even exist a disenfranchised Other when the contact zone extends into the distant past. Given du Chaillu's central thesis in *The Viking Age*, the absence of images indicating a Celtic civilisation is logical, but from an ethnic point of view, he participates in silencing an indigenous, British pre-history as well as a surviving Scots-Gaelic, Irish and Welsh culture that was considerably under pressure in the late nineteenth century. In the book, Scotland, Ireland and Wales mainly appear in the saga extracts as enemies or as defeated by the Vikings, or obliquely, as the previously mentioned "original and effete inhabitants" (du Chaillu 1889, vol. 1, 3) of the British Isles. As actual, historical contact zones, encounters in Scotland, Wales and the Isle of Man were indeed mostly characterised by hostility, while commercial exchange dominated relations in Ireland, where intermarriages and alliances between Vikings and the original population took place from the early ninth century onwards (Maier 2017, 160, 182). Trading with Ireland is mentioned in some saga quotations (1889, vol. 2, 146, 209, 392), but it seems to have been more important for du Chaillu to downplay any connections between the Vikings and an earlier British population and instead show Norse exchange with prominent civilisations such as Greece, Rome, Byzantium and the Arab world. Although the Tjängvide stone embodies evidence to the contrary, no Celtic artefacts are explicitly included or acknowledged in the illustrations.

Narrative pictures as well as stories like du Chaillu's *Ivar the Viking* inevitably stage an encounter with imagined or reconstructed people and societies, but as Amédée Forestier's Iron Age illustrations reveal, attempts to recreate pre-historic life risks projecting present-day value systems onto previous eras. In such contexts, the representation or description of ancient items serves internal credibility by concealing interpretative biases. As a result, the staged encounter involves only the segment of a past society that the collected items are marshalled to support. Unlike Forestier, du Chaillu does not integrate Viking women in the Victorian gender system but tends to overvalue their relative status in pre-Christian times. While acknowledging the gender hierarchies brought in by Christianity (1889, vol. 2, 1), he mainly concentrates on high-born women in his text, which leads him to draw far-reaching conclusions about women's conditions in general. He locates the chivalric spirit in the north and argues that when the attitudes spread to other parts of Europe, "the opinion only of women of higher rank was valued," whereas among "the earlier tribes of the North all were respected" (du Chaillu 1889, vol. 2, 2). Such statements are supported by pictures of women's belongings that almost exclusively indicate high rank, even though they are tempered by the saga extracts where bondswomen, servants, concubines and slaves appear. It is inaccurate to conclude that women are silenced in du Chaillu's history since they figure prominently in the quoted passages, but considered as a collection, *The Viking Age* provides a distorted version of female life. While women have a noticeable textual presence in the work, their visual presence is faint.

Illustrations of a spindle and shuttle are included (du Chaillu 1889, vol. 2, 368), and du Chaillu expresses admiration for Viking Age skill in weaving (1889, vol. 1, 4), but on the whole, mundane objects are absent from both text and illustrations. He mentions that women's graves mainly contain objects like combs, pins, needles, buttons, jewellery and ornaments (1889, vol. 1, 249), but only the most decorated or valuable pieces are represented. Instead, du Chaillu foregrounds the elevated position of women with the help of ornamented fibulas (1889, vol. 1, 257, vol. 2, 342), glass vessels (1889, vol. 1, 280) and keys (1889, vol. 2, 362, 478). To a certain extent this could be explained by the fact that everyday items are less likely to be preserved for posterity, but it also reproduces nineteenth-century museum practices where the concept of a rich Viking culture was invented. In addition, the association between women and objects of beautification is a gendered convention that needs to be supported by osteological evidence to count as fact (Whitehouse 2017, 224). If our "encounters with the past are limited by the relics we find in our archives" as Adriana Craciun and Mary Terrall phrase it (2019, 16), it is important to understand the composition of those archives and how they are made meaningful. Women's belongings in *The Viking Age* have been subjected to a succession of selection biases which means that very few examples that can be connected to ordinary women's lives and culture are included.

The "Victorian Viking craze" (Wawn 2000, 8) was closely bound up with a widespread fear of degeneration that, in turn, led to a celebration of masculine virtues. Viking subject matter in art and literature thus primarily represented male Vikings and showed them as "rugged, muscular and armoured" (Marshall 2017, 8). Such would have been the images that circulated in England as well as in Scandinavia at the time, supported by the artefacts on display at museums and represented in scholarly literature. It is only logical, then, that both in the text, and overwhelmingly in the illustrations, du Chaillu privileges stories and items that support the image of a hardy, masculine culture. Nevertheless, by also including Viking Age women in this framework he partially upsets the contemporary desire for unambiguous gender identities. The fact that daggers have been found in female Bronze Age graves causes him to speculate "that the women had been warriors" (1889, vol. 1, 92), and amazons are mentioned in the very brief chapter about women's occupations (1889, vol. 2, 362–68). Insights from twenty-first-century gendered archaeology however suggest that the warrior designation should be understood as temporary, regardless of the warrior's sex. Like beautification items, weaponry grave goods should consequently not be taken as a conclusive gender identification of the buried individual (Whitehouse 2017, 224, 228). While Fedir Androshchuk concurs with the view that Viking Age women's lives may not have been as restricted as sometimes claimed, he also states that "there is still no firm evidence to confirm the existence of female warriors" in the period (2018, 58). The fierce women in the sagas may have been described precisely because they were exceptional (Raffield et al. 2017, 188). To some extent, such a realisation can be deduced from how du Chaillu organises his collection. Although he uses saga evidence to suggest that women were actively engaged in war and plunder, and even mobilises Joan of Arc as a latter-day support of the idea (du Chaillu 1889, vol. 2, 367–68, 368 n. 1), his material evidence rather contradicts the notion: the most richly illustrated chapter in volume 2 of *The Viking Age* is the chapter about weapons where none is associated with a woman (1889, vol. 2, 65–101).

5 Conclusion

Curiosities become curious when they are removed from their original context and use, and as Kate Hill points out, there is a connection between early touristic souvenir collection and archaeological collection (2016, 180). By the token of their antiquity the represented Viking Age artefacts in du Chaillu's book could be regarded as curiosities, but his agenda-driven manner of selection transforms them into a collection, albeit not one based on "objective study rather than subjective judgement" (Hill 2016, 177). The narratives embodied in the objects as well as in their previous collections have shifted and occasionally been lost. Gathered from a number of disparate collections in museums, scholarly treatises and archaeological sites, du Chaillu's collection forms a personalised assemblage framed to build up an image of a highly civilised warrior culture with artistic and cultural ambitions. Yet, given the prominence of the illustrations, the work is driven by the images as much as by du Chaillu's thesis, as recognised by the *Edinburgh Review* (1891, 332) commentator: "The text illustrates the figures, rather than the figures the text". While the illustrations in du Chaillu's travelogues provide authenticity for the exotic or alien phenomena described, the sheer quantity of similar objects from the Viking period paradoxically returns the artefacts from the exceptional to the everyday. The mass of examples adds plurality and sometimes even suggests a counter-narrative where the Viking warrior is replaced by a rural, peaceful population, or, as the Tjängvide stone suggests, a different version of history where influences have travelled from the west to the north rather than the other way round. Such more multifaceted narratives are embedded in the over-abundance of artefact images and saga extracts in *The Viking Age* even if they are not acknowledged in the narrative text.

Today it is a truism that an archaeological collection is not only the result of what was there but also reflects the purpose of the collector. The representational power of images is equally questioned and the limitations of different processes of mediation are recognised. The illustrations and excerpts in *The Viking Age* instead demonstrate a belief in objects and their representation as an unmediated access to the past. Although the images lack the fundamental characteristic of an actual collections as the assembly of physical samples, they carry the signs of an antiquarian desire to amass things. When *The Viking Age* is considered as a history, the sense-making processes become central and questions regarding du Chaillu's failings as a historian and the validity of his arguments come to the fore. Understood as a collection, the work in both its production and reception primarily illuminates late nineteenth-century trust in the deceptive objectivity of artefacts. A remediating factor is that, after organising his source materials into subsidiary collections or chapters, du Chaillu largely allows the items to speak for themselves. As a history, *The Viking Age* cannot be separated from du Chaillu's genealogical claims, and primarily becomes an example of the imperialist principles and gender anxieties of his time. As a collection, this totalising vision can be suspended, inviting reinterpretation and allowing the work to be read for its absences as well as its inclusions.

Bibliography

Androshchuk, Fedir. 2018. "Female Viking Revisited." *Viking and Medieval Scandinavia* 14: 47–60.

Axelsson, Bodil, Fiona R. Cameron, Katherine Hauptman & Sheenagh Pietrobruno. 2022. "Conclusion." In *Museum Digitisations and Emerging Curatorial Agencies Online: Vikings in the Digital Age*, edited by Bodil Axelsson, Fiona R. Cameron, Katherine Hauptman & Sheenagh Pietrobruno, 121–27. Cham: Palgrave Macmillan.

Boast, Robin. 2011. "Neocolonial Collaboration: Museum as Contact Zone Revisited." *Museum Anthropology* 34 (1): 56–70.

Bugge, Alexander. 1905. *Vesterlandenes indflydelse paa Nordboernes og saerlig Nordmaendenes ydre kultur, levesaet og samfundsforhold: 1. Vikingestiden*. Christiania: Dybwad.

du Chaillu, Paul. 1889. *The Viking Age: The Early History, Manners, and Customs of the Ancestors of the English-Speaking Nations. Illustrated from the Antiquities Discovered in Mounds, Cairns, and Bogs as well as from the Ancient Sagas and Eddas. With 1366 Illustrations and Map.* 2 vols. London: John Murray.

Clifford, James. 1997. "Museums as Contact Zones." In *Routes: Travel and Translation in the Late Twentieth Century*, edited by James Clifford, 188–219. Cambridge: Harvard University Press.

Craciun, Adriana & Mary Terrall. 2019. "Introduction." In *Curious Encounters: Voyaging, Collecting, and Making Knowledge in the Long Eighteenth Century*, edited by Adriana Craciun & Mary Terrall, 3–18. Toronto: University of Toronto Press.

The Daily News. 1889. "The Viking Age." 24 October.

The East and South Devon Advertiser. 1896. "Art and Literature." 22 February.

Edinburgh Review. 1891. "Scandinavian Antiquities." Unsigned review of *The Viking Age*, by Paul du Chaillu & Five Other Works of Northern Pre-History. *Edinburgh Review* 173 (April): 331–59.

Edmonds, Mary K. 1998. "Paul Belloni du Chaillu." In *Dictionary of Literary Biography vol. 189: American Travel Writers 1850–1915*, edited by James Schramer & Donald Ross, 109–31. Detroit: Gale Research Inc.

Ellis, Caitlin. 2021. "Remembering the Vikings: Ancestry, Cultural Memory and Geographical Variation." *History Compass* 19 (4): e12652. https://doi.org/10.1111/hic3.12652.

Findlen, Paula. 2020. "Objects of History: The Past Materialized." *History and Theory* 59 (2): 270–82.

Glasgow Herald. 1885. No title. 30 October.

The Globe. 1893. "The Library Table." 23 October.

Hill, Kate. 2016. "Souvenirs: Narrating Overseas Violence in the Late Nineteenth Century." In *Britain and the Narration of Travel in the Nineteenth Century: Texts, Images, Objects*, edited by Kate Hill, 175–89. London and New York: Routledge.

The Illustrated London News. 1890. "New Books." January 4.

The Illustrated London News. 1891. "The Scandinavian Exhibition." 25 July.

Jansson, Sven B. F. & Elias Wessén. 1962. *Sveriges runinskrifter vol. 11: Gotlands runinskrifter part 1*. Stockholm: KVHAA.

Johannesson, Lena. 1982. *Xylografi och pressbild: bidrag till trägravyrens och till den svenska bildjournalistikens historia*. Diss. Uppsala and Stockholm. Stockholm: Nordiska museet.

Johannesson, Lena. 2018. "Tekla Vilhelmina Nordström." Kvinnobiografiskt lexikon. https://www.skbl.se/en/article/TeklaNordstrom. Accessed 27 February 2024.

Maier, Bernhard. 2017. *The Celts: A History from Earliest Times to the Present*. Translated by Kevin Windle. Edinburgh: Edinburgh University Press. *ProQuest Ebook Central*, https://ebookcentral.proquest.com/lib/umeaub-ebooks/detail.action?docID=6994883.

Marshall, Nancy Rose. 2017. "Victorian Imag(in)ing of the Pagan Pyre." *19: Interdisciplinary Studies in the Long Nineteenth Century* 25: 1–37. https://doi.org/10.16995/ntn.795.

McCook, Stuart. 1996. "'It May Be Truth, But It Is Not Evidence': Paul du Chaillu and the Legitimation of Evidence in the Field Sciences." *OSIRIS*, 2nd series 11: 177–97.

Miller, Peter N. 2017. *History and Its Objects: Antiquarianism and Material Culture since 1500*. Ithaca and London: Cornell University Press.

Momigliano, Arnaldo. 1950. "Ancient History and the Antiquarian." *Journal of the Warburg and Courtauld Institutes* 13 (3–4): 285–315.

Moser, Stephanie & Sam Smiles. 2005. "Introduction: The Image in Question." In *Envisioning the Past: Archaeology and the Image*, edited by Sam Smiles & Stephanie Moser, 1–12. Malden, MA and Oxford: Blackwell Publishing.

Nonconformist and Independent. 1889. "The Viking Age." 7 November.

Phillips, James E. 2005. "'To Make the Dry Bones Live': Amédée Forestier's Glastonbury Lake Village." In *Envisioning the Past: Archaeology and the Image*, edited by Sam Smiles & Stephanie Moser, 72–91. Malden, MA and Oxford: Blackwell Publishing.

Pietrobruno, Sheenagh. 2021. "Curating Viking Objects Through Customized Scrolling: How Search Engines Personalize Historical Narratives." *Convergence: The International Journal of Research into New Media Technologies* 27 (5): 1393–417.

Poulter, Emma K. 2014. "Hybridity – Objects as Contact Zones: A Critical Analysis of Objects in the West African Collections at the Manchester Museum." *Museum Worlds: Advances in Research* 2: 25–41.

Pratt, Mary Louise. 1991. "Arts of the Contact Zone." *MLA Profession* (1991): 33–40.

Raffield, Ben, Neil Price & Mark Collard. 2017. "Polygyny, Concubinage, and the Social Lives of Women in Viking-Age Scandinavia." *Viking and Medieval Scandinavia* 13: 165–209.

Resløkken, Åmund Norum. 2022. "Objects, Traditions, and the Hermeneutics of Viking Heritage." *Nordic Museology* 2: 56–72.

Science. 1890. "M. du Chaillu on The Viking Age." *Science* 15 (365): 74, rpt from *The Times* January 7, 1890.

Sigurðsson, Jón Viðar. 2022. *Scandinavia in the Time of the Vikings*. Translated by Thea Kveiland. Ithaca: Cornell University Press.

St. James's Gazette. 1889. "The Viking Age: Mr. du Chaillu on the Origin of the English." 25 October.

"Tjängvidestenen." http://svenskuppslagsbok.se/79929/tjangvidestenen/ Accessed 27 February 2024.

Wawn, Andrew. 2000. *The Vikings and the Victorians: Inventing the Old North in Nineteenth-Century Britain*. Cambridge: Boydell and Brewer.

Whitehouse, Laura M. 2017. "The Weapons Make the Man? A Re-Analysis of Early Anglo-Saxon Weapon Burial and Interpretation." In *Archaeologies of Gender and Violence*, edited by Bo Jensen & Uroš Matić. Oxford: Oxbow Books.

OBJECT: Stave church

FIGURE 6.1 Borgund church, 1866
PHOTO: UNKNOWN/THE NATIONAL LIBRARY OF NORWAY

CHAPTER 6

Architecture on the Move: Collecting and Connecting Stave Churches in Nineteenth-Century Norway

Sine Halkjelsvik Bjordal

Abstract

This chapter investigates how the collecting of Norwegian stave churches "on paper" by nineteenth-century artists and art historians was based on and contributed to the development of specific views of the North. Its starting point is a collection of illustrations of the stave churches of Borgund, Urnes and Heddal published by the Norwegian painter J.C. Dahl (1837). After Dahl's publication, the illustrations of and knowledge about the churches circulated in Europe through English, German and French art historical publications, all of them interpreting the "peculiar" architecture from the North in their own ways, based on both nationalist sentiments and universalist ideas about the historical evolution of art. The chapter argues that the many possibilities for differing art historical interpretations of the stave churches made them the perfect objects for a developing field of European art history searching for missing "links" in national and universal histories of art development.

Keywords

architectural history – art history – cultural history – history of texts – circulation of knowledge – stave churches – J.C. Dahl

On 15 November 1843, The Oxford Society for Promoting the Study of Gothic Architecture (also known as the Oxford Architectural Society) held their second meeting of the autumn term. The proceedings were in the same order as always, starting with the reading of the minutes of the last meeting, and then "any matter of business", which this time concerned two letters about a Church to be erected on the island of Colaba in India. After this, it was time

 | DOI:10.1163/9789004749665_007

for the usual paper for the evening to be read. According to the written report from the meeting, Mr. Patterson[1] of Trinity College presented this evening "a series of lithographs, illustrative of a peculiar style of wood architecture of very ancient date". The collection of lithographs consisted of "exterior and interior views, ground plans, etc., of three Churches at Hitterdal,[2] Urnes, and Borgund, in Norway". In addition to presenting the collection of drawings, Mr. Patterson read a translation of some remarks published together with the plates, before a discussion of the particular churches followed (Oxford Society 1843, 10).

We will return to the Oxford Society and to what the construction of a church on Colaba and some ancient wooden churches in Norway had to do with this. But what kind of buildings were these churches recorded on lithographic plates? What sort of collection was this? The plates had been donated to the Society by Patterson, but how he had gotten hold of them is unknown. What is known, however, is that the lithographs and the remarks about them were part of a German work entitled *Denkmale einer sehr ausgebildeten Holzbaukunst aus den frühesten Jahrhunderten in den innern Landschaften Norwegens* (hereafter: *Denkmale*),[3] published by the Norwegian painter J.C. Dahl (1788–1857), who at this point was a professor at the renowned art academy in Dresden. The work had been published in three booklets in Dresden six years earlier. Depicted and described in the booklets were three examples of what is today considered, along with the Viking ships, Norway's most significant contribution to the world cultural heritage: the Norwegian stave churches.

The object this chapter takes as its starting point is the most famous – and probably most often depicted – of these churches: the church of Borgund in Lærdal in Sogn, situated on the travel route stretching from the innermost part of the long Sognefjord and over the Filefjell mountain pass, historically one of the main routes in southern Norway. In the first half of the nineteenth century, this church started to attract travellers of different sorts, but it had long been known as a peculiarity among those with antiquarian and historical interests. When Joseph Paul Gaimard (1793–1858) and his fellow travellers set out for the *La Recherche* expedition to Norway and the Arctic in 1838, 1839, and 1840 – an ambitious French expedition that resulted in 26 scientific volumes and five

1 James Laird Patterson (1822–1902), later Auxiliary Bishop of Westminster.

2 Now called Heddal.

3 Translation: "Monuments of a highly developed wooden architecture from the earliest centuries in the interior landscapes of Norway". All translations from Danish, French and German sources in the chapter are done by the author, with good help from ChatGTP 4o, unless otherwise specified.

FIGURE 6.2 Exterior of the church of Borgund, from J.C. Dahl's *Denkmale,* 1837
PHOTO: THE NATIONAL LIBRARY OF NORWAY

collections of lithographic plates – the travel route did *not* include a detour to Filefjell for the draftsmen following the expedition. However, three plates showing the exterior of the church of Borgund from two sides, a ground plan, and two of the portals, were published in *Voyages de la Commission scientifique du Nord en Scandinavie. 1, Danemark, Norvège, Spitzberg. Atlas Historique et Pittoresque* in 1852. The French artist, Auguste Mayer (1805–1890), who participated in the expedition, must have either seen the church on some other occasion, or he had seen other drawings of it.[4] A small detail on one of the plates suggests that he might in fact have had access to Dahl's *Denkmale.* On the plate showing the ground plan and the two portals, the ruler displaying the scale says "Echelle de 5 métres" (scale five metres). On the drawing of the ground plan in *Denkmale*, the exact same scale is drawn, but here it says "5 ellen", five ells. The lack of accurate scaling indicates that the measurements are not Mayer's own.

This is hardly evidence of anything, but it leads us to what will be the main focus of this chapter: the very specific paper collection of stave churches

4 It has not been possible to discover whether Mayer actually went to Borgund. He followed the expedition in 1838, and travelled by land from Christiania (Oslo) to Trondheim, but this route did not pass by Borgund (Knutsen 2002, 43–46, 82).

published by Dahl in 1837 and how the church of Borgund and its fellow specimens in Hitterdal and Urnes contributed, through this collection, to the shaping of a specific form of cultural knowledge about the North – while at the same time themselves being shaped as a particular type of object. As will be shown below, the stave churches played a key role in the romantic view of Scandinavia, a view that predominated among the general public in the nineteenth century (Barton 2007, 28). This chapter, however, argues that there is also another, parallel, story to be told about the stave churches in the nineteenth century and the role they played as objects from the North. This is a story about art history, architectural knowledge, and the then emerging field of architectural history which strove to collect all the world's past and contemporary architecture on paper, while at the same time searching for historical examples that could be used as models for new, national architectural styles. For both these purposes, it will be shown, Dahl's collected stave churches fitted perfectly.

1 The Land of the Stave Churches?

In 1837, when Dahl's *Denkmale* was published, and in 1843 when Patterson found it appropriate to present these buildings to a society promoting the study of gothic architecture, the stave churches were scarcely known outside of Norway. And within Norway, only a small group of painters and scholars had begun to realize that the peculiar-looking, small, and run-down old wooden churches in remote areas of the country could, and should, be considered valuable historical monuments (Bjordal 2021). Dahl himself had left Norway many years earlier to get his education abroad. Based in Dresden, he still travelled back to Norway several times, the first journey in 1826. Some friends of Dahl, the painter Johannes Flintoe (1787–1870) and the cartographer Gerhard Munthe (1795–1896), made sure Dahl would see several stave churches on this tour by preparing a route for him that would take him past several of the most remarkable churches. At a later journey to Norway, in 1834, Dahl experienced that many of the stave churches he had seen in 1826 had since been demolished. He then realized that it was, as he writes in *Denkmale*, "urgent", to make the most remarkable of the churches known by means of a publication, "because the spirit of alteration and renovation is even now at work upon them and threatens them with […] destruction" (Dahl 1837, foreword).[5] Two years later, Dahl

5 Translated quotes from *Denkmale* are to a large degree taken from the Oxford Society (1843).

sent the architect and painter Franz Wilhelm Schiertz (1813–1887) on a mission to collect a number of Norway's medieval stave churches on paper.

As Edward Kaufman has shown, architectural travel was at its peak around 1850, after having started to become popular around a century earlier. Architects, many of them British, travelled to gather knowledge about buildings in other places that were not already represented in books, which at this point applied to most of the world's architecture. This was the same time as "a significant reorientation in the nature of architectural knowledge" started to take place – it was now no longer only the Roman Renaissance tradition that impacted on the practice and theory of architecture, but a whole array of architectural traditions. An important part of the reorientation was moving the architectural gaze from the south of Europe, with the classical architecture of Greece and Rome, to the North (Kaufmann 1989, 59).

Buildings can, however, also move in several ways, and this was particularly so of the stave churches. First of all, stave churches are themselves mobile as they are seemingly built as "construction sets". In the 1880s three stave churches were actually moved from their original locations to be restored and re-erected in Christiania, Bergen and Trondheim, respectively, and then exhibited and understood as "exceptional exemplars" of Norway's medieval wooden architecture (Bjordal 2021, 367–69). The most common way of collecting stave churches in the nineteenth century, however, was not to move the buildings, but to collect them on mobile paper.[6] In an article on "travelling" images of Norway in the illustrated press, Iver Tangen Stensrud (2021) shows how circulating images of, among other things, stave churches in various magazines contributed to the reproduction of well-established stereotypes of Norway and the Norwegian people in the second half of the nineteenth century. These stereotypes can be summarized by H. Arnold Barton's three aspects of the typical romantic view of Norway: *Majestic Norway, Norway the land of heroes,* and – especially relevant for the stave churches – *Picturesque Norway,*

> the land of proud, timbered farmsteads purportedly dating back to Viking times and tiny crofts on steep, rocky mountainsides, of simple sæters, on high-lying pastures with rosy-cheeked milkmaids who sang as they went about their chores, of ancient stave churches, of colorful local folk dress, elaborate peasant weddings, athletic dances, and Hardanger fiddles. (Barton 2007, 30)

6 This can lead the reader to think of Bruno Latour's (1986) concept of "immutable mobiles", a concept the author finds very interesting and good to "think with". However, it has not been my intention to fully employ Latour's concept here.

The stave churches were thus moving also in a third way: they affected the people looking at them, producing curiosity, excitement and awe. And when it comes to romantic views of Norway and the tourist travel boom – especially from English speaking countries – in the second part of the nineteenth century, a fourth aspect could therefore be added, to the view on the stave churches particularly: The *peculiar and curious* Norway. This view can be found in a range of travel descriptions and illustrated magazines. For instance, in the article "A Trip to Norway" in the journal *All Around the World* we can read about Borgund church, "this curiosity of art", that

> the pointed roof, the sculptured water-spouts, and the grotesque ornaments [...] have a decided Burmese or Chinese physiognomy. Smaller than the church of Hitterdal, this edifice appears also to be more ancient, and a covered gallery, with columns blackened by time, runs all around it [...] everything breathes that venerable perfume of times long gone by, and of which the traces are every day disappearing more and more. (Ainsworth 1866, 59)

In another format and type of text, Mattieu W. William's *Through Norway with Ladies*, the following can be read about the churches of Borgund and Hitterdal:

> Of course we visited the queer old wooden toy church of Borgund, with its nave 39 feet long, and its incomprehensible surrounding narrow passage. Wooden beams, wooden walls, wooden tiles – wooden throughout. This and its twin sister at Hitterdal [...] are about the oddest specimens of ecclesiastical architecture in existence. (Williams 1877, 319–20)

Even in the American *Ballou's Pictorial Drawing-Room Companion* from 1857, an image of the church of Borgund can be found with the following description: "The curious pile of buildings delineated in this picturesque sketch, is certainly an architectural anomaly. It appears to have been erected after the model of one of those card houses which children are delighted to pile story upon story".[7] Others again point to the similarities between Borgund and "a Chinese pagoda" (Johnston 1884, 62; Gaimard 1852, 2), and, according to Bayard Taylor, also writing about Borgund: "[a] more grotesque affair could hardly be found in Christendom; it could only be matched among the monstrosities of Chinese art" (Taylor 1864, 371).

7 *Ballou's Pictorial Drawing-Room Companion*, 1857, XIII (20): 309.

But the picturesque and peculiar are not the only perspectives on stave churches in the travel writing of the time. Sometimes a seemingly different view 'sneaks' into the texts, often alongside the others. We can read about the "German Romanesque" architecture of Borgund church (Williams 1877, 320), that Borgund and Hitterdal churches are "Byzantine-Gothic, *on dit*" (Pritchett 1879, 27), and about the "picturesque variety which is peculiar to Gothic architecture" found in Borgund (Forester 1850, 177). The afore-mentioned Taylor describes how the architecture of Borgund shows "the Byzantine style in the rounded choir and the arched galleries along the sides, the Gothic in the windows and pointed gables, and the horned ornaments on the roof suggest the pagan temples of the ante-Christian period" (Taylor 1864, 371; see also Fjågesund & Symes 2003, 250–52).

Some of the travellers were probably familiar with the art historical terms in circulation at the time, especially later in the century, and could themselves interpret the architectural shapes they saw. But they might also have heard or read about the Norwegian stave churches, and they could have *seen* them on paper, before studying them in real life. Both John Murray's *Hand-Book for Northern Europe* from 1849, and Thomas Forester's *Norway in 1848 and 1849*, recommend Dahl's *Denkmale* to those interested in expanding their knowledge about the stave churches (Murray 1849, 164; Forrester 1850, 179).

These short references to Dahl, together with the scattered art historical interpretations of the stave churches, are small pointers to understandings of the stave churches that go beyond the typical picturesque and curious. They are cues leading to other publications than the typical travelogue and to other forms of knowledge shaped by and shaping the stave churches. Before the boom of travellers coming to Norway with their romanticizing gazes in the second half of the nineteenth century, the stave churches had themselves travelled, from Norway to Germany, France, England, and other places. Here, they were welcomed by art historians, archaeologists and architects not as curiosities or romantic objects, but as valuable historical sources and architecture worth studying. And it all started with Dahl's small paper collection of the three churches of Borgund, Hitterdal and Urnes, published as *Denkmale* in 1837.

2 Collecting Stave Churches: *Denkmale einer sehr ausgebildeten Holzbaukunst*

Before expanding on the field of art/architectural history in the middle of the nineteenth century and the reception of *Denkmale*, a more elaborate description and explanation of Dahl's work should be given. The work consisted of 21 plates, and the drawings showed, as seen in the above quote from the Oxford Society's *Proceedings*, "exterior and interior views, ground plans, etc." from all the churches, "etc.", meaning details such as portals, columns, and capitals with elaborate ornamentation, in addition to a chair and a baptismal font from two other churches in Telemark. Each plate was accompanied by a short foreword and afterword, written by Dahl, who introduced the work with the following paragraph:

> Rarely has architecture in wood reached a pitch of development beyond such as necessity has dictated, and attained to a higher degree of beauty and decoration, and yet more rarely that the monuments of such a style have remained to us from a remote period of antiquity. Hence it is that some of the Churches of my own country, Norway, which are of very great antiquity, deserve some attention. (Dahl 1837, unnumbered)

He then claims that one "should have kept these old, beautiful, truly national[8] forms, which are, I must say, intimately interwoven with religious concepts, and adapted them to a larger scale according to the demands of the time" (Dahl 1837, unnumbered). Here, he is referring to the common practice of his time of tearing down old, cramped churches and substituting them with modern, more spacious churches. In most of the short text, Dahl presents his thoughts on the different architectural elements of the churches. He writes that the churches are "characteristic", but continues by claiming that the Nordic building style[9] "seems to have been a combination of several originals". He sees elements of Latin Christian architecture, for instance, in the squared abacuses of the capitals. In other ornaments he sees "old, pagan patterns", which might, Dahl conjectures, have been formed after foreign oriental patterns. In the ground plans, on the other hand, "Byzantine ideas" can be identified. This, Dahl argues, can easily be explained by the Varangian trading routes (Dahl 1837, foreword). On each plate, he also gives a description of the ornaments seen, like "predominantly Byzantine style", or "old pagan-Nordic motives mixed with the Oriental-Byzantine" (Dahl 1837, tab. IV and V). Especially the capitals of the columns in

8 It is not clear what Dahl means by "national" here, but he probably means that the architecture is connected to a specific people, the Norwegians. Bressani (2023) writes about the emergence of "race-thinking" in the field of architectural history 1770–1850 and points to 'nation' and 'people' as some of the terms assuming that certain traits and cultural dispositions pertained to certain human communities.

9 "Style" was at this point not a very well defined term. See Hvattum (2023) for more about the development of the concept in the eighteenth and nineteenth centuries.

the church of Urnes show Byzantine influence, Dahl claims, as they remind him of motives and decorations in "the well-known bible of Charles the Bald (which is of the ninth century), and of the same style of ornaments which abound in Greek manuscripts of the same period" (Dahl 1837, afterword).

Post-nineteenth century scholars have pointed out all the mistakes Dahl made in his interpretations of the stave churches and the strangeness of his reasoning. For Dahl's contemporaries, however, *Denkmale* was a serious work that contributed with important knowledge, even if – as will soon be seen – the stave churches and their origins could be interpreted in a variety of ways. But from where did Dahl himself get his knowledge? A first important source of information for Dahl had been the Norwegian clergyman and polyhistor Andreas Faye (1802–1869). Faye had himself published an article on stave churches in the Norwegian *Skilling-Magazin* in 1836, which he sent to Dahl. He also, on Dahl's request, provided him with historical information about the churches. When *Denkmale* was published, however, there were few traces of Faye and his historical knowledge. Instead, Dahl had apparently received some help from the renowned German art historian Carl Friedrich von Rumohr (1785–1843) when finishing the publication. In the archival material left after Dahl died, several drafts of the foreword to *Denkmale* can be found, and in these drafts Rumohr's corrections and suggestions can be seen.[10] Dahl's biographer, Andreas Aubert (1920, 218), assumes that most of the art-historical reflections in *Denkmale* were strongly influenced by Rumohr. Rumohr had, at this point, a specific interest in wood as a building material, and this probably explains his fascination with the medieval wooden churches Dahl had found in the interior landscapes of Norway (Wexelsen 1973, 82).

Faye was very enthusiastic about Dahl's work, even if Dahl did not choose to use the information sent to him from Norway. In a letter to Dahl in 1837, Faye heartily exclaims: "I eagerly await the sight of the mentioned Work from Your Hand. Who would have thought that Norway should possess Buildings remarkable enough to be presented to the Eyes of Europe by Your Masterly Hand?"[11] In Europe, Dahl found himself in close proximity to many contemporary discussions on art, architecture, antiquities, and (cultural) history. This influenced his view on the stave churches, but it certainly also influenced the

10 Some of the drafts can be found in the archives after Dahl at KODE (Bergen Kunstmuseum), others are located at the National Library of Norway, Håndskriftsamlingen, Ms.fol. 1882:h.

11 Letter from Faye to Dahl, 24.05.1837, University of Bergen Library, Spesialsamlinger, Ms. 539 L.

stave churches themselves, when they started circulating in the form of Dahl's (and Schiertz's) paper collections.

3 Art, Culture, Architecture: Architectural Histories in the Nineteenth Century

Architectural history as an academic practice and field in the first half of the nineteenth century needs some explication. In fact, architectural history was part of the not very well defined field of art history, which was also something quite different from what it is now. Art history was at this point a very broad field, including and interacting with what we today would call architectural history and certain forms of archaeology. In addition, the history of art was in many ways seen as an "appendage" to a much broader field of cultural history (Karlholm 1996, 43).

An important figure in the history of art history and for its development in the nineteenth century, was Johann Joachim Winckelmann (1717–1768). Architectural historian Mari Hvattum claims that Winckelmann's *Geschichte der Kunst des Altherthums* (1764) is "the single most important contribution to the modern notion of style in architecture", as "[s]eeping from historiography into architectural practice, Winckelmann's style became a means not only of understanding history, but of making it" (Hvattum 2023, 65). What Winckelmann did in his study of Greek art opened up for the relativization of art expressions, even though he himself operated within a classicistic art historical paradigm. He also developed an idea about growth and decline in artistic development. In this way, the study of art – architecture included – could reveal a specific culture's developmental level. Winckelmann, Hvattum argues, thus "wrote the history of culture by means of art" (2023, 69).

Another influential thinker was the German philosopher Johann Gottfried van Herder (1744–1843), who also was inspired by Winckelmann. Unfortunately, there is no room here for a thorough explanation of Herder's thoughts, but important in this context is Herder's idea of the world and humanity seen as a whole and the consequences this had for the writing of histories of art and architecture. As the art historian Dan Karlholm has shown, the main goal of the art-historical research in the middle decades of the nineteenth century was nothing less than writing a "cultural history of the whole humankind" (Karlholm 1996, 46–47). But the field of architectural history did not take an interest in collecting knowledge of the past just to write a world history of cultural development; it was also intensely interested in the present, and, not least, the future: Implicit in architectural historical endeavours at this time lay an ambition of actively encouraging the development of a national art and architecture. In this way, architectural publications, often richly illustrated, could also be seen as pattern collections, providing historical models that could be used in the construction of present and future styles (Karlholm 2004, 33).

Inspired by Herder, Winckelmann and other Enlightenment thinkers, architectural historians aimed at collecting knowledge about art historical developments "all over the world in all ages". This ambition received its foremost expression in a brand new genre of publication: the art-historical handbook, which soon developed into the more specific world architectural history survey. These works can be seen as collections of architectural styles, and they acquired their contents from a range of other texts, such as monographs of buildings, architectural travel reports, and archaeological surveys – all providing a pile of new knowledge about the history of art and architecture. The architectural historians thus moved away from classical aesthetics and traditional architectural publications, and studied "the beauty and variety of architecture, drawing on new theories and concepts from other disciplines such as natural history, geography, history, geology and ethnography" (Brouwer 2023, 19–20).

Dahl's acquaintance Rumohr (together with Franz Theodor Kugler (1808–1858), a professor at the Academy of Arts in Berlin from 1833) is often credited with having introduced 'the historical paradigm' in the field of art history. This entails an understanding of art history precisely as a form of cultural history, where human artistic creations – in a broad sense – played an important role. Of particular interest was how art and architecture could be read as historical 'documents', from which knowledge about cultural encounters and exchanges could be gained. The source of this knowledge was visual similarities; if something looked similar to another thing, the two were probably connected in some way or another. In a letter to Faye about *Denkmale* in 1838 Dahl himself writes: "In the work, I have confined myself solely to *the historical that expresses itself in the buildings* – and I have juxtaposed these with renowned works in the art of building and with motifs found in manuscripts and choir books from the ninth century".[12]

With all this in mind, Dahl's reasoning about Varangians and Charles the Bald's bible seems a bit less strange. In addition, it is no wonder that the remarkable and characteristic medieval churches in the peripheral North were of considerable interest to art historians/architectural historians/archaeologists

12 Letter to Andreas Faye from J. C. Dahl, April 11, 1838. Andreas Faye's archive, Riksarkivet in Norway, PA15/F/Fi/L0034/0004, original italicization.

of art in the middle of the nineteenth century. As the French archaeologist Pierre-Victor[13] put it:

> It is important to find out how these Scandinavian monuments are connected to the monuments of other countries. The archeology of art must side with natural history and other sciences and spread its investigations all over the world. By the example of Scandinavia, we see how the nations being the farthest away from each other very often still are connected in many ways. They form a sort of chain around the world, a chain where one must follow all links to get a view of the whole circle, so that a complete archeological system can be obtained. (Pierre-Victor 1841a, 65)

Pierre-Victor writes about the archaeology of art but his interests and thoughts were concurrent with architectural historians at the time. The juxtaposition with natural history is accurate not only when it comes to spreading the investigations of objects of art all over the world, but also to the systematization and categorization of objects. There was no Linnean taxonomic system of architectural styles, but the many world architectural history surveys can be seen as attempts at organizing and systematizing objects of art, finding a system that could illuminate the many cultural and historical connections between different cultural expressions. The question, then, is: to whom, where, and – not least – to what period(s), could the stave churches be connected?

4 Connecting Stave Churches: Byzantine, Romanesque, or Maybe Old Egyptian?

The first receptions – and connections – of Dahl's stave churches came in form of reviews of Dahl's work, that is articles where *Denkmale* was presented and discussed as a work relevant to art historians and architectural history. These reviews probably contributed to the first spreading of Dahl's work in the art historical milieu, especially in German speaking areas, but also to other places, like France and England. Already in 1837, Professor Franz Kugler commented on the first booklet of *Denkmale* in his own journal *Museum*. As seen above, Dahl had called attention to Byzantine and old Nordic style elements in the churches. Kugler, on his side, was more interested in the *Germanic* that expressed itself in the buildings. For instance, he points to the ornaments on

13 The name is a pseudonym, he was also called Pierre-Victor Lerebours or Pierre-Simon Lerebours.

one of the portals (plate V) and claims that these most likely express a "nationell germanische Gefühlsweise" – a national Germanic sensation – as the style of the portals is very similar to examples from German art of the twelfth century (Kugler 1837, 519–21).

Another German professor who presented *Denkmale* to a broader scholarly public was the same Rumohr who had helped Dahl finish his work. Rumohr writes that there are many churches that seem to have been possible models for the stave churches, for instance some churches from the eleventh and twelfth century in the Duchy of Schleswig. The church of Hitterdal, however, is very divergent from the other two, Rumohr argues. The model here must most likely have been ancient Egyptian (*altägyptischen*), even if it is hard to say how this can have happened, Rumohr admits, since there are no historical sources that can shed light on the cultural transfer (Rumohr 1840, 347). When discussing the specific parts of the churches, Rumohr concurs with Dahl and the idea about pagan forms, for instance as seen in the *nicht kunstlosen* dragon windings (Rumohr 1840, 345). Other decorations, however, probably came to Norway from Greek, rather than Roman antiquity, Rumohr writes, and he wonders if the decoration could have been mediated through Western Europe, or maybe even the Byzantine empire, as Dahl suggests. Still, Dahl is mistaken, Rumohr claims, about the Bible of Charles the Bald as an example of the Byzantine style – here, Dahl should have taken into consideration the distinctive French-German school of art under the Carolingian dynasty (Rumohr 1840).

It might seem a bit odd to compare architecture with illuminated bibles. But in the eighteenth and nineteenth century many art historians had an interest in illuminations, or miniatures, and the visual culture of the Middle Ages (Worm 2009, 259). In an article in *Kunstblatt* from 1841, Kugler writes: "[T]he study of illuminations must be categorically described as utterly fundamental for medieval art in general, since from the study of their progress and development conclusions may be drawn for all other branches of art" (Kugler cited and transl. in Worm 2009, 261). Similarities between art expressions could thus be found across different formats, and the stave churches were particularly open for interpretation. The manifold architectural details of unknown origin exhibited forms and patterns that could be compared to a variety of forms and patterns in art and architecture from other places and times. It all depended on the eyes of the beholders and the knowledge they already had: *What does this look like? Where have I seen something like this before?*

Dahl's *Denkmale* received some attention in France as well. Another review of the work can be found in a "report" written by Pierre-Victor, who had studied the plates in *Denkmale* together with the architect M. Thoillet. Based on Thoillet's comments, Pierre-Victor concludes that the architectural construction seems to be something specific to the Nordic countries. This is mostly expressed in the intricate ornamentation, which "seems to be inspired by the belief in the Odin mythology" (Pierre-Victor 1841b, 10). He still sees several similarities with French architecture and art, for instance images in the well-known illuminated twelfth century French Bible, *La Bible de Souvigny* (Pierre-Victor 1841b, 17).

After the initial reception in the art-historical field, Dahl's *Denkmale* must have caught the attention of several key figures in European art history, since the stave churches did not stop their circulation here. As cultural objects showing both recognizable and unfamiliar forms they fitted perfectly into the brand new genre of world art/architectural history surveys which attempted to present and organize all the past and contemporary architecture of the world. One of the first to publish this kind of universal art historical handbook was – again – Kugler, with his *Handbuch der Kunstgeschichte* published in 1842. This was a work that would become the standard for many of the art-historical handbooks and architectural histories following in its wake, with Kugler's division of the world's art history into four main periods – art in its earliest developmental stage; Classic art; Romantic (Medieval) art; and Modern art (Schwarzer 1995, 25) – especially influential. The 'problem' with the ambition of universality, however, is its basis in an understanding of wholeness. Every object of art is seen as connected to something else: it has an origin, but it is also possibly itself an origin to something else – forming "a chain around the world". Much of the effort in the field of architectural history in the 1840s and the following decades thus centred around filling the 'gaps' in the chain. Where in this chain could the Northern stave churches fit?

Kugler places the stave churches in §7, "Monuments in the Scandinavian countries and North America", in a chapter titled "Art of the Romanesque style", in the third part of the book, "The History of Romantic Art". The ground plans and the constructions of the churches of Borgund, Hitterdal and Urnes point most of all, Kugler writes, to the Byzantine rather than the Romanesque. This pertains also to the small "arcades" in the churches of Hitterdal and Borgund, and it can easily be explained by the Varangian exchanges with Russia and Constantinople from the tenth century. The ornaments, however, which in the church of Urnes especially seem to be very old, remind Kugler of old Anglo-Saxon miniatures, and according to him they can thus be considered an expression of the "original Germanic aesthetic sense" (Kugler 1842, 479–80).

The German art historian Karl Schnaase (1798–1875), in his *Geschichte der bildenden Künste* (1842–1879), refers to *Denkmale* and writes several pages about the stave churches in the volume on "Das eigentliche Mittelalter", in a chapter on "England, nebst [together with] Irland und Scandinavien". He mostly describes the churches, but he also discusses the theory about Byzantine influence, which he does not find very convincing. In the architectural details, Schnaase sees a Western Romanesque style. The carvings, he admits, are the most peculiar, and he discusses whether they are modelled after Irish or Anglo-Saxon shapes, although he can certainly also see some "local" elements, which are similar to shapes that can be found in art by all Germanic peoples (Schnaase 1854, 446–47).

Wilhelm Lübke (1826–1893), professor first of architecture and then of art history, categorized the stave churches as Christian Medieval architecture, Romanesque style, in his work from 1855. He writes that the characteristic architecture must be understood as a "transformation" (*Umwandlung*) of the Romanesque architecture used in other countries based on local building materials and traditions. After having given an explanation of the characteristics of the churches of Borgund, Urnes and Hitterdal, Lübke goes on to argue that the structure and ground plan of the churches has often wrongly been interpreted as based on Byzantine models. Rather, Lübke claims, the forms are a result of local needs and the materials used. He is especially amazed by the carvings, with their arabesques and their "strangely fantastical character" (*seltsam phantastischem Charakter*), which he – like others – thinks are reminiscent of the illuminations in old manuscripts. "Thus", he writes, "even in the most remote North, dominated by a significantly different building material, an echo of the powerful formative laws that shaped the entire Christian architecture of the West in that epoch, can be heard" (Lübke 1855, 276–78).

Yet another way of interpreting the three churches can be found in James Fergusson's (1808–1886) *The Illustrated Handbook of Architecture* that came in two volumes in 1855. In Fergusson's work the stave churches have their place in volume II, *Part II. Christian Architecture, Book IX. Gothic Architecture in Northern Europe*. However, Fergusson emphasizes the remarkable appearance of the churches, which he finds "very unlike anything in stone architecture". Hitterdal, he writes, is in fact "more like a Chinese pagoda, or some strange creation of the South Sea islanders". The same goes for Borgund, "even more fantastic in its design, and with strange varced pinnacles at its angles, which give it a very Chinese aspect" (Fergusson 1855, 934).

As seen above, the Chinese appearance was a common interpretation of the stave churches also in travel writing.[14] When it comes to Fergusson, it is worth mentioning that he actually spent several years in India, working as an indigo planter. In this period he made several tours of the Indian subcontinent, and

14 The publisher of Fergusson's work was the same Murray who published series of very popular travel guides books. In, for instance, Murray (1869) one finds a reference to Fergusson's *Handbook* in the section on the church of Borgund (p. 110).

he later published an extensive work on Indian and Asian architecture (Chopra 2023, 132–33). This can explain Fergusson's emphasis on the Asian aspect of the stave churches. The practice of collecting architecture on paper in a colonial context was – and is – not unproblematic; Fergusson's architectural history "was not only dependent on colonial conquest – on its mapping, networks, and resources – but also simultaneously exploited native knowledge and denied native intellect and expertise" (Chopra 2023, 134).

Fergusson's colonial connections and mindset bring up another problematic aspect of world art/architectural history surveys, namely the alleged 'universalness' of the publications. The art/architectural historical handbooks in the 1840s and later were universal in at least two ways: they were books meant not only for professionals but also for the general public and, in addition, they were *all-encompassing*, presenting the architectural styles "of all ages and countries" (Karlholm 1996, 64; Brouwer 2023, 20). The latter was of course not entirely true, as "the survey texts included only those styles that were deemed worthy of study from the perspective of Europe, or more precisely Europe's centers of power: Germany, Britain, and France". This led to the promotion of Western traditions and the suppressing and marginalizing of others (Brouwer 2023, 20).

The stave churches were not marginalized, but their building material – wood – made them rare, and they were thus sometimes hard to categorize and interpret for architectural historians trying to organize their collections of world architecture. The churches did, however, undoubtedly display forms that gave them a place among the monumental Western architecture. Edward Augustus Freeman (1823–1892), who wrote the first systematic world architectural history survey in the English language, apparently had a problem with the categorization and inclusion of Borgund, Hitterdal and Urnes in his existing system of architecture. But he found a solution. At the end of the chapter on Norman architecture Freeman put in a (long!) note, starting with the following explanation:

> As the series of Romanesque styles terminates here, I will add a brief account of an interesting class of churches which exhibit the treatment of that style when applied to an unwonted material. These are the wooden churches of Norway, which exhibit an infinity of rich Runic, and sometimes distinctively Romanesque and Byzantine carving. (Freeman 1849, 251)

Freeman was a "rather odd figure in the annals of anglophone architectural historiography", often seen as being in opposition to the leading critics of his time, like the well-known English art historian John Ruskin (1819–1900) and the so-called ecclesiologists[15] (Bremner 2023, 277). Freeman wanted his work to be an intervention in the heated debates then raging about the relevance of Gothic revival architecture in modern Britain. In 1841 he became an active member of the Oxford Society for Promoting the Study of Gothic Architecture. It is not unlikely that it was here that Freeman had become acquainted with Dahl's stave churches; the long note on the stave churches in his architectural history ends with the following: "See a fuller description [of the churches] in the Proceedings of the Oxford Architectural Society for Michaelmas Term, 1843, communicated by the Rev. J. L. Patterson" (Freeman 1849, 251). It is plausible that Freeman was, in fact, present when Patterson presented Dahl's *Denkmale* on 15 November.

5 Conclusion: Collections, Connections, Pasts and Futures

When the Oxford Society for Promoting the Study of Gothic Architecture was founded in 1839, members agreed that the object of the Society should be to "collect Books, Prints, and Drawings; Models of the forms of Arches, Vaults, & c.; Casts of Mouldings and Details" (Oxford Society 1839, unnumbered). This was completely in line with the art/architectural historical ambition of using the past to build the future, and the society came to actively encourage the use of the Gothic Revival style in new church architecture (Prout 1989, 379). Some of the prints collected by the Society were the stave churches drawn by Schiertz in Norway in the summer of 1837 and published by Dahl in Dresden the same year. When presented to the society by Mr. Patterson, the churches were interpreted as a form of gothic architecture: "We have here all the elements of a really fine Church, great loftiness, sufficient length, divided into Nave, Transept, and Chancel; and breadth, divided into Nave and aisles, with a clerestory over" (Oxford Society 1843, 15).

That the stave churches were understood as an expression of gothic style is not surprising. What might be a bit more surprising is part of the discussion that followed after Mr. Patterson had presented *Denkmale*. Mr. Patterson remarked in conclusion that his reason for presenting these lithographs to the society was a hope that "they might afford some useful hints for the erection of similar Churches in countries where the same materials and no others were readily to be found". More concretely he mentioned the British colonies of

15 Members of The Cambridge Camden Society, from 1845 known as the Ecclesiological Society, the most important architectural society in England at the time.

Newfoundland and New Zealand. Following this, the Chairman, the Rector of Exeter College, agreed that:

> these examples of ancient wooden Churches are of great importance at the present time and, rude, mutilated, and patched, as they are, there is still much in them worthy [of] the attention of a clever architect, who might from the materials and ideas here furnished, supply a great desideratum for many of our colonies. (Oxford Society 1843, 15–16)

He "trusted that the subject would not be suffered to drop, but that some competent architect would come forward and carry out the idea" (Oxford Society 1843, 16).

As has been shown in this chapter, Dahl's collection of stave churches circulated widely, partly by itself, but also to a large degree with the help of an emerging field of architectural history and a new genre of publication, which had the ambition of collecting and connecting architecture from "all" places in all times. Although the examples shown in this chapter are partly taken out of their larger textual and academic context, the ideas they convey still show us something important: that in the inner landscapes of Norway, in the middle of the nineteenth century, it was not merely that some picturesque and curious wooden churches, representing something distinctively Norwegian, existed. Rather, for nineteenth century architectural historians these buildings, through their structural design, ground plans and ornamentations, were sources to important knowledge about cultural connections and histories of influences and mobility – in addition to providing inspiration for new (church) architecture. By being collected on paper, the church of Borgund and the other medieval wooden churches could also be connected to a larger – allegedly universal – history of art and cultural exchange, and to possible futures. When discussing imaginings and interpretations of the North it is thus important to remember that not all these imaginings and interpretations emphasized what was characteristic, distinctive and unique about the North. For some, the most important task was to find similarities and commonalities connecting the North to the rest of the world.

Bibliography

Ainsworth, William Francis. 1866. "A Trip to Norway." *All Around the World*, vol. 4: 1–64.

Aubert, Andreas. 1920 [1893]. *Maleren Johan Christian Dahl. Et stykke av forrige aarhundredes kunst- og kulturhistorie*. Kristiania: Aschehoug & Co.

Barton, Arnold, H. 2007. "The Discovery of Norway Abroad, 1760–1905." *Scandinavian Studies*, 79 (1 Spring): 25–40.

Bjordal, Sine Halkjelsvik. 2021. *'Om denne haves intet mærkværdigt' En tekst- og kunnskapshistorisk studie av stavkirkene på 1700- og 1800-tallet*. PhD dissertation. Oslo: University of Oslo.

Bremner, G. A. 2023. "An Aryan Descent: Race, Religion, and Universal Civilization in E. A. Freeman's A History of Architecture (1849)." In *Narrating the Globe. The Emergence of World Histories of Architecture, 1790–1900*, edited by Petra Brouwer, Martin Bressani & Christopher Drew Armstrong, 277–94. Cambridge, Massachusetts: The MIT Press.

Bressani, Martin. 2023. "Finding True North: Nation, Race, and Architecture in Germany 1770–1850." In *Narrating the Globe: The Emergence of World Histories of Architecture, 1790–1900*, edited by Petra Brouwer, Martin Bressani & Christopher Drew Armstrong, 99–130. Cambridge, Massachusetts: The MIT Press.

Brouwer, Petra. 2023. "A New Genre: World Architectural History Surveys." In *Narrating the Globe: The Emergence of World Histories of Architecture, 1790–1900*, edited by Petra Brouwer, Martin Bressani & Christopher Drew Armstrong, 19–66. Cambridge, Massachusetts: The MIT Press.

Brouwer, Petra, Martin Bressani & Christopher Drew Armstrong. 2023. *Narrating the Globe: The Emergence of World Histories of Architecture, 1790–1900*. Cambridge, Massachusetts: The MIT Press.

Chopra, Preeti. 2023. "The West is the Best: World History of Architecture as Imperial Form of Knowledge." In *Narrating the Globe: The Emergence of World Histories of Architecture, 1790–1900*, edited by Petra Brouwer, Martin Bressani & Christopher Drew Armstrong, 131–58. Cambridge, Massachusetts: The MIT Press.

Dahl, Johan Christian Claussen. 1837. *Denkmale einer sehr ausgebildeten Holzbaukunst aus den frühesten Jahrhunderten in den innern Landschaften Norwegens*. Dresden: J.C.C. Dahl.

Dietrichson, Lorentz. 1890. "Vangs stavekirke i Schlesien." *Aftenposten* 22 May 1890.

Fergusson, James. 1855. *The Illustrated Handbook of Architecture*. 2 Volumes. London: John Murray.

Fjågesund, Peter & Ruth A. Symes. 2003. *The Northern Utopia: British Perceptions of Norway in the Nineteenth Century*. Amsterdam: Rodopi.

Forester, Thomas. 1850. *Norway in 1848 and 1849: Containing Rambles Among the Fjelds and Fjords of the Central and Western Districts*. London: Longman, Brown, Green, and Longmans.

Gaimard, Paul M. 1852. *Voyages de la Commission scientifique du Nord, en Scandinavie, en Laponie, au Spitzberg et aux Feröe : pendant les années 1838, 1839 et 1840 sur la corvette la Recherche*. Volume 1: Atlas historique et pittoresque. Paris: Arthus Bertrand.

Hvattum, Mari. 2023. *Style and Solitude: The History of an Architectural Problem*. Cambridge, Massachusetts: The MIT Press.

Johnston, J. 1884. *How we Spent our Summer Holiday in Norway*. Reprint from *The Bolton Weekly Guardian*. Bolton: George Winterburn.

Karlholm, Dan. 1996. *Handböckernas konsthistoria. Om skapandet av 'allmän konsthistoria' i Tyskland under 1800-tallet*. Stockholm/Stehag: Brutus Östlings Bokförlag Symposion.

Karlholm, Dan. 2004. *Art of Illusion: The Representation of Art History in Nineteenth-Century Germany and Beyond*. Bern: Peter Lang.

Kaufman, Edward. 1989. "Architecture and Travel in the Age of British Eclecticism." In *Architecture and its Image. Four Centuries of Architectural Representation*, edited by Eva Blau & Edward Kaufman, 58–85. Montreal: Centre Canadien d'Architecture.

Knutsen, Niels M. 2002. *La Recherche. En ekspedisjon mot nord*. Tromsø: Angelica Forlag.

Kugler, Franz. 1837. "Denkmale einer sehr ausgebildeten Holzbaukunst aus den frühesten Jahrunderten in den innern Landschaften Norwegens. Herausgegeben von J. C. C. Dahl". Review. *Museum* 39: 517–521.

Kugler, Franz. 1842. *Handbuch der Kunstgeschichte*. Stuttgart: Ebner & Seubert.

Latour, Bruno. 1986. "Visualisation and Cognition: Drawing Things Together." In *Knowledge and Society: Studies in the Sociology of Culture Past and Present*, edited by H. Kuklick, 1–40. New York: Elsevier.

Lübke, Wilhelm. 1855. *Geschichte der Architektur von den ältesten Zeiten bis auf die Gegenwart*. Leipzig: Verlag von Emil Graul.

Murray, John. 1849. *Hand-book for Northern Europe*. London: John Murray.

Murray, John. 1858. *A Hand-book for Travellers in Denmark, Norway, Sweden, and Iceland*. 3rd revised edition. London: John Murray.

Murray, John. 1869. *A Knapsack Guide to Norway*. London: John Murray.

Oxford Society. 1839. *The Rules and Proceedings of the Oxford Society for Promoting the Study of Gothic Architecture*. Oxford: The Society.

Oxford Society. 1843. *Proceedings of the Oxford Society for Promoting the Study of Gothic Architecture: For Michaelmas Term*. Oxford: The Society.

Prout, David. 1989. "'The Oxford Society for Promoting the Study of Gothic Architecture' and 'The Oxford Architectural Society', 1839–1860." *Oxoniensia* LIV: 379–91.

Pierre-Victor. 1841a. *Coup d'oeil sur les antiquités skandinaves ou aperçu général des diverses sortes de monuments archéologiques de la Suède, du Danemark et de la Norvège*. Paris: Challamel.

Pierre-Victor. 1841b. *Rapport Fait à la Société libre des Beaux-arts, par Pierre-Victor, membrecorrespondant, sur d'anciennes constructions en bois sculpté de l'intérieur de la Norvège*. Paris: Imprimerie de Ducessois.

Pritchett, Robert Taylor. 1879. "Gamle Norge." *Rambles and Scrambles in Norway.* London: Virtue & Co.

Rumohr, Carl Friedrich von. 1840. Review of *Denkmale.* In *Münchner Jahrbücher für bildende Kunst*, 2, edited by Rudolf Marggraff, 344–47. Leipzig: Wilhelm Engelmann, Rudolph Weigel.

Schnaase, Carl. 1854. *Das eigentliche Mittelalter*. Bd. 4, Abth. 1. Geschichte der bildenden Künste im Mittelalter; Bd. 2, Abth. 2. Geschichte der bildenden Künste. Düsseldorf: Buddeus. https://doi.org/10.11588/diglit.1293#0003.

Schwarzer, Mitchell. 1995. "Origins of the Art History Survey Text." *Art Journal. Rethinking the Introductory Art History Survey*, 54 (3): 24–29.

Stensrud, Iver Tangen. 2021. "'Europe is Becoming Dreadfully Used Up': Travelling Images of Norway in the Illustrated Press." In *Nordic Travels,* edited by Janicke Kaasa, Jakob Lothe & Ulrike Spring, 121–40. Oslo: Novus.

Tylor, Bayard. 1864. *Northern Travel: Summer and Winter Pictures: Sweden, Denmark and Lapland.* New York: G.P. Putnam.

Wexelsen, Einar. 1973. *J.C. Dahl. Forståelse og innsats for bevaring av norske fortidsminnesmerker*. Thesis. Bergen: University of Bergen.

Williams, W. Mattieu. 1877. *Through Norway with Ladies.* London: Edward Stanford.

Worm, Andrea. 2009. "The Study of Medieval Illuminated Manuscripts in German Scholarship *ca.*1750–1850." In *Medieval Art and Architecture after the Middle Ages*, edited by Janet T. Marquardt & Alyce A. Jordan, 246–73. Newcastle upon Tyne: Cambridge Scholars Publishing.

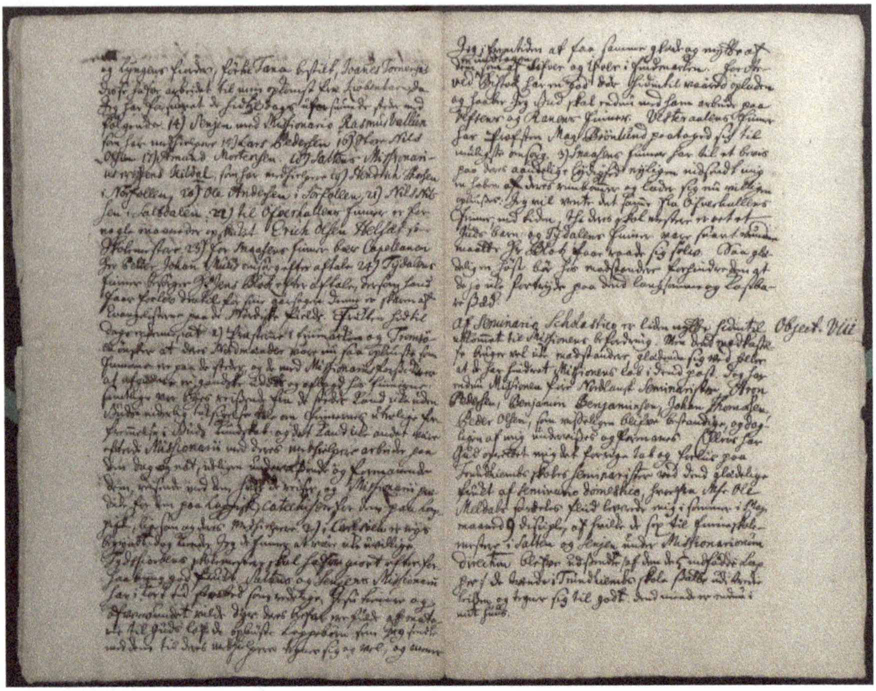

FIGURE 7.1 Extract from the Mission Collegium's archive

CHAPTER 7

The Archive of the Mission Collegium: Twice repatriated, Once Destroyed?

Harald Lindbach

Abstract

The article addresses the problems that arise when archives are fragmented and divided, which is common for minorities and indigenous peoples living on different sides of national borders. One of the most important archives for Sámi history is the Missionary Collegium archive, which covers much of the eighteenth century. Today, part of the archive is located at the Sámi archive in Guovdageino, and another part is at the National Archives in Copenhagen. Can digitization help to recreate the archive as it once was? Do the different national digitization platforms pose obstacles?

Keywords

archives – Sámi – history – digitization – eighteenth century

Any archivist will be happy to tell you about how archives are not collections, and how they differ from collections. Archives are by-products of work processes; the creation of the archive is a necessity for obtaining the goals of the creator of the archive. For the creator the archive is the set of tools that assures that the goal is achieved in the best possible way.

Although there are distinct differences between a collection and an archive, the archive can also be viewed as a collection: the information contained and kept in the archives are collected to serve certain purposes. The purpose of the information collected is related to and dependent on the tasks the organization is set up to achieve. A tax office and a health centre will be asking you for different information, but both institutions will collect and keep information

 | DOI:10.1163/9789004749665_008

as a necessary part of the workflow. The information in the archives obviously depend on the purpose of the creator.

To rule and exploit remote or colonized areas states need to collect large amounts of information. The presence of minorities and Indigenous peoples in the archives of the dominant culture may seem to undergo a three-part process (Lindbach 2019, 197–98). In the first phase, the government collects information, it maps the population and the natural resources and obtains information about geographical and cultural conditions. In the next phase, the integration of the local population into the dominant culture takes place. The church and legal system have been developed at this stage and are being used by the population. Now the minority's own voice emerges to some extent in the archival material, but mainly in the language and on the premises of the dominant culture. This process does not necessarily happen at the same time everywhere, but is related to proximity to institutions, integration in trade and business, and size ratio between minority and majority population. The archive of the Mission Collegium partially documents this phase, which in the Sámi area in northern Scandinavia can be said to characterize the period from the end of the eighteenth century to the beginning of the twentieth century.

The third phase occurs in connection with the development of organizations and institutions for and by minorities and indigenous peoples. Archiving takes place to a greater extent on the minorities' own premises, and often also in their own language, but document management is still intimately connected with the dominant culture's institutions. There is a constant flow of information both ways. Economically and administratively, there is also often a relationship of dependence between the state's and the minority's organizations and institutions.

In what may be described as the post-colonial period repatriation has been an important part of the reconciliation process, in Norway manifested through the work of The Truth and Reconciliation Commission, which was set up by the parliament in 2018. Objects of cultural and historic importance have been returned to the place of origin. In numerous more serious cases skeletons, skulls and other bodily remains have been returned and been given a proper burial according to local religious practice. However, not all colonial remains are equally suitable for repatriation.

In what follows, I will focus on the archive of the Danish-Norwegian Mission Collegium, and especially the part of the archive that was handed over to Norway in the 1820s and covered the mission among the Sámi in the eighteenth century. As an archivist specializing in the archives of the northern region of Norway, I have been involved in the decision to move this part of the archive from the National Archive of Norway in Oslo to the Sámi Archive in Guovdageino (Kautokeino). The transfer of the archive was obviously influenced by the idea of repatriation, of returning objects of high historical and cultural value to the place of origin, but also by a general principle that archives should, to the greatest extent possible, be accessible to the group that the archive describes, the group that is likely to use the material most diligently.

1 The Creation of the Collegium

In 1714 king Frederik IV of Denmark created *Collegium de cursu evangelii promovendo*, normally referred to as the Mission Collegium. First and foremost, the collegium was intended as a tool for the Christianization of the people of India (Steen 1954, 102). Tarangambadi, or Trankebar as it is commonly referred to in the Danish and Norwegian source material, was a Danish colony from 1620 to 1845 when it became part of British India. The mission in Trankebar became operational already in 1705 when the missionaries Bartholomeus Ziegenbalg and Henrik Plütschau arrived in India (Uglem 1979, 12). Both were educated in Halle, a stronghold of pietism.

The Mission Collegium was given the responsibility for the mission in India in 1715. Already the following year the collegium was also given responsibility for missions among the Sámi in Norway. Mission districts were established from Snåsa in the south to Eastern Finnmark in the north. The mission in Greenland, a Danish colony, was also administrated by the Mission Collegium from 1737.

The Mission collegium functioned as a collegium until 1771. What was the origin of the name, and which primarily categorized the collegium system introduced in the late seventeenth century, was that the various collegiums consisted of an expert management group who were formally equal. The collegium had the right of presentation, which meant that they had the right to put matters directly before the king. After 1771, the institution was subordinate to the Danish chancellery (Stoa 1989, 14).

From 1380 to 1814 Norway was united with the kingdom of Denmark. After the loss of independence when the national council was deposited in 1536 Norway simply became a part of the Danish kingdom. This meant that most archives of national importance were created and kept in the capital, in Copenhagen. This was also where the administration of the mission collegium was situated.

Although the main task for the Mission was the Christianization of the East Indies a circular regarding other possible mission assignments was issued in 1715. A group of pietistic clergymen known as The Seven Stars (Syvstjernen) had for a period tried to influence the Danish authorities with proposals of stronger church discipline, better theological education and stronger emphasis on the catechism and piety. The group replied to the circular with a suggestion for mission among the Sámi people of Norway and the authorities were positive (Steen 1954, 108–09). In 1716 the former vicar and member of The Seven Stars Thomas von Westen was given the task of leading the mission in Norway, a position he held until his death in 1727.

A reason for the urgency in setting up a mission among the Sámi in Norway may have been the great Nordic war (1700–1723), where Sweden and Denmark where on opposite sides. Missionary activities among the Sámi could contribute to consolidating the Danish-Norwegian right to supremacy in the northernmost areas of Scandinavia, a proper border between Norway and Sweden in the northern areas was not determined until 1751.

Another Norwegian cleric, the Vicar Hans Egede at Vågan in Lofoten, had already sent several proposals regarding mission among the people of Greenland. This would take a little longer, but in 1721 the financial foundations where in order and Hans Egede could sail off to Greenland where he was to lead the mission for the next 15 years.

2 The Transfer and the Split

Denmark suffered defeat in the Napoleonic Wars and as a result lost Norway which declared its independence and adopted its own constitution on 17 May 1814. However later that year Norway was joined with Sweden in a personal union. After the break with Denmark archives that specifically dealt with Norwegian affairs were transferred to Christiania (Oslo). In 1845, there were an estimated 2,500 shelf meters in the new Norwegian national archive at Akershus fortress that housed the national archive at the time. About half of this were archives that had been transferred to Norway in the 1820s due to Denmark's loss of Norway (Herstad 2008, 187).

Among the archives transferred to Norway were parts of the Mission Collegium. Six shelf meters that covers the period from 1688–1811 were transferred to the national archive of Norway. The transfer thus contains some archival material that predates the creation of the collegium in January 1714. This material has most likely originally been transferred from the Danish Chancellery to the Mission Collegium. Parts of the archive are unfortunately lost. In 1795, letters received in the period 1715–1734 were destroyed in a fire in Copenhagen (Steen 1954, 49).

The part of the mission collegium's archive that is currently kept at the Danish National Archives consists of the archive of the Missions collegium and the Directorate for Waisenhuset (an orphanage) which was co-located with the mission, which covers the period 1714–1859. The archive consists of a total of 17 volumes and 82 packages. Most of this covers matters that specifically concern the mission in Greenland or the East Indian colonies. Material relating to Waisenhuset is also incorporated in the archive since when it was established in 1727, it was subordinated to the Mission Collegium (Kornerup 1943, 152). However, important source material related to the mission among the Sámi can also be found in Copenhagen. A recent example is a letter by the missionary Thomas von Westen dated 1718 that was discovered by the researchers Liv Helene Willumsen and Randi Hege Skjelmo (Willumsen & Skjelmo 2017). The letter sheds new light on the work of the mission. The letter particularly describes the working and living conditions for the local missionaries and possibilities and plans for future improvements.

3 The Transferred Archive

The part of the Mission Collegium's archive that was transferred to Norway consists of a copybook for the Finnmark mission which covers the years 1784–1802 but is missing the year 1796. A copybook is a collection of all outgoing correspondence arranged chronologically to ensure retrieval. Furthermore, there are drafts for outgoing letters for the period 1795–1802 which is missing the year 1799, as well as a journal of incoming letters for the period 1715–1788. The archive also contains an archive series with royal rescripts dealing with the mission and covering the period 1700–1802. A long series of 29 archive boxes with incoming letters, make up the main part of the archive, and cover the period from 1725–1802. The archive also contains an archive series with cases arranged by subject. This series covers the period from 1708 to 1800. Among other things, it contains accounts of the missionary calls, and matters concerning the schools that the mission set up. Finally, the archive contains a total of 30 serial numbers with accounting material for the period from 1708 to 1811. This series also contains accounting material for Seminarium Lapponicum, a seminarium that was set up in Trondheim in 1752 to educate teachers and missionaries in the Sámi language. The school was closed in 1774 after the passing of Professor Knud Leem who himself had a background as a missionary and priest in Sámi areas and wrote several treatises on the Sámi language.

The archive contains a series of royal rescripts. A rescript was a letter to the official or officials who had to comply with the order given by the king and was often used when there was no general legislation that covered everyone, but an injunction aimed at a specific official or a group of officials. The royal instruction regarding the introduction of the mission is found among these documents. Here it is stated that the mission in addition to the Sámi also included the Rufinns (Rye-Finns), a term normally used to describe Skogfinnene (the Wood Finns), a group of Finns that in the seventeenth century immigrated from Finland to the remote coniferous forest area between Sweden and Norway in the south. However, the term may also describe the Kvens of northern Norway. The Kvens is a people closely related to the Finns of Finland, that traditionally inhabited the Torne Valley area, between what is now Finland and Sweden and settled in the northernmost part of Norway at least from the seventeenth century on. The Kvens do appear in the archival material as teachers, pupils and members of various congregations covered by the mission.

The royal rescripts obviously reflect the views of the highest administrative level. It is nevertheless natural to see the rescripts as reactions to the information that comes in from priests and missionaries, and that they thereby reflect the reality and the practical problems that face the missionary collegium and the church's employees at the various places of work. In the archive series containing the royal rescripts we find a rescript from 1728 which is a royal response to a complaint the missionary Simon Kildal has submitted on behalf of all missionaries to Bishop Eiler Hagerup (Mission Collegium Da L0001/0003). The points of complaint, which are reproduced in the rescript, are that the Sámi oppose the activities of the missionaries and the schoolmasters and make threats towards them. It is further complained that despite the king's order of 27 September 1726 against selling liquor to the Sámi, this trade is still ongoing.

However, the series does contain more than royal rescripts. Several cases that never reach such a high administrative level can also be found in this part of the archive. In 1760 the Sámi youth Henrich Mortensen from Skjervøy – north in Troms county – sends a humble request to the king for financial support to fulfil studies so that he may become a priest and work among his own (Misjonskollegiet Da L001/003). The letter has three attachments, the first is a recommendation from the priest in Skjervøy, Cornelius Duns. Here it says that Mortensen was born in Skjervøy parish. He is "born of Sámi parents" (fød af finne forældre) and "has been a light for the youth of his people" (har været et lys for det folkes ungdom).[1] The other two attachments are a prayer, which

1 All translations by the author.

can either be a copy or written by Mortensen himself, as well as a well-crafted handwritten copy of the medieval Latin hymn "Puer Natus in Betlehem".

However, after Mortensen spent half a year waiting in Copenhagen, the reply that he received was not positive. He was to be sent to Bergen as soon as possible and then back to his place of birth where he may be given some training by his local vicar so that he could be of use to his fellowmen, not as a vicar, but as catechist or teacher (Misjonskollegiet Da L0001/0003).

The treatment of Mortensen was in stark contrast to the attitude of the Swedish church which already early on recruited Sámi priest, but in accordance with the development of the mission on the Norwegian side where local knowledge and Sámi language were gradually ignored by the authorities toward the end of the century. This is reflected in the copy books of the collegium that covers the period from 1784 to 1802 (Misjonskollegiet DA L0001). The entries largely provide information related to administrative and financial dispositions.

On 27 June 1789, a letter was sent from the Mission Collegium to Bishop Schønheyder. The Dane Johan Christian Schønheyder, who was bishop in Nidaros (Trondheim) from 1788 to 1803, was an avowed opponent of using the Sámi language in connection with school and mission activities (Steen 1954, 276). The letter points to the uncertainty that has become apparent on whether to use Sámi language or not. It states that the different Sámi languages and dialects make it difficult to develop books that can be useful everywhere, and it asks for advice on how to promote the Danish language among the Sámi.

Ten years later, the mission collegium's language policy has clearly turned in the direction of a hostile attitude towards the Sámi language. In connection with Johan Christopher Lohman being appointed as a missionary in Skjervøy in a letter of 1 June 1799 (Jnr. 25/99) it is indicated that the letter of vocation instructs him to ensure that the Danish language is more and more introduced among the old and young Sámi, and that this is considered in the schools.

The series of incoming letters consists of a total of 29 archive boxes and consists of letters received by the Mission Collegium. There is thus a large mismatch between the outgoing and incoming correspondence in this archive. An actual series of outgoing correspondence, i.e., copy books, exists only for the years 1784 to 1802. The incoming correspondence deals partly with financial and practical problems relating to the work of the mission. The series does also contain some rapports from the local missionaries on the conditions in the local Sámi congregations. These may shed some light on living conditions and customs in various areas. However, a lot of material first and foremost contains information on more practical problems in the various mission areas and congregations, often with the emphasis on economy. Much correspondence concerns checking that allocated funds have been used correctly, that buildings have been improved and that this has been done in the prescribed manner.

The introduction of the missionaries and schoolmasters to the established congregations often caused problems for the local vicars. Many of the priesthoods were initially in a difficult financial situation and took a large part of its income from religious services to the local community. When the missionaries where introduced parts of this income had to be divided, leading of course to dissatisfaction and correspondence. This contradictory relationship was particularly visible in the first half of the eighteenth century, but never completely disappeared.

4 The Principle of Provenance

As we have seen, the mission's archive was split up and distributed between two different states. Based on modern archival principles, it is inconceivable that this would be resolved in the same way in our time.

The combination of the eighteenth century's scientific focus on classification systems associated with the biologist Carl von Linné, the zoologist Georges Cuvier and others, and the emergence of the first real depository institutions came to have an impact on the way archives were catalogued and arranged. The pertinence principle is in many ways a logical consequence of its time. Especially in the latter half of the eighteenth century, arrangement forms were developed in which subject-wise arrangement principles were applied. This led to archives being routinely broken up and reconstructed so that material dealing with the same subject was combined and archived in subject based systems. This arrangement principle allowed for very subjective assessments and was also extremely labour-intensive. The most important negative consequence was that the internal logic of the individual archives was destroyed, which subsequently made reconstruction of proceedings very difficult or impossible. In practice, archive organization according to pertinence meant that individual archives were turned into collections of individual documents from different archives.

The Dutch handbook "Handleiding voor het Ordenen en Beschrijven van Archieven" from 1898 is often regarded as the cornerstone of modern archival theory and methodology. Although in many ways it appears as a clear break with previous methods and theory, it is also the result of a development that had taken place over time. In 1891, the world's first professional archive organization (Vereniging van Archivarissen) was founded in the Netherlands. Both the organization's member magazine and the organization's annual member seminar became a forum for discussions related to methods for organizing and cataloging archives (Horsman, Ketelaar & Thomassen 2003, 254).

In the Dutch manual or handbook from 1898, the principle of provenance is established. The principle is two-fold: external provenance sets requirements to keep each individual archive from each individual creator for themselves, while internal provenance aims to preserve the internal order of the individual archive. External provenance had its origin in the French term *respect des fonds*, which was introduced in the regional archives in France in 1841 (Marthinsen 2012, 38). Internal provenance had its origins in a Prussian regulation from 1881, that acquired that archival documents should retain the order in which the material received during the proceedings. However, it was the Dutch manual that provided an archival scientific basis for the principle. The handbook gained enormous importance within the archival profession.

5 Documentation of Transformation

On 14 January 1814, a peace treaty between the Danish and Swedish kings was concluded in Kiel. The peace agreement in Kiel threw Norway into a series of events that led to a separate Norwegian constitution and a loose union with Sweden. The treaty also stated that archives should be transferred from the government offices in Copenhagen. This resulted in negotiations about the extent of the transfers in 1819. It turned out that the Danes wanted to get rid of larger quantities of Norwegian archives than the Norwegian authorities had planned for. This resulted in a lack of depository capacity and probably also in a certain degree to lack of control.

The transfer thus took place before the principle of provenance was generally established as a foundation for archival science. The result is that the archive was divided according to geography and topic. If one studies the archive as it appears in the Norwegian depository and catalogue today one will easily get the impression that it solely covers the Sámi areas in Norway, and that the mission's main activities were aimed at the Sámi, and to a certain extent also the Kvens of Finnmark and Troms. The collegiums' top decision-making level often appears as invisible in the material and the connection between the various mission areas is lost. The archival material as it is presented today can therefore easily lead researchers to draw simplified and hasty conclusions. This will of course also apply to those parts of the archive which are still in the Danish National Archives in Copenhagen. For more than one hundred years both archives have been incomplete, a situation that obviously has made research more difficult.

In 2021 the Norwegian part of the archives was moved to the Sámi Archives in Guovdageino (Kautokeino). The idea behind the move was to place archival material that is important for Sámi culture and history in an institution that specializes in the field. It is of course also important that the archives are situated in the heart of the Sámi homeland and are co-located with the Sámi college.

Sámi Archives began as a collaborative project in 1988 to secure source material for Sámi history. A foundation was established in 1995 and ten years later management responsibility was transferred to the National Archives and today the archive is a separate institution within the Norwegian National Archives with national responsibility for collecting and mediating Sámi archives. In addition to the archive of the Mission Collegium the archive also holds the Norwegian copy of Lappekodisillen, an addendum to the border treaty of 1751 between Denmark-Norway and Sweden that amongst other things secures the right of the Sámi to cross the borders freely with their reindeer herds. Lappekodisillen is often referred to as the Magna Carta of the Sámi. Apart from these archives, the institution's holdings mostly consist of private and public material from modern Sámi society. The two archives from the eighteenth century thus widen the scope of the Sámi archives considerably.

Although, as we have seen, the archive of the Mission Collegium mainly documents the administrative and economical sides of the mission, as more or less a byproduct there is a considerably amount of information about the transformation of Sámi society during the time of the Collegium that can be deducted from the archive.

The introduction of missionaries to the different Sámi areas also meant that schoolmasters were assigned, normally two to three teachers per missionary (Steen 1954, 140–41). In the early years, the mission placed a strong emphasis on teaching and spreading Christianity in the Sámi language. However, by the middle of the eighteenth century this was challenged by priests in districts where there was a lack of missionaries were solely using Norwegian or Danish (Lindbach 2019, 93–94).

Knowledge of Christianity, knowledge of the official state languages and, eventually, reading skills, become important factors for incorporating the Sámi into the state community in the eighteenth century. These are factors that trigger or exclude from citizenship rights. Although these requirements can eventually be described as general, they applied to all residents, there is little doubt that the requirements were more difficult to achieve for populations with their own language and their own religious tradition and culture, this making it difficult for them to receive full citizen rights.

The Lutheran confirmation that was introduced in 1736 meant that literacy became even more important. Everyone should know the Holy prayer, the creed, the ten commandments and other central texts by heart. To be able to go to communion, get married or act as godfather at a child's baptism, one had to satisfy certain requirements. This also applied to those who had to take an oath in connection with proceedings, which could cause problems for those who were not found knowledgeable enough to take an oath in a court case. This seems to have particularly affected the Sámi, a result that, although not apparent in the missions' own archives can be seen in the local court records (Lindbach 2019, 112–14).

However, the arrangement with schoolmasters in the districts seems to have had great effect, in a report from 1759 Bishop Gunnerus says he is so satisfied with the arrangement that he proposes to extend it to several districts (Lindbach 2019, 187). There are obviously several stages in mastering the alphabet. For most people at the time, the art of reading was primarily an aid to learning the necessary frameworks of faith by heart. For this, a basic ability to spell through text will be sufficient. In the eighteenth century, there were far fewer people who mastered the art of writing than there were people with elementary reading skills. Still, some went further and found the knowledge of writing quite useful (Lindbach 2021, 286). We know that writing has been used as a mean of communicating between reindeer herders when one part was situated by the sea in Norway and the other part was still situated on the Swedish side of the border (Lindbach 2019, 122).

6 The Archive, Long Divided, Must Unite

All in all, the archive of the Mission Collegium is extremely important as source material for Sámi history. It is the only archive from the eighteenth century that focuses mainly on subjects relating to the Sámi. When used together with other source material such as episcopal archives and court records a broader understanding of both the mission itself and the Sámi societies of the time may be uncovered.

Today the Norwegian branch of the archives in the depositories of the Sámi archives are being digitized. This is a slow process due to technical challenges such as various formats and questions related to metadata supply, cataloguing and presentation. Completed in digital form, the archive will constitute a treasure chest for researchers and will undoubtedly contribute to bringing to light previously unknown elements of Sámi history.

Digitization will also, in the longer term, make it possible to reconstruct the original and actual order of the archive. To achieve this, it is a prerequisite that the Danish National Archives also digitizes its part of the archive. Still there remain problems related to the fact that the various national archives have different presentation platforms. In other words, it may be necessary to present a complete archive from the Mission Collegium from both the Norwegian and Danish sides.

A reconstructed archive will shed light on the connection between the various series and archive parts that now remain hidden. To have full access to the archive at the present, one needs to visit both the Sámi Archives in Guovdageino and the National Archive in Copenhagen. In addition, one needs a fairly advanced understanding of archival theory and practice in the eighteenth century to get a full understanding of how the pieces were once joined together.

The gains of a reconstructed archive would be immense. The administrative context will become clear, researchers and others will be able to follow the communication back and forth from central to local levels. The connection between the various mission fields will be easier to uncover. Recreating the original order in the archive will make it possible to see the connection between the Sámi areas, Greenland, and the mission in the Danish colonies in the East Indies and in the 1790s in the Danish slave islands in the West Indies.

Bibliography

Herstad, John. 2008. *"Archivaren – Gud bevar'en": Henrik Wergeland som riksarkivar.* Oslo: Universitetsforlaget.

Horsman, Peter, Eric Ketelaar & Theo Thomassen. 2003 "New Respect for the Old Order: The Context of the Dutch Manual." *American Archivist* 66 (2): 249–70.

Kornerup, Bjørn. 1943. *Danske Kancelli og de dermed beslægtede institutioner.* København: Heitzels.

Lindbach, Harald. 2021. "Gateway to the Sámi Past: The Sámi Hidden in Archives from 18th Century Scandinavia." In *Library and Information Studies for Arctic Social Sciences and Humanities*, edited by Spencer Acadia & Marte Fjellestad, 278–93. London and New York: Routledge.

Lindbach, Harald. 2019. *Minoritetspolitiske dokumentasjonsstrategier i Nordområdene på 1700–tallet.* Tromsø: Universitetet i Tromsø.

Marthinsen, Jørgen H. 2012. *Arkivteori – en innføring.* Riksarkivarens skriftserie 36. Oslo: Universitetsforlaget.

Steen, Adolf. 1954. *Samenes kristning og finnemisjonen til 1888.* Oslo: Egede

Stoa, Nils Johan. 1989. *Arkivmateriale fra Dansk-Norske sivile sentralinstitusjoner før 1814*. Oslo: Universitetsforlaget.

Uglem, Olav. 1979. *Norsk misjonshistorie*. Oslo: Lunde.

Willumsen, Liv Helene & Randi Hege Skjelmo. 2017. *A Letter by the Missionary Thomas von Westen*. Trondheim: The Royal Norwegian Society of Sciences and Letters.

FIGURE 8.1 John Linnell (1792–1882), after Edward Price (active *c.*1823–1854), View Across the Fiord from Herrinsholmen, circa 1826–1827 from series Seven Views in Norway, etching on paper, UCL Art Museum, LDUCS-1456
REPRODUCED WITH PERMISSION FROM UCL ART MUSEUM

CHAPTER 8

Collecting, Re-collecting and Connecting with the North: Teaching and Promoting Scandinavian Studies through UCL's Collections

Elettra Carbone and Jesper Hansen

Abstract

In this chapter we examine and theorise pedagogical approaches to the use of archival fragments in language teaching and learning. Embracing the findings of several reports on university collections showing how these are spaces with the potential to bring teaching, research and public engagement together, we discuss how UCL's Collections have been used in the past eight years to promote and enhance the visibility of Scandinavian Studies in the UK and raise the profile of foreign language learning. We begin by presenting in detail a teaching project we carried out in 2015/2016, which initially led to a series of joint Danish and Norwegian language classes designed around some of the Nordic fragments we unearthed in UCL's Collections. This project formed the basis of several further outputs, including an outreach and widening participation project for the promotion of multilingualism and language awareness which culminated with the showcase *Not Just Words: Learning languages through art and objects* (January 2023–September 2024). Ultimately, we argue that the showcase, which featured five Nordic languages as part of its video tasters designed around items from the UCL Art Collections, championed a new approach to collection-based language teaching which emphasises the indissoluble link between language and culture.

Keywords

object-based learning – foreign language teaching and learning – Scandinavian Studies – Nordic Studies – university collections – pedagogy – teaching-based research

The copy of *View Across the Fiord from Herrinsholmen* featured here can be found today as part of the over 10,000 artworks that make up the UCL Art

 | DOI:10.1163/9789004749665_009

Collections. Together with the Scandinavian Studies library collections and other papers and artworks linked to the Nordic region, this is only one of the many fragments or "microhistories of 'displaced' things and persons" that represent a piece of Nordic cultural history in the UCL Collections (Greenblatt 2010, 16–17). University collections like UCL's constitute a particular type of archive which, historically, has gone through different stages: an expansion phase from the eighteenth century to the early twentieth century where collection-based research and teaching were widely recognised as indispensable; a period of marginalisation during the twentieth century where that connection between collections, research and teaching was partially lost as university collections were seen as old-fashioned and lacking relevance to contemporary university curricula; finally, a period of revival from the end of the twentieth century where the university collections were rediscovered as material heritage with cultural significance and able to initiate new lines of enquiry in teaching and research (Ludwig and Weber 2013, 652–53; Hannan, Duhs & Chatterjee 2013, 162). According to a national survey from 2000, there are roughly 400 higher education museums, galleries and collections in the UK (Arnold-Forster & Mirchandani 2001, 48) and several studies have highlighted in the last decade how university collections can be a valuable resource to bring teaching, research and external engagement together (Kelly 2001, 197).

In this chapter, we explore what "collecting the North" means in the context of UCL's university collections, reflecting on how – as highlighted in the anthology *Managing University Museums* on the role and organisation of this specific type of museum – university collections are "national and international resources" that "can authenticate theoretical teaching by providing original artefacts for practical study that enrich the learning experience" and "be a place of innovative thinking" which can become "an integral part of academic programmes" (Kelly 2001, 10–11). What role do the Nordic items in the UCL collections play in supporting Scandinavian Studies teaching at UCL, the first institution in the UK where the academic study of Scandinavian Studies was established in 1917? While the items linked to the Nordic region might not all have become part of UCL Collections due to a conscious attempt to create a discipline-specific teaching collection, we will argue that they can today all support the teaching, promotion and advancement of the study of the languages and culture of this area. University collections can be a particularly important resource for "less widely taught languages", a category to which the Nordic languages belong in the UK since UCL and the University of Edinburgh are the only two institutions left with a full provision of modern mainland Scandinavian languages at degree level (Garton 2017, 98; Uebel, Kraemer & Giupponi 2024, 1–2).

In this chapter, we will present two teaching projects: a Scandinavian Studies project on which we collaborated in 2015–2016 and which resulted in a pop-up exhibition called *What we know about Scandinavia*; and an ongoing widening participation and public engagement[1] project which began in 2020 and resulted in the showcase *Not Just Words: Learning languages through art and objects*, co-curated by Elettra Carbone and Andrea Fredericksen (then Curator of UCL Art Museum).[2] The project from 2015–2016 consisted of three joint Danish and Norwegian language object-based language sessions taught in UCL Art Museum and designed around relevant items from UCL's Collections. This pilot led to a series of joint Danish, Norwegian and Swedish language classes which are still running to date. The second project, which began in 2020, was born out of a collaboration between several language tutors in the School of European Languages, Culture and Society (SELCS) at UCL and then UCL Art Museum's Curator Andrea Fredericksen. It resulted in a series of nine language video tasters initially aimed at Year 12 pupils, namely students normally between 16 and 18 of age who are reaching the end of the schooling system. The tasters were launched in the summer of 2020 and presented an aspect of one of the languages taught at UCL taking as a starting point an item from UCL's Art Collections. Created as part of SELCS's outreach and widening participation strategy to promote foreign languages to pupils during lockdown, this series, called the *Language and Culture Show and Tell*, showcased four Scandinavian languages: Danish, Norwegian, Swedish and Icelandic.[3] Finnish joined the series a year later as the series continued – and still continues – to expand, with Faroese, which was previously taught at UCL, added in 2024. While the series now counts a total of 18 video tasters, in this chapter we will focus on presenting and analysing the UCL Nordic contributions.

Ultimately, this chapter demonstrates how a "local" Scandinavian Studies teaching project had an impact on exploring more broadly the potential that

1 Widening participation is defined as efforts "ensuring that students from disadvantaged backgrounds can access higher education" and be supported in their studies (Atkins & Ebdon 2014, 6). Public engagement, on the other hand, is defined as the "ways in which the activity and benefits of higher education and research can be shared with the public [and communities]" (Research England 2020, 12).

2 Thanks are extended to the UCL Library, Culture, Collections and Open Science (LCCOS) and particularly to the UCL Art Museum team: Dr Andrea Fredericksen, who, as mentioned in the article, has been instrumental to these projects, but also Dr Nina Pearlman (then Head of UCL Art Collections) and Lucy Waitt (then Curatorial and Collections Assistant, UCL Art Museum), whose unwavering support throughout the project has been key to the success of the exhibition *Not Just Words*.

3 Further details on this project and all video tasters can be accessed here: https://www.ucl.ac.uk/european-languages-culture/language-and-culture-show-and-tell.

the fragments held in university museums, libraries and archives hold for the development of new language pedagogies and the promotion of language teaching and learning (Boddington, Boys & Speight 2013, 4; Hooper-Greenhill 1999; Flaherty & Kovach 2023, 2).[4]

1 Re-collecting the North: Promoting Inter-Scandinavian Communication

Hannan, Duhs and Chatterjee present the results of a UCL survey conducted in 2010–2011 which suggests some pedagogical benefits of Object-based Learning (OBL) and shows how this has been embedded in UCL's teaching and learning strategies to introduce a more "active and experiential learning" in a greater variety of disciplines. They argue that OBL can "encourage students to form their own connections between their existing knowledge and new concepts", and university museums, which can provide teachers and students with original local resources, "can play an important role in the development of a new and improved approach" to teaching and learning (Hannan, Duhs & Chatterjee 2013, 161, 166). The study emphasises UCL's institutional support of OBL, quoting a teaching and learning strategy document where it is stated that: "UCL has an outstanding resource in its Museums and Collections and departments are strongly encouraged to exploit the potential of object-based learning in their teaching" (Hannan, Duhs & Chatterjee 2013, 161). As part of this institutional push to find new approaches to a more research-based and student-centred education, we initiated conversations with Andrea Fredericksen (then Curator at UCL Art Museum) and Nicholas Grindle (Associate Professor (Teaching) UCL Arena). Following these we began to pilot a new teaching approach to a series of joint Danish and Norwegian language classes that would take place in UCL Art Museum and introduce collection-based teaching as a language learning methodology in the UCL Scandinavian Studies curriculum. In this section we will explore the initial aims of the project we carried out in 2015–2016 and present the methodology and immediate outcomes, which led, among other things, to what we will refer to as a "re-collection of the North" in the UCL Collections.

4 Note that while most institutions in the UK continue to use "Scandinavian Studies" in their name, the teaching and, particularly, the research within this field has a broader Nordic scope. "Nordic Studies" has therefore become increasingly recognised as a more representative name for this discipline.

Several studies have already addressed the benefits of engaging with museum collections in teaching and outreach emphasising how collections and museums can become a space for the research and teaching of disciplines that do not necessarily have a close connection with them (Wagner et al. 2015; Sederberg 2013; Schwarts 2008; Hooper-Greenhill 2000). In addition, some studies have explored their applications to the curriculum design of language courses at all levels, from *ab initio* to advanced (Flaherty & Kovach 2023). In their article on English language teaching, Shehzad and Awan highlight some of the advantages of using museums and art collections for this specific purpose arguing that thanks to museums foreign language students can develop at the same time their cultural and linguistic sensibility (Shehzad & Awan 2010). Presenting the experiences of how the Department of French and Italian at Oberlin College redesigned the French language curriculum in collaboration with Allen Memorial Art Museum, Leelah and Murphy maintain that "[s]tudent engagements with artworks and with each other in the museum setting are spontaneous and authentic" and support the students' "language acquisition, cultural competence, historical awareness, and interdisciplinary collaboration" (Leelah & Murphy 2023, 25–26). Finally, teaching in a museum space offers the possibility to break away from the formal classroom learning, providing "an alternative to the structured methods of learning often used in a classroom" (De Feo & White 2023, 48).

These perspectives summarise what we wanted to achieve when we began working on this language teaching project in collaboration with UCL Art Museum in 2015: to introduce research-based learning in the Danish and Norwegian classroom using materials already at UCL, helping students to create connections between language and culture (Fung 2017); and to create a new space for "real-life" language use breaking away from the classroom space and from the topics traditionally covered in textbooks. Moreover, by designing our sessions as "joint" Danish and Norwegian sessions, we also wanted to encourage students to develop what studies refer to as "inter-Scandinavian communication", namely the practice among Danish, Norwegian and Swedish speakers of communicating with each other using their own respective languages (Braunmüller 2002, 3–5). Stressing the linguistic connections between the three mainland Scandinavian languages is something that has long been part of Scandinavian Studies' strategy, not only because the discipline of Scandinavian Studies itself relies on the close collaboration of colleagues working across the Nordic region but also because, from the students' perspective, there can be career advantages in navigating all the Scandinavian languages. Barfod argues that the degree to which inter-Scandinavian communication is successful can vary according to the individual's linguistic resources and experiences but also to the context in which the communication takes place (Barfod 2019, 34–38). The joint classes in UCL Art Museum therefore had the aim to create a context for inter-Scandinavian communication where students could use their own language of study while collaborating on a series of joint tasks and, possibly more importantly, to build a positive attitude towards inter-Scandinavian communication (Barfod 2019, 52).

In order to find the primary materials for our classes, rather than turning to the Scandinavian Library collections – the only collections specifically defined as 'Scandinavian', and thus subject-specific – inspired by the session in UCL Art Museum, we set out to look for "new" items with a Nordic connection. In our search we were guided by what other scholars have referred to as the principle of "serendipity" in research, understood as "an unexpected rupture, an opportunity, fortunate circumstances, and discoveries" (Daly Goggin & Goggin 2018, 4; Hoeflich 2007, 814, 825). As a methodological principle, it recognises that research is often far from linear and can be "messy" thus requiring "a prepared mind" but also an "open mind" (Daly Goggin & Goggin 2018, 3, 6). As a pedagogical approach, "serendipity" recognises the importance of "curiosity" as a legitimate research strategy that encourages the "(re)discovering of the inherent pleasures in browsing" (Lewis Gaillet 2018, 67).

We began by compiling our own collection of items searching first the UCL Art Collections and then UCL Special Collections' catalogues using generic words relating to the Nordic region. This process could be defined as "re-collecting the North", where the concept of "re-collecting" addresses both the practice of "remembering" or "re-discovering" unknown or forgotten connections between the Nordic region and UCL's institutional history and archives (connections that were not directly linked to the history of the discipline) and the act of "gathering" or "re-assembling" an inventory of resources from previously unrelated papers and materials in the archives using the lens of Nordic Studies.[5] Simply using as search terms words such as Scandinavia, Scandinavian, Denmark, Norway, Sweden, Danish, Norwegian and Swedish returned a number of interesting hits in both collections, and we compiled an initial list of items, taking also into account which items students could feasibly engage with given the time constraints of the course.[6]

Some of our more notable findings include seven engravings of Norwegian landscapes, one of which is featured at the beginning of this chapter. Part of UCL Art Museum's collections, these engravings were made by the British

5 For the concept of "re-collecting" we are using the definitions provided by Jeffers 2016, 147.

6 Samples of the items we collated can be seen in the online exhibition Nordic Fragments: https://nordicfragments.org/.

painter and engraver John Linnell, The Elder (1792–1882) after drawings by the British artist Edward Price (1800–1885), today only remembered for his journey to Norway in 1826 which remains immortalised in his illustrated journal *Norway. Views of Wild Scenery* (1834). While the engravings by Linnell did not feature in the book, they still clearly represented moments in Price's travel narrative and were connected to the history of British travel to Norway in the early nineteenth century (Fjågesund & Symes 2003).

In our search for items connected to the Nordic region, we found more examples of British travel representations of Scandinavia. Another example of this emerged in UCL Special Collections as we unearthed the writings on Denmark and Sweden by the British Dante scholar Henry Clarke Barlow (1806–1876) compiled during his journey to Scandinavia in the Summer of 1856. These documents were part of the Barlow Papers, which became part of UCL's Collections when Barlow bequeathed them to UCL in 1876. In addition to his library, which included several editions of the Divine Comedy, and "a fund for the endowment of public lectures on Dante", these contained papers such as manuscripts and notes relating to his published and unpublished work on Dante or other topics of his interest as well as papers relating to his journeys (Flaxman Gallery 1995, 1). For the teaching project, rather than focusing on the rather lengthy travel diary, we presented the students with the representation of Scandinavia in two unpublished essays on Copenhagen, "Revelations of a Writing-Case: At Copenhagen" and "Revelations of a Writing-Case: Leaving Copenhagen" where Barlow, collecting his impressions of the Danish capital, comments on the positive societal effects of easier access to culture and education (Carbone 2019).

More recent examples of British impressions of Scandinavian societies emerged in the papers relating to the post-war Scandinavian journeys of former UCL staff members such as J. B. S. Haldane (1892–1964), who was Professor of Genetics at UCL from 1933 to 1937 and of Biometry (Biostatistics) from 1937 to 1957, and Hugh T. N. Gaitskell (1906–1963), who was Head of the Department of Political Economy at UCL between 1938 and 1939 and went on to become the Leader of the Labour Party from 1955 to 1963. Both visited Denmark and Sweden, in 1946 and 1954 respectively, and while the papers documenting their journeys focus on different disciplinary interests (genetics and biochemistry on the one hand and welfare and international relations in post-war Europe on the other), they both cast light on the importance of international relations and collaborations, particularly in post-war Europe.

Based on the initial list we compiled, we designed three joint Danish and Norwegian classes for each language year-group (basic, intermediate and advanced) featuring some of these items. A more comprehensive list of the

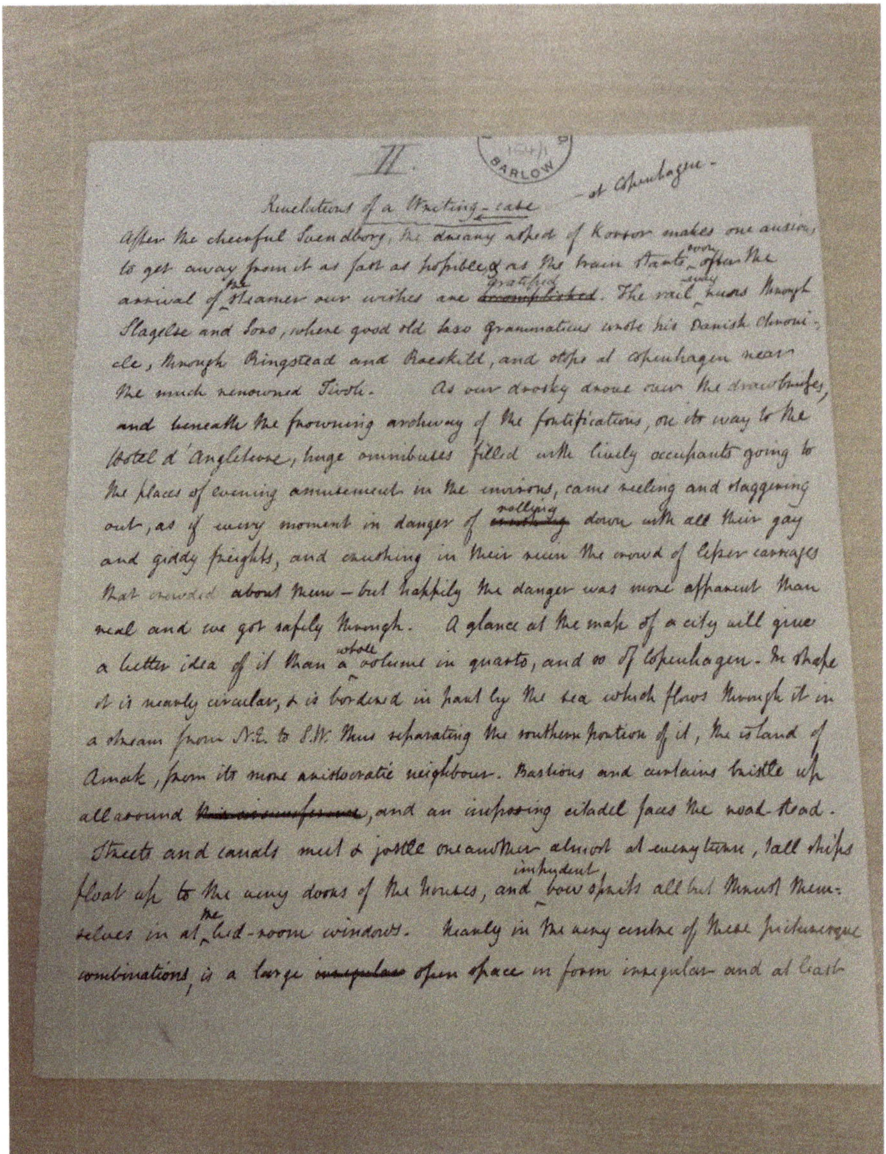

II.

Revelations of a Writing-case – at Copenhagen –

After the cheerful Svendborg, the dreary aspect of Korsor makes one anxious to get away from it as fast as possible, & as the train starts soon after the arrival of the steamer our wishes are gratified. The rail runs through Slagelse and Soro, where good old Saxo Grammaticus wrote his Danish Chronicle, through Ringstead and Roeskild, and stops at Copenhagen near the much renowned Tivoli. As our drosky drove over the drawbridges, and beneath the frowning archway of the fortifications, on its way to the Hotel d'Angleterre, huge omnibuses filled with lively occupants going to the places of evening amusement in the environs, came reeling and staggering out, as if every moment in danger of rolling down with all their gay and giddy freights, and crushing in their ruin the crowd of lesser carriages that crowded about them – but happily the danger was more apparent than real and we got safely through. A glance at the map of a city will give a better idea of it than a whole volume in quarto, and so of Copenhagen. In shape it is nearly circular, & is bordered in part by the sea which flows through it in a stream from N.E. to S.W. thus separating the southern portion of it, the island of Amak, from its more aristocratic neighbour. Bastions and curtains bristle up all around, and an imposing citadel faces the road-stead. Streets and canals meet & jostle one another almost at every turn, tall ships float up to the very doors of the houses, and impudent bow sprits all but thrust themselves in at the bed-room windows. Nearly in the very centre of these picturesque combinations, is a large open space in form irregular and at least

FIGURE 8.2 First page of Henry Clark Barlow's "Revelations of a Writing-Case: At Copenhagen" (1856). Unpublished manuscript. UCL Special Collections, UCL Archives, London, Barlow Papers, BARLOW/133-269/154-164/154

items we had uncovered was distributed to the students who were required to complete an assessment linked to these sessions, which required them to carry out a detailed description of an object of their choice and some basic independent research on its history and cultural significance. For each group the teaching sessions and the assessment were adjusted to reflect the courses' learning outcomes, and thus the students' expected linguistic level. Finally, we decided to include in the project an element of public engagement, signing up to contribute to the programme of the 2016 UCL Festival of Culture – a public-facing series of events showcasing aspects of UCL research through workshops, exhibitions, talks and performances which ran until 2019 – with a pop-up exhibition featuring the result of the students' research.

While not all the items we used to design our sessions belonged to the Art Collections, all the joint sessions were taught in UCL Art Museum. This was partially due to the fact that the museum is one of the spaces on campus where the consultation and handling of collection materials is permitted, and partially because one of the aims of the project was to move the teaching to a non-classroom space. As Sederberg points out, "a museum-based approach integrates aesthetic and intellectual goals, learner-centred teaching models, and interactive and experiential learning" allowing teachers to develop models for content-based instruction (CBI), namely approaches that attempt to integrate "the teaching of language and culture, and to making interdisciplinary teaching possible" (Sederberg 2013, 252). While the language classes remained "language driven", meaning that the focus was on language instruction, the sessions were thematically based, drawing, where possible, on the skills and knowledge acquired through other subjects the students may be taking as part of their degrees (Cenoz 2015, 11–12). Given that for the majority of Scandinavian Studies students, studying the language is part of a language and culture degree, our students were used to this type of content and language integrated learning (CLIL). The museum sessions and tasks allowed them to describe historical objects, while articulating their reflections on their cultural significance, on the role of the museum (and particularly of university collections) and the intercultural contacts between the UK and the Nordic region.

For first-year students we began with vocabulary building tasks (such as the creation of conceptual maps in Danish and Norwegian linked to the word 'museum') allowing students to slowly build the vocabulary they would need to interact with the space, learning words such as 'maleri' (painting), 'skulptur' (sculpture), 'gjenstand/genstand' (object) and 'landskap/landskab' (landscape). Students were also asked to translate a short text on the history of UCL Art Museum and its collections, which we provided in both Danish and Norwegian. Given that for most students this was their first encounter with the space, they learnt to relate to it in their languages of study rather than English, navigating rather challenging expressions, such as "tverrfaglig undervisning og forskning" (interdisciplinary teaching and learning) while finding out about the museum's functions.

Tasks introducing more technical and space-specific vocabulary and expressions were combined with others encouraging the students' use of day-to-day vocabulary and creativity. Students were asked, for instance, to make up stories inspired by the objects and images on display. In addition to the above-mentioned engravings of Norwegian landscapes, which allowed students to make up stories using words relating to Nordic landscapes – such as 'fjell' (mountain), 'fjord' (fiord), 'får' (sheep) – these also included other artworks representing historical figures. The stories inspired by the seventeenth-century engraving of Jens Juel (1631–1700), Baron van Juling, Danish diplomat and statesman, "wearing full wig, ceremonial clothing", by Gerard Valck, after Wallerant Vailliant, featured many creative narratives around the words 'aristokrat' (aristocrat) and 'parykk' (wig). Students also learnt to tackle detailed descriptions of a selection of objects as well as of the museum space, and concluded in the last session with an oral presentation of an item of their choice from the collections.

The topics chosen for second and fourth-year students were the same but what varied was the number of tasks, the level of the vocabulary used and the extent of the focus on the discussion of abstract topics. We began with a session on defining the museum as institution and its functions, relating that to the history of UCL Art Museum and other collections. The sessions allowed students to present (in Danish and Norwegian) their own previous experiences with museums and collections and reflect on different types of collections across time. We then moved on to the analysis of different items from the collections, including the Linnell engravings of Norwegian landscapes combined with extracts from Price's travel book on Norway and samples of Barlow's descriptions of Copenhagen from 1856. These formed the basis of discussions around what attracted British travellers to Scandinavia in the nineteenth century. We concluded the sessions with a discussion on the steps involved in the organisation of an exhibition, planning for the pop-exhibition in May 2016.

While most of the collection materials the students worked with were in English, all discussions were conducted in Danish and Norwegian. UCL Art Museum therefore became a "pedagogical laboratory" or, more precisely, an inter-Scandinavian laboratory where students could experiment with new content (Flaherty & Kovach 2023, 2; León 2023, 146) while coming to grips with and practicing this type of communication. This is a function that, as we will return to at the end of this section, it has retained to date.

The first concrete outcome of the sessions was a compulsory assessed project which became part of the students' language portfolio. Towards the second half of Term 2, students were required to select an object in the UCL Collections with a Nordic connection, conduct their own research and write a short individual project. In order to complete this task all students were asked to attend a training session on the handling and use of archival materials run by UCL Special Collections staff. The projects were also the basis of the text for the catalogue of the pop-up exhibition *What we know about Scandinavia*. On 25 May 2016 the students became curators for a day as the objects they selected for their projects featured in a pop-up exhibition at UCL Art Museum. The exhibition, which attracted 104 visitors in less than 4 hours, allowed the students to take ownership of the project and work together as a group to create a coherent narrative about UCL and Scandinavia. They took our work on "re-collecting" the collections to another level by re-curating the exhibits, a process that allowed the students to intensify the links between university collections and curricula instruction (Schwartz 2008, 42).

The students were responsible for preparing all the exhibition materials (captions and mini-catalogue) in dual language (English and Norwegian or Danish) and took turns to invigilate the exhibition and interact with visitors. Besides fostering student agency in research and writing, this final activity made the students ambassadors of their own languages and programme (Rodgers 2015, 154). The exhibition, which, as mentioned above, was featured as one of the events of the UCL Festival of Culture, presented a piece of British-Nordic history to the academic and non-academic public, while demonstrating to the public, as well as to the students, the connections between "what we have" and "what we do" at UCL (Cuno 1990, 42). This also drew the visitors' attention to the existence and history of Scandinavian Studies as a discipline at UCL.

The evaluation of the project gave very positive results. The questionnaire we devised asked students to rate two aspects: the joint classes with Danish and Norwegian students and, more specifically, the UCL Art Museum teaching. Student satisfaction was very high in relation to both aspects. Students seemed to perceive the joint museum classes as a way to learn about cultural as well as linguistic similarities and differences between Danish and Norwegian. Given that the cultural content of the joint classes is provided by the objects in UCL Special Collections and their contexts, it is also not surprising that students were equally positive about the setting of the classes and the content of their projects as they explained it provided them with a change of learning environment, the use of language skills in another context outside the ordinary classroom and the opportunity to familiarise themselves with specific vocabulary and to discuss objects present during the class. On the whole students stated that the project had allowed them to explore a different topic and style of writing which is not normally associated with language learning and pushed them out of their comfort zone.

Following the success of the project carried out in 2016, the Scandinavian Studies Programme introduced three joint classes in UCL Art Museum at first year level as part of the language curriculum. The sessions, which have included Swedish since 2016, prepare students for an object-based assessed presentation at the end of the academic year. In 2019, a final-year joint session on ninetieth-century teaching materials on the Scandinavian languages (samples of which include nineteenth-century phrasebooks, grammars and dictionaries and are part of UCL's Scandinavian Rare Books collection) was also introduced.[7] This session fits with one of the learning outcomes for the course which requires students to reflect on foreign language learning and teaching and produce, as part of their assessment, a foreign language learning resource (mostly in the form of a textbook chapter) in their language of study (Danish, Norwegian or Swedish).

2 Connecting with the North: Promoting Nordic Studies in the UK

In addition to the curriculum changes to the study of the Scandinavian languages, the collection and object-based approach was, during the COVID-19 pandemic, at the core of a project promoting not only the Nordic languages but foreign language study in general, multilingualism and language awareness. This consisted of an outreach and widening participation project called the *Language and Culture Show and Tell* which resulted in the creation of a series of freely accessible language video tasters showcasing some of the languages offered at UCL and created around an object from UCL's Art Collections. At a time when many of our outreach and widening participation activities had moved online and visiting the institution in person was not possible due to the pandemic, UCL's Collections offered an opportunity to establish a connection with UCL as an institution and a physical space, while recreating the experience of learning through objects in museum spaces digitally to make the collections and our approach to learning as accessible as possible. In other words, we wanted to create the experience of what Bautista and Balsamo refer to as the "mobile museum", namely "a multimedia learning experience based on museum collections" (Bautista & Balsamo 2013, 57).

While they were created within a university environment, the *Language and Culture Show and Tell* materials were originally designed with students

7 For the organisation of these sessions, we would like to credit our colleagues Anne Grydehøj, Annika Lindskog, and Jelena Ćalić.

in Year 12 in mind, meaning that they were tailored from the beginning to be accessible to a broader audience. Given that the videos were initially intended as asynchronous content for the Y12 Widening Participation Language and Culture Residential Summer School organised by SELCS (School of European Languages, Culture and Society) in 2020, an invitation to contribute to the project was initially sent to tutors teaching languages offered in this department.[8] For each language the participating tutor was asked to produce the following: a short video which took an object from the UCL Art Collections as its starting point and a worksheet based on the content of the video. The series, which initially counted 9 languages (Danish, Dutch, French, Icelandic, Italian, Norwegian, Portuguese, Spanish and Swedish) in 2020, expanded to 13 (with the addition of Arabic, Finnish, German and Mandarin) in 2021 and now counts 16 as Ancient Greek, Hebrew and Romanian joined in 2023 with one more language (Faroese) added in 2024.

In January 2023 the video tasters became the centre of a temporary showcase hosted by UCL Art Museum, *Not Just Words: Learning languages through art and objects* where the objects featured in the video tasters were exhibited alongside iPads showing the videos. Visitors could therefore engage with the content in the physical space but, as all video tasters were freely available online, they could decide whether to experience these in the museum space or access them from elsewhere thanks to the QR code in the room. The "mobile museum" thus became the "distributed museum", a "modern museum" that "adapts its traditional functions to the new cultural environment of the digital age" demonstrating "the creative capacity of the museum" offering "new learning experiences" (Bautista & Balsamo 2013, 57–58, 68).

All the modern Nordic languages currently offered at UCL (Danish, Finnish, Icelandic, Norwegian and Swedish) are represented in the *Language and Culture Show and Tell* series and in the showcase, and we will therefore focus on presenting these materials and their use of collections considering not only which linguistic and cultural aspects relating to the Nordic region these tasters address but also reflecting further on the process of "re-collecting" the North in the UCL Collections. While linked by their overall aim to promote interest in the Nordic languages and cultures, the five video tasters can be divided into two categories in relation to how they use the art object on which they centre. The Danish, Norwegian and Swedish tasters use artworks that are closely connected to the language, culture and history of these countries while the Icelandic and Finnish ones find their own, unique way to connect with the North.

8 We would like to thank The Sutton Trust and UCL Access and Widening Participation for their generous funding and support for the creation of these resources in connection with the Y12 WP Residential Summer Schools online in 2020 and 2021.

FIGURE 8.3 Section of the exhibition *Not Just Words: Learning languages through art and objects*, (January 2023–September 2024), UCL Art Museum, London
PHOTOGRAPHER: KIRSTEN HOLST. REPRODUCED WITH PERMISSION FROM UCL ART MUSEUM

The Danish video taster by Anne Grydehøj focuses on the topic of language and language contacts using a *Portrait of George, Prince of Denmark*, by John Smith, after Godfrey Kneller (mezzotint on paper, circa 1704–1708, UCL Art Museum, LDUCS-2041). Initially, the portrait of Prince George of Denmark (1653–1708) is used as an image that can be described to demonstrate the linguistic influences that other languages, such as French, had on Danish: words such as 'portræt' (portrait), 'prins' (prince) and 'paryk' (wig) have in fact all entered the Danish language via French. The image of Prince George is also used as a prompt for a more historical introduction on the status of different languages in Denmark in the sixteenth and seventeenth centuries when French and German were the languages of culture, commerce, the military and of the aristocracy, and Danish was still at the bottom of the language hierarchy. On the whole, the taster encourages the learner to reflect on language contacts creating parallels between Danish language history and more recent phenomena which have seen the introduction of English words in the Danish language as well as the addition of the word "hygge" in the English language in 2016 (Peterson & Beers Fägersten 2024; Breunig & Kallestrup 2020, 159). By focusing on contact-induced processes, the Danish taster tries to establish a linguistic common ground of similarities between Danish and English (primarily), a strategy used also by the Norwegian and Swedish tasters (Kaivapalu & Martin 2014, 284–87).

Taking inspiration from *View Across the Fiord from Herrinsholmen*, one of the engravings of Norwegian landscape by John Linnell, after Edward Price (circa 1826–1827 from the series *Seven Views in Norway*, etching on paper, UCL Art Museum LDUCS-1456), the Norwegian taster by Elettra Carbone explores the role that Norwegian nature and landscape have played in attracting British travellers to the country from the early nineteenth century onwards. Placing the object in its historical context, the taster explains how the engraving is representative of this as Scandinavia became more fashionable both as a destination and as the topic of travel literature with most travel accounts about the region being written by British travellers (Fjågesund 2014, 283–84). The taster employs the artwork not only to cast light on cultural contacts between the countries but also linguistic ones using vocabulary drawn from the engraving. It demonstrates, for instance, how words such as 'fjord' (fiord) and 'sjø' (sea) are recognisable in Norwegian for English speakers while also highlighting some lexical differences – false friends like the word 'sky' which in Norwegian means 'cloud' – and grammatical differences, such as the presence in Norwegian of grammatical gender.

The Norwegian taster shares its focus on nature and landscape with the Swedish one but, while both draw on the historical context of the artwork, they use it in different ways. The Swedish taster by Annika Lindskog reflects on how nature and geography have played significant roles in shaping Swedish national identity and values, taking the figure represented in the artwork *Portrait of Charles Linnaeus* by Charles Edward Wagstaff, after Lorens Pasch (etching and engraving, circa 1810–1840, UCL Art Museum, LDUCS-4533) as its starting point. The taster begins by examining the role eighteenth-century scientist and botanist Carl Linnaeus (1707–1778), who was given the task to document Swedish geography, natural resources and people, played in the construction of Swedish national identity, particularly in relation to the country's nature and natural resources (Koerner 1999). It then moves on to show the legacies of his promoting Swedish nature which can be found in modern-day Sweden. While the linguistic element of the taster is inspired by the artwork, unlike the Danish and Norwegian tasters, it is independent of it and focuses on presenting examples of the "presence" of nature in Swedish culture. This includes examples of how personal names derive from natural elements (like in the case of the surname 'Ljungberg' meaning 'heather' ('ljung') and 'mountain' ('berg'), and the name 'Björn' meaning 'bear').

The Icelandic and Finnish video tasters take a different approach compared to the Danish, Norwegian and Swedish ones, because no artwork directly connected to these countries was identified in the collections. Both use objects that might be associated with British culture – namely a map of sixteenth-century London and a Turner painting – but nonetheless manage to connect these to aspects of Icelandic and Finnish culture. Ultimately these are examples of how the North can not only be "re-collected" in the UCL Collections but also "connected", promoting what Sederberg highlights as the ultimate aim of Museum pedagogy, namely that of making connections across contexts and disciplines, promoting not only the acquisition of knowledge but also curiosity and supporting ways in which the objects can be "read" (Sederberg 2013, 253–54).

The Icelandic taster by Helga Luthersdottir juxtaposes a sixteenth-century birds-eye map of London, *Plan of London, from Civitates Orbis Terrarum (volume 1, plate 1)* by Franz Hogenburg, published by Georg Braun (etching and engraving with hand-colouring on paper, 1572, UCL Art Museum, LDUCS-4787) with maps of Iceland from a similar time period, exploring the linguistic strategies employed by Jón Ólafsson (*c.*1594–1679) to describe a space and cultural context entirely foreign to him as he visited London in the summer of 1615 and wrote about it in his autobiography *Æfisaga Jóns Ólafssonar Indíafara: samin af honum sjálfum (1661)* (*The Life of the Icelander Jón Ólafsson, Traveller to India, Written by Himself*). Coming from a rural country with no cities, Ólafsson needs to find words to picture London's urban configuration. In the taster Luthersdottir illustrates how Ólafsson uses various terms to describe places where people live together such as 'pláss' (place), 'borg' (city), 'byggð' (township), and 'þorp' (village) and yet, as she argues, none of these terms "seems to carry a fixed connotation of population […] but rather his terms simply mean 'not a farm' – that is, not Iceland" (Lúthersdóttir 2020). By connecting images of sixteenth-century Iceland and London, the taster gives an example of what Agar refers to as different "languacultures", a concept that emphasises the connections between language and culture (Agar 1994, 20). "Languaculture" also defines culture as what happens "when you encounter differences, become aware of something in yourself, and work to figure out why the differences appeared", differences that can lead to changes or change potentials in linguistic practices (Risager 2006, 110–11).

The Finnish taster by Riitta-Liisa Valijärvi introduces Finland and Finnish language through an artwork by JMW Turner, namely *The First Steamer on the Lake of Lucerne* (drawing, circa 1841, UCL Art Museum, LDUCS-2176). The choice of artwork takes inspiration from Finland's geography. Internationally known as "the country of a thousand lakes" because of its 188,000 lakes, the image is suggestive of this prominent Finnish natural feature which, unsurprisingly, is prominent in both personal and place names. Drawing on its presentation of Finland's geography and landscape, the taster presents a grammatical aspect linked to the description of space, namely the local grammatical cases in Finnish. These are defined as grammatical endings attached to words which roughly have the same meaning as English prepositions such as "in", "on", "at", "from" and "into." While the word lake features as an example, like in the Swedish taster, the artwork is not the centre of the taster but the inspiration for it. It provides the learner with a "familiar" visual dimension, a "connection" that can bridge the gap created by the grammatical differences between Finnish and English.

3 Towards an Object-Based Methodology for the Promotion of Foreign Language Learning

The teaching projects presented in this chapter all have one aspect in common as artworks and objects – whether directly connected to the Nordic region or not – function as "learning portals" through which to examine the different ways in which knowledge is constructed (Boddington, Boys & Speight 2013, 4, 174–75). Objects from the collections are placed in their historical and cultural contexts to allow learners to explore their connections with the Nordic region and/or in new contexts where the connection with the Nordic region is created through creativity and associations. Both strategies allow the teacher and the learner to explore the benefits of creativity and serendipity as a teaching and learning methodology but also as a research strategy, one that encourages staff and student researchers to make full use of the valuable "local" resources offered by university collections and museum spaces.

While the engagement with subject-specific collections – such as the Scandinavian Studies library collections – can be a straightforward and effective way to introduce collection-based teaching in the language classroom, university collections can often offer more than what has been "collected" specifically to support a discipline. By "re-collecting" the UCL Collections through the lenses of Nordic Studies, we were able to present our Scandinavian Studies students with previously unexplored materials with clear connections to the Nordic region which we – and they – were able to "read" and make sense of thanks to their subject-specific knowledge while also developing new language and research skills beyond the traditional classroom context. Moreover, the *Language and Culture Show and Tell* shows how this museum-based pedagogical approach to language learning can be employed to engage non-academic audiences particularly, but not exclusively, to demonstrate the way language and culture are intertwined. In this way, collection objects are used to create connections that bridge the familiar and unfamiliar. Ultimately, as Sederberg points out, "incorporating strategies from museum learning" into teaching "is one way to place culture at the core of language instruction, and to provide learning experiences for our students that will stay with them after the semester is over, helping them think more critically and mindfully about the connections between language and culture, and gain skills to read the objects of cultural archives" (Sederberg 2013, 260).

Finally, it is important to point out that the impact of our engagement with UCL's institutional collections has gone much beyond teaching, outreach and public engagement. Conceived as an example of "research-based education" in 2015, it has also developed into a case of what could be called "teaching-based research", namely research that came directly out of the exploration of new teaching practices. The 2015 research project not only provided the first "fragments" of Carbone's research on British representations of Scandinavia emerging from the UCL Collections; it also encouraged us to consider a different approach to research, one that takes its starting point from serendipity and contingency. As Carbone's work progresses to unearth new materials, these will not only be included in her research project but may also feed back again into teaching and new research-based projects. The link between research-based education and teaching-based research is therefore a dynamic one and one that, in this case, also reminds us of the important role that university archives and access to them play in the production of culture and dissemination of education, particularly when supporting less commonly taught disciplines such as Nordic Studies.

Bibliography

Agar, Michael H. 1994. *Language Shock: Understanding the Culture of Conversation.* New York: William Morrow.

Arnold-Forste, Kate & Sophia Mirchandani. 2001. "Collections in the United Kingdom." In *Managing University Museums Organisation for Economic Co-Operation and Development*, edited by Organisation for Economic Co-operation and Development, 47–54. Paris: OECD Publishing.

Atkins, Madeleine & Les Ebdon. 2014. "National Strategy for Access and Student Success in Higher Education." Department for Business Innovation and Skills.

Barfod, Sonja. 2019. "Interskandinaviske interaktioner i et historisk og nutidigt perspektiv." *Globe* 8. https://doi.org/10.5278/ojs.globe.v8i0.3089.

Braunmuller, Kurt. 2002. "Semicommunication and Accommodation: Observations from the Linguistic Situation in Scandinavia." *International Journal of Applied Linguistics* 12 (1): 1–23. https://doi.org/10.1111/1473-4192.00022.

Breunig, Malene & Shona Kallestrup. 2021. "Translating Hygge: A Danish Design Myth and Its Anglophone Appropriation." *Journal of Design History* 33 (2): 158–74. https://doi.org/10.1093/jdh/epz056.

Carbone, Elettra. 2019. "The Cheerful Danes: Henry Clarke Barlow's 'Revelation of a Writing-Case' (1856) on Copenhagen." *Scandinavica* 58 (2): 153–73.

Cenoz, Jasone. 2015. "Contents-Based Instruction and Content and Language Integrated Learning: The Same or Different?" *Language, Culture, and Curriculum* 28 (1): 8–24. https://doi.org/10.1080/07908318.2014.1000922.

Cuno, James. 1990. "The Harvard University Art Museums as a Catalyst for Teaching and Scholarship." *Director's Report – Harvard University. Art Museums*, 1990/1991: 41–44.

Daly Goggin, Maureen & Peter N. Goggin. 2018. "Stumbling into Wisdom in Rhetoric, Writing, and Literacy Research: An Introduction." In *Serendipity in Rhetoric, Writing, and Literacy Research*, edited by Maureen Daly Goggin & Peter N. Goggin. Chicago: Utah State University Press.

De Feo, Daniele & Veronica White. 2023. "Benvenuti al Museo!: Italian Classes at the Princeton University Art Museum." In *Language Learning in Academic Museums: New Paradigms for Cultural Study, Language Acquisition, and Campus Engagement*, edited by Heather Flaherty & Jodi Kovach, 1st ed., 45–64. Blue Ridge Summit: Rowman & Littlefield Publishers, Incorporated.

Ewbank, Inga Stina. 1999. "Introduction: 'Anglo-Scandinavian Cross-Currents'." In *Anglo-Scandinavian Cross-Currents*, edited by Inga Stina Ewbank, Olav Lausund & Bjørn. J. Tysdahl, 11–27. Norvik Press Series a No. 17. Norwich: Norvik Press.

Fjågesund, Peter. 2014. *The Dream of the North: A Cultural History to 1920*. Boston: BRILL. https://doi.org/10.1163/9789401210829.

Fjågesund, Peter & Ruth A. Symes. 2003. *The Northern Utopia: British Perceptions of Norway in the Nineteenth Century*. 1st ed. Vol. 10. Boston: BRILL. https://doi.org/10.1163/9789004485013.

Flaherty, Heather & Jodi Kovach. 2023a. "Introduction." In *Language Learning in Academic Museums: New Paradigms for Cultural Study, Language Acquisition, and Campus Engagement*, 1st ed., 1–7. Blue Ridge Summit: Rowman & Littlefield Publishers, Incorporated.

Flaxman Gallery. 1995. *H. C. Barlow and His Place in Dante Studies*. London: Printed by Central Services, University College London.

Fung, Dilly. 2017. *A Connected Curriculum for Higher Education*, 1st ed. London: UCL Press. https://doi.org/10.14324/111.9781911576358.

Greenblatt, Greenblatt, Stephen. 2010. "Cultural Mobility: An Introduction." In *Cultural Mobility: A Manifesto*, edited by Stephen Greenblatt, 1–23. Cambridge: Cambridge University Press.

Hannan, Leonie, Rosalind Duhs & Helen Chatterejee. 2013. "Object-Based Learning: A Powerful Pedagogy for Higher Education." In *Museums and Higher Education Working Together: Challenges and Opportunities* edited by Anne Boddington, Jos Boys & Catherine Speight, 159–68. Farnham: Ashgate.

Hoeflich, Michael H. 2007. "Serendipity in the Stacks, Fortuity in the Archives." *Law Library Journal* 99 (4): 813–63.

Hooper-Greenhill, Eilean. 1999. *The Educational Role of the Museum*, edited by Eilean Hooper-Greenhill, 2nd ed. Leicester Readers in Museum Studies. London: Routledge.

Hooper-Greenhill, Eilean. 2000. "Changing Values in the Art Museum: Rethinking Communication and Learning." *International Journal of Heritage Studies: IJHS* 6 (1): 9–31. https://doi.org/10.1080/135272500363715.

Jeffers, Alison. 2016. "Recollecting and Re-Collecting: The Ethical Challenges of Social Archiving in Post-Conflict Northern Ireland." In *Theatre History and Historiography. Ethics, Evidence and Truth*, edited by Claire Cochrane & Jo Robinson, 147–62. London: Palgrave Macmillan.

Kaivapaulu, Annekatrin & Maisa Martin. 2014. "Symmetry of Similarity: Definition, Perception, Measurement: Finnish and Estonian Noun Morphology as a Testing Ground." In *Language Contacts at the Crossroads of Disciplines*, edited by Heli Paulasto, Helka Riionheimo, Lea Meriläinen & Maria Kok, 1st ed., 283–318. Newcastle-upon-Tyne: Cambridge Scholars Publishing.

Kelly, Melanie. 2001a. "Conclusions." In *Managing University Museums Organisation for Economic Co-Operation and Development*, edited by Organisation for Economic Co-operation and Development, 197–200. Paris: OECD Publishing.

Kelly, Melanie. 2001b. "Introduction." In *Managing University Museums Organisation for Economic Co-Operation and Development*, edited by Organisation for Economic Co-operation and Development, 7–18. Paris: OECD Publishing.

King, Lyndel. 2001. "University Museums in the 21st Century: Opening Address." In *Managing University Museums Organisation for Economic Co-Operation and Development*, edited by Organisation for Economic Co-operation and Development, 19–28. Paris: OECD Publishing.

Leelah, Preea & Libby Murphy. 2023. "Curating the Curriculum: Museums as Language Labs for Authentic Target-Language Conversations." In *Language Learning in Academic Museums: New Paradigms for Cultural Study, Language Acquisition, and Campus Engagement*, edited by Heather Flaherty & Jodi Kovach, 1st ed., 25–44. Blue Ridge Summit: Rowman & Littlefield Publishers, Incorporated.

León, Rosamaría. 2023. "Writing with Your Eyes: Multiliteracies and Community-Based Learning in an Art Museum." In *Language Learning in Academic Museums: New Paradigms for Cultural Study, Language Acquisition, and Campus Engagement*, edited by Heather Flaherty & Jodi Kovach, 1st ed., 141–58. Blue Ridge Summit: Rowman & Littlefield Publishers, Incorporated.

Lewis Gaillet, Lynèe. 2018. "Fortuitous Happenstance: Serendipity in Archival Research." In *Serendipity in Rhetoric, Writing, and Literacy Research*, edited by Maureen Daly Goggin & Peter N. Goggin, 59–69. Chicago: Utah State University Press.

Ludwig, David & Cornelia Weber. 2013. "A Rediscovery of Scientific Collections as Material Heritage? The Case of University Collections in Germany." *Studies in History and Philosophy of Science. Part A* 44 (4): 652–59. https://doi.org/10.1016/j.shpsa.2013.07.012.

Lúthersdóttir, Helga Hlaðgerður, dir. 2020. *Language and Culture Show and Tell: Icelandic.* https://www.ucl.ac.uk/european-languages-culture/language-and-culture-show-and-tell.

Peterson, Elizabeth & Kristy Beers Fägersten. 2024. *English in the Nordic Countries: Connections, Tensions, and Everyday Realities*, edited by Kristy Beers Fägersten & Elizabeth Peterson, 1st ed., vol. 1. Routledge. https://doi.org/10.4324/9781003272687.

Research England. 2020. "Knowledge Exchange Framework: Clustering and Narrative Templates: Report Detailing the Clustering Arrangements and Providing Narrative Statement Templates for the First Iteration of the Knowledge Exchange Framework." Research England.

Risager, Karen. 2006. *Language and Culture: Global Flows and Local Complexity / Karen Risager.* Languages for Intercultural Communication and Education. Bristol; Multilingual Matters. https://doi.org/10.21832/9781853598609.

Rodgers, Susan. 2015. "Toward a Pedagogy for Faculty and Student Co-Responsibility in Curating College Museum Exhibitions." *Arts and Humanities in Higher Education* 14 (2): 150–65. https://doi.org/10.1177/1474022214554111.

Schwartz, John Pedro. 2008. "Object Lessons: Teaching Multiliteracies through the Museum." *College English* 71 (1): 27–47. https://doi.org/10.2307/25472303.

Sederberg, Kathryn. 2013. "Bringing the Museum into the Classroom, and the Class into the Museum: An Approach for Content-Based Instruction." *Die Unterrichtspraxis* 46 (2): 251–62. https://doi.org/10.1111/tger.10144.

Shehzad, Wasima & Muhammad Safeer Awan. 2010. "Culturally Distant Discoursal and Linguistic Patterns Experienced and Learned through Museums." *International Journal of Arts and Humanities* 38 (38).

Smith Bautista, Susanna & Anne Balsamo. 2013. "Understanding the Distributed Museum: Mapping the Spaces of Museology in Contemporary Culture." In *Museums and Higher Education Working Together: Challenges and Opportunities*, edited by Anne Boddington, Jos Boys & Catherine Speight. Farnham: Ashgate.

Uebel, Emily Heidrich, Angelika Kraemer & Luca Giupponi, eds. 2024. *Sharing Less Commonly Taught Languages in Higher Education: Collaboration and Innovation.* Taylor & Francis. https://doi.org/10.4324/9781003349631.

Uebel, Emily Heidrich, Angelika Kraemer & Luca Giupponi. 2024a. "Sharing Less Commonly Taught Languages in the 21st Century." In *Sharing Less Commonly Taught Languages in Higher Education*, 1st ed., 1:1–10. Routledge. https://doi.org/10.4324/9781003349631-1.

Wagner, Laura, Shari R. Speer, Leslie C. Moore, Elizabeth A. McCullough, Kiwako Ito, Cynthia G. Clopper & Kathryn Campbell-Kibler. 2015. "Linguistics in a Science Museum: Integrating Research, Teaching, and Outreach at the Language Sciences Research Lab: Linguistics in a Science Museum." *Language and Linguistics Compass* 9: 420–31. https://doi.org/10.1111/lnc3.12164.

OBJECT: Writing tools

FIGURE 9.1 Sighvatur Grímsson's writing desk and note from his son Gísli. The National and University Library of Iceland (NULI). Turn, make horizontal
PHOTOGRAPH BY DAVÍÐ ÓLAFSSON

CHAPTER 9

Written on a Nineteenth-Century Laptop: Things in Literacy Practices

Davið Ólafsson

Abstract

Texts are essentially materially conditioned. First, they would not be available without, or beyond, their physical representations. Furthermore, their production, storage and consumption are intricately tied to other objects that function as both means and motive for literacy practices. In this study, I delve into the literacy practices of ordinary individuals in nineteenth-century Iceland, examining their interactions with objects tied to the creation and use of texts. By exploring this historical context, I aim to uncover the ways in which everyday people engaged with literacy, highlighting the significance of various items related to the production and consumption of texts during this period.

Keywords

literacy – manuscript culture – material culture – the North – microhistory – diaries – Fridtjof Nansen

1 Manuscript Culture and Its Materiality

The manuscript collection at The National and University Library of Iceland (NULI) comprises approximately 15,000 handwritten documents, the majority of which date back to the early modern and modern periods (seventeenth to twentieth century). As an archive, these miscellaneous documents depict a vibrant and diverse scribal culture, encompassing both learned individuals and numerous popular scribes (Ólafsson 2009; 2017). The manuscript collection contains only a handful of artifacts associated with the creation and/or use of handwritten materials. Among them is a box the size of a book that one

 | DOI:10.1163/9789004749665_010

of those popular scribes utilized as a writing surface. Accompanying this item was a letter written by his son when it was donated to the National Library in 1930. In the letter, he stated that his father had conducted all his writings on this desk, from the very first one up to 1890, whether he was at home or away (NULI, Gísli Sighvatsson 14 March 1930). In the case of this specific scribe, this small desk or box was just one of several objects linked to his literacy practices, among the various entities that were considered 'vibrant' and imbued with 'agency' (Bennet 2010).

Various recent studies in the fields of cultural history, literary studies and philology have over the last three decades revealed and mapped out a vast and varied field of cultural practices identified as post-medieval manuscript culture (see for example Eggertsdóttir & Driscoll 2017; Magnússon & Ólafsson 2017). These studies have exposed the production, circulation, and consumption of handwritten texts as one of the most abounding features of the cultural history of nineteenth-century Iceland (Ólafsson 2009; 2013). This form of literary practice had been enduring since medieval times but became much more widespread and diverse in the nineteenth century. Manuscripts as phenomena are principally identified by their text-bearing qualities, not by their *content* (literary or non-literary, historical, or fictional, religious, or secular, vernacular, or learned), their *materiality* (vellum, paper, bark, cloth, etc.) or *format* (scrolls or in codex). They are *text-bearing objects*. Like all other objects, the text-bearing ones do not 'live' or 'happen' in isolation and without relations, but emerge through complex systems of production, transmission, and reception. And they have *historicity* and *historiography*. Throughout the last half-a-century or so, text-bearing objects – books, manuscripts, documents, etc. – have become subjects within humanities and social sciences, under different banners and within (or sometimes across) different disciplines.

Moreover, these entities possess materiality. Traditionally, objects and texts have been seen as distinct categories. In fields like archaeology and art history, objects serve as crucial physical artifacts offering insights into the past, while historians and literary scholars have predominantly focused on textual analysis. However, these apparent dichotomies are seldom precise or beneficial. In the past three decades, numerous scholarly endeavours have sought to investigate the materiality of texts. This chapter delves into the material conditions of manuscript culture among ordinary people in nineteenth-century Iceland.

In recent decades, the study of textual material in the humanities has undergone notable transformations, one of which is characterized by a material turn manifesting in various forms, under diverse banners, and across different disciplines. A prominent strand of this development is evident in the field of book history, which gained prominence in the 1980s through the influential works of scholars like Roger Chartier and Robert Darnton (Chartier 1989). Originally stemming from the French Annales movement, book history evolved into a distinct interdisciplinary branch of scholarship, notably delineated in Darnton's essay "What is the History of Books?" (Darnton 1982). In his seminal paper, first published in 1982, Darnton introduced his conceptualization of the "communication circuit". This model aimed to comprehensively capture and analyse the multifaceted aspects of book production, dissemination, and consumption. Addressing the titular question of his essay, Darnton proposed a fusion of analytical bibliography, sociology of knowledge, literary history, and cultural history, envisioning the emergence of a new scholarly discipline. The communication circuit delineated the interaction among key participants involved in the journey of books from authors to readers, including printers, shippers, and booksellers. According to Darnton, book history encompasses every phase of this process and its entirety, encompassing variations across space and time and its interconnectedness with economic, social, political, and cultural systems. This perspective does not imply that every student of book history must exhaustively explore every link of the communication circuit or its embedded systems. Rather, the circuit serves as a map delineating the expansive realm of modern book history.

My academic background stems, on the one hand, from the field of book history as it developed in the 1980s and, on the other hand, from new philology (Nichols 1990) and post-medieval manuscript studies that emerged in the 1990s (see Love 1993). In recent years, I have increasingly focused on the material aspects of these practices. (Ólafsson 2023; 2025; Magnússon & Ólafsson 2023). Like most other fields within the humanities, traditional approaches within book history have centred around human agency, often relegating non-human entities to a subordinate position. This foundational perspective underwent substantial revision with a growing emphasis on the role of things and their material conditions. One influential advocate of this critique was the French scholar Bruno Latour, who called for a re-evaluation of the agency attributed to objects (Latour 1988, 142). Consequently, actions, activities, and events came to be seen as emerging from the dynamic interplay between human and non-human actors. Over subsequent decades, this perspective was further developed under the banners of Actor-Network Theory (ANT) (Law & Hassard 1999; Latour 2005) and Object-Oriented Ontology (OOO) (Harman 2018), both of which underscore the interconnectedness and interplay between human and non-human actors.

2 An Arctic Explorer in the Westfjords

On 1 August 1900 Icelandic peasant farmer Sighvatur Grímsson noted, rater indifferently, in his diary: "Dr. Fridtjof Nansen visited me and two other important gentlemen"[1] (Lbs 2375 I 4to, 1 August 1900). He had noted the day before that the Norwegian explorer had arrived at the fjord on a steamship. Nansen had organized an expedition during the summer for fisheries research under the supervision of Norwegian naturalist Dr Johan Hjort. The ship was first supposed to investigate the seabed and fish life around Norway and then proceed under the northeast coasts of Iceland for research. When the research around Iceland was completed, the ship was to head to Jan Mayen and then to Tromsø. The reason for the expedition sailing into Dýrafjörður was undoubtedly to meet Norwegian whaler Lauritz Jacob Berg. Berg had been engaged in whaling and whale processing since 1890 at Framnes in Dýrafjörður, adjacent to Sighvatur's farm.

The welcome extended to the esteemed group was described by a young Icelander, who had been a playmate of Berg's children in their childhood years:

> Once, the Norwegian oceanographic research ship Mikael [sic] Sars arrived at Dýrafjörður. Scientists and ship crew members visited Framnes, and as part of the celebrations, they were mounted on horses. They rode out along with the ladies and their daughters. In this group were professors Johan Hjort, Hákon Gran, and Fridtjof Nansen, who was already world-famous for his exploratory voyages across the Arctic Ocean. (Ólafsson 1973, 6)[2]

Like Sighvatur, Nansen was also a diarist, at least during expeditions. Excerpt from his notes from this expedition to Jan Mayen and Iceland were published in a book called *Frilufts-liv,* first published in 1916. On 31 July Nansen noted: "I visited an old bard, Sighvatur Grímsson, Berg's nearest neighbour, and his steward or tenant farmer. He is 60 years old and did not look like a poet at first glance" (Nansen 1916, 120–21).[3] After a small chat in front of the farm Sighvatur invited

---

1 "Dr. Friðþjófr Nansen heimsókti mig og 2 stór herrar aðrir". [All translations are mine, DÓ]

2 "Einhverju sinni kom norska hafrannsóknaskipið Mikael Sars á Dýrafjörð. Vísindamenn og skipstjórnarmenn komu í heimsókn að Framnesi, og til hátíðabirgða voru settir undir þá hestar og riðu þeir og frúin og dæturnar með þeim út á Mýramelinn. I þessum hópi voru prófessorarnir Johan Hjort, Hákon Gran og Fridtjiof Nansen, sem þegar þá var heimsfrægur fyrir könnunarferðir um Norðuríshafið".

3 "Jeg besøkte en gammel skald, Sigvathor [sic] Grímsson, Bergs nermese nabo, og hans forprakter eller husmand. Hann er 60 år, og så just ikke ut som dikter av det ytre".

FIGURE 9.2 The farm of Höfði around 1900. National Museum of Iceland

Nansen to come inside. They passed through the low, crooked, and dark hallway – and Nansen immediately made mental associations with his travels in Greenland – "almost like a passageway to an Eskimo hut" (Nansen 1916, 121).[4]

Then they entered a room so low that Nansen could not stand upright. There he found books piled in bundles along the walls, mostly unbound. Finally, they climbed a ladder to the second floor, which was a common area beneath the sloping ceiling: a living room, workspace, and bedroom for him and the entire family. There were his tools, pen, and ink, on a table by the window. And here, as far as Nansen understood, did Sighvatur spend most of the winter, and whenever he was free from his farm work (Nansen 1916, 122).

It is probable that Nansen's curiosity and awareness of this writing practice served as the primary motivation for his visit to Sighvatur's humble abode. The lower floor was stacked with books and manuscripts but in the attic, he had mainly kept books and manuscripts that he needed to use for his writing tasks. There, Sighvatur unveiled his Opus Magnum, his most ambitious project, a work both relentless and enduring:

> Then he brought out his manuscripts: six hefty volumes about Icelandic ministers, written finely and densely, – and there were also several

4 "nesten som en husgang til en eskimo-hytte."

> additional volumes. I cannot judge the content, but the sheer volume of the work lies in the writing alone. – And this was done without hope of ever seeing it in print. (Nansen 1916, 122)[5]

The work Nansen refers to here was Sighvatur's lifelong endeavour, known as *Prestaæfir á Íslandi* (Ministers' Lives in Iceland). He started writing these biographical entries on every serving church minister in Icelandic history around 1870 and continued until his death in 1930. The ultimate edition of this work is a substantial collection of around 15,000 pages, encompassed within 22 volumes held within the manuscript collection of The National Library of Iceland (Lbs 2358–2373 4to). Despite its magnitude – and perhaps because of it – it never saw print during his lifetime and to this day, only small sections of it have been published. The explorer of the Nordic periphery was seemingly astonished by the juxtaposition of this highly literate peasant: "Truly, I felt admiration for this old man whom I met in the cramped space. – Whom did he work for then? Not for fame – it must have been for the future – for humanity?" (Nansen 1916, 122).[6] At the turn of the twentieth century, he viewed Iceland as rural and underdeveloped, both culturally and industrially. When he first encountered the farmer at Höfði he was cutting hay with a small scythe, which required him to harvest in a stooped or squatting position. What he encountered were obsolete farming methods and literacy practices of a bygone era: "This is Iceland! Burdened under memories, mostly crippling, it lives in the past, forgetting the present, using the small scythe" (Nansen 1916, 123).[7]

3 A Peasant Scribe

Sighvatur Grímsson (1840–1930) – who adopted the additional name Borgfirðingur (literally "one from Borgarfjörður") – was one of those ordinary folks in Iceland that engaged intensely in the vernacular literary practices of the era (Ólafsson 2009). Throughout most of his long life of 90 years, he wrote, read, bought, sold, presented, and received handwritten material of various

5 "Så dro han frem sine manuskripter: seks svære binn om Islans prester, fint og tett skrevet,— og så kom desuten flere binn med tillegg. Om inhollet kan jeg ikke dømme; men et kjempearbeide er bare skrivingen. – Og det er gjort uten håb om noensinne å se det i trykken."

6 "Sandelig, jeg fikk beundring for denne gamle mannen, som jeg sto der i det trange rum.— For hvem arbeidet da han? Ikke for berømmelsen—det måtte da være eftertidens.—For menneskeheten?"

7 "Det er Islann! Tynget unner minner, mest til forkrøbling, lever det i fortiden,—glemmer nutiden,—og bruker stuttorvet."

genres. He was associated with several "scribal communities" in his lifetime, both local and far reaching (Ólafsson 2013). Born into a family of poor fishermen and peasant farmers in a small village in western Iceland in 1840 Sighvatur's enthusiasm was leaning towards literature and history from an early age despite being deprived of all formal education. Having lost both his father and later his mother during adolescence, Sighvatur Grímsson left his hometown of Skipaskagi (now Akranes) at the age of nineteen. He served for a decade as a farmhand and fisherman before becoming a tenant farmer in districts of mixed farming and maritime activity, first in the commune of Kaldrananeshreppur in the county of Strandasýsla for few years and later at the farm Höfði in Dýrafjörður for nearly 60 years (Ólafsson 2009).

Within this mundane and outwardly typical life course, Sighvatur lived a parallel life as a cultural institution of his own, an astonishingly industrious scribe and lay scholar, a collector and distributor of printed as well as handwritten books, and a local "poet laureate". The manuscripts he composed, copied, or compiled throughout his life and then bequeathed to the National and University Library of Iceland (NULI), totalling around 200 items. Among his largest and most abiding projects was his diary of nearly seven decades (Lbs 2374–2377 4to, NULI). Albeit being a tenant farmer throughout most of his life, it was books and manuscripts that were his passion and as for the diary this is thoroughly confirmed. Sighvatur Grímsson's diaries give us access to a detailed and extensive account of his literary activities throughout his adult life. Initially, entries related to reading and writing were scarce and scattered, but they soon evolved into a nearly comprehensive record of his book acquisitions, reading habits, and writing endeavours. His writings and transcriptions varied widely, encompassing everything from single letters to multi-volume chronicles, lineage charts, occasional poetry, and the exchange of manuscripts and printed books among individuals, families, and households. His diaries also reveal the intricate connections between literary interactions and various forms of exchange – be it food, goods, favours, or goodwill – highlighting Sighvatur's high level of professionalism and expertise in his chosen pursuits.

In this chapter I explore the material conditions related to the creation, distribution, and conservation of text-bearing objects. The emphasis will be on the instruments and procedures, living conditions, and environmental influences during the period which Sighvatur resides with his family at Klúka in Kaldrananeshreppur, 1869–1873 (see Ólafsson 2009, 160–80; 2022; 2024). Despite this being decades before Nansen's visit, it can be argued that these manuscripts and testimonies provide insight into a vibrant manuscript culture among the Icelandic common people during its peak in the latter part of the nineteenth century.

The vast quantity of texts copied by Sighvatur over the term of four years for himself as well as a variety of receivers demanded access to paper, pens and ink as well as some minimum spatial and material settings. In this chapter I will explore the materiality of this endeavour, primarily the attainment of paper via Sighvatur Grímsson's writings, primarily the entries in his diaries during this period. Despite the briefness of the daily entries, they provide evidence of a vigorous exchange of manuscripts and printed material between Sighvatur Grímsson and various members of the community. This includes individuals, families and households, men, women and adolescents, affluent farmers, poor lodgers, and farmhands. A close study of his diaries gives evidence of a vast interest in reading in most households, fuelled and fulfilled by Sighvatur's dynamic input.

4 Literacy Artifacts

In his description of Sighvatur Grímsson's living conditions Fridtjof Nansen mentions various things and settings related to reading, and in particular writing, of texts. There are books in stacks along the walls of the narrow tunnel and lower floor, leading to a second-floor common room (*baðstofa*) under a sloping ceiling. That space served as a living room, bedroom for the entire family and workspace for the ever-active scribe and scholar. There he finds his workstation with pen and ink, sitting on a table by the window (Nansen 1916, 122).

In a short autobiography he wrote (using the third person) in his early fifties Sighvatur describes his childhood encounters with literacy and its materiality (Lbs 3623 8vo; Grímsson 1965). There he describes how he, at the age of seven, learned to recognize the letters of the alphabet and to connect them into syllables and words at his mother's knee. After Sighvatur had learned to read printed and handwritten texts, he felt a strong desire to write began to awaken within him. Yet there was no means for it, and no one would offer help or encouragement. Nevertheless, he started to produce ink from various materials, depending on what was available at the time, sometimes using indigo, soot, light smoke, and others. At times, he managed to procure materials of this kind from others. With these, he collected envelopes from letters and discarded paper, sometimes blank. On occasion, he had to fashion a pen from feathers, while at other times, he obtained steel pens, mostly used. He would then eagerly sharpen and use them. Sighvatur obtained writing samples from various people, including an alphabet of capital letters from a local scribe. (Lbs 3623 8vo, [2–3]). Among his manuscripts, preserved in the manuscript department of the National Library, are numerous examples of both direct and indirect outcomes of these adolescent writings (see Ólafsson 2009, 124–39). This pursuit of writing instruments persisted throughout his adulthood, requiring him to resourcefully procure paper, ink, and pencils to satiate his bibliomania and meet the local demand for handwritten books.

4.1 *Pens*

The pens that Nansen saw at Sighvatur's desk were probably a combination of a steel nib (Icel., *penni/pennaoddur*), and a wooden shaft (Icel., *pennastöng*). Across Europe steel-point nibs started to replace quill (feathers) as the main writing tool in the seventeenth century and became common around the turn of the nineteenth century. Such instruments had become popular in Iceland around or before mid-century.

There are several mentions of pens in Sighvatur Grímsson's diary 1869–1873. The first comes from March 1869 when Sighvatur composed a stanza where he addresses his pen as a "crooked chap" and claims that "everyone is haunted by the fear of your spear" (Lbs 2374 I 4to: 27 May 1869; Lbs 2325 8vo II, 101). An entry in the diary on 20 August 1871 provides a clear insight into the material aspects of his writing practices. On that day he arrived at the farm Kleifar to return a handwritten copy of a certain *rímur*-cycle (ballad). According to his journals Sighvatur had made at least two transcripts from it, one for a fellow farmer and the second for himself. Upon the return he notes that he had obtained "12 pens for ink" (Lbs 2374 I 4to: 20 August 1871). The number of nibs acquired by Sighvatur suggests that they were not very durable and that they were usually sold by the dozen. Similarly in October 1872 we learn that Sighvatur sent his neighbour to a nearby trading station to buy some tobacco and pens. Four days later the neighbouring farmer returned with 24 pens, or two dozen (Lbs 2374 I 4to: 17 and 21 October 1872). This is further supported by import statistics from the mid-nineteenth century, which quantify pens by the dozen. Yearly import to Iceland in the 1860s was ranging between just over 300 dozen up to more than 1000 dozen or 12.000 nibs, an ample supply to cater to the requirements of this predominantly rural society of around fifty thousand individuals (*Skýrslur* 1870, 89, 618, 856; *Skýrslur* 1875, 132, 388).

From Kleifar, Sighvatur proceeded directly to the Hella farm, where he acquired around 30 nibs and a shaft (Lbs 2374 I 4to: 20 August 1871). At this point, Sighvatur had written extensively for the Hella household, indicating that these items might have been used as partial compensation for his scribal services in the preceding weeks and months. A similar scenario likely occurred in July 1872 when Sighvatur arrived at farm Sandnes, where he gathered various goods, probably as a compensation for commissioned writing tasks. Along

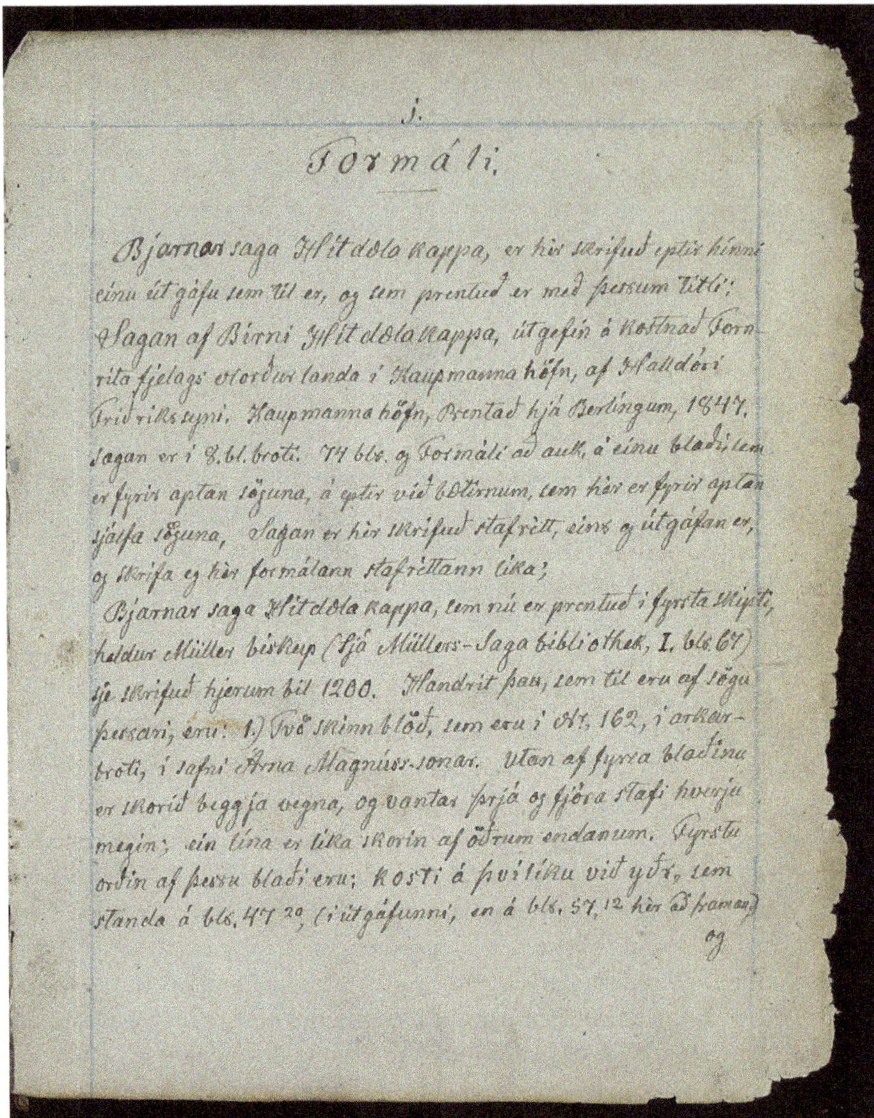
1.

Formáli.

Bjarnar saga Hitdæla kappa, er hér skrifuð eptir hinni
einu útgáfu sem til er, og sem prentuð er með þessum titli:
Sagan af Birni Hítdælakappa, útgefin á kostnað Forn-
rita fjelags Norðurlanda í Kaupmannahöfn, af Halldóri
Friðrikssyni. Kaupmannahöfn, Prentað hjá Berlingum, 1847.
Sagan er í 8.bl.broti. 74 bls. og Formáli að auk, á einu blaði, sem
er fyrir aptan söguna, á eptir viðbætirnum, sem hér er fyrir aptan
sjálfa söguna, Sagan er hér skrifuð stafrétt, eins og útgáfan er,
og skrifa eg hér formálann stafréttann líka;
Bjarnar saga Hítdæla kappa, sem nú er prentuð í fyrsta skipti,
heldur Müller biskup (Sjá Müllers-Saga bibliothek, I. bls. 67)
sje skrifuð hjerumbil 1200. Handrit þau, sem til eru af sögu
þessari, eru: 1.) Tvö skinnblöð, sem eru í Nr. 162, í arkar-
broti, í safni Árna Magnússonar. Utan af fyrra blaðinu
er skorið beggja vegna, og vantar þrjá og fjóra stafi hvorju
megin; ein lína er líka skorin af öðrum endanum. Fyrstu
orðin af þessu blaði eru: kosti á þvílíku við ydr, sem
standa á bls. 47 20, (í útgáfunni, en á bls. 57, 12 hér að framan)
og

FIGURE 9.3 Handwritten compilation of Icelandic Family Sagas, penned by Sighvatur Grímsson 1880–1881. Main text, enclosed by light blue borders. The National and University Library of Iceland (NULI). Lbs 2329 4to. Sögubók, 1880–1881. Handrit.is

with some food and four needles, he obtained two pen nibs and a borrowed jar of ink (Lbs 2374 I 4to: 9 July 1872). These examples illustrate a significant link between the availability of steel pen nibs and the creation of handwritten materials, which played a crucial role in the literacy practices of nineteenth-century

Iceland. Additionally, examples can be found of Sighvatur acquiring pencils, in one instance a coloured pencil. (Lbs 2374 I 4to: 27 July 1872). It is reasonable to assume that these implements were used for marking the margins of handwritten books, much like what we observe here below.

4.2 *Paper*

A second essential component for scribal practices was of course paper. While Sighvatur actively collected printed materials, it is reasonable to assume that the books observed by Nansen, stacked along the walls, were both print material and manuscripts. At his death, Sighvatur bequeathed a collection of approximately 2000 printed books, booklets, and periodicals to a local communal library, Sýslubókasafn Vestur-Ísafjarðarsýslu ("Aldaminning" 1940). His over 200 handwritten volumes, however, were handed over to the National Library of Iceland in accordance with a 1906 agreement which stipulated that the library paid him an annual pension in exchange for the manuscripts upon his death. These handwritten books were, naturally, penned on paper that had to be procured either by the scribe or the individual commissioning the work. Paper in nineteenth-century Iceland was primarily imported, the vast majority from Denmark. During the second half of the century all foreign trade was restricted to either residential merchants located at certified trading posts or non-resident merchants, who were permitted to trade from their ships at authorized locations. With these merchants, farmers, who had butter, tallow, wool and meat, and fishermen with their catch, could trade their goods for imported commodities like grain, beans, sugar, coffee, tobacco, and paper (see Ólafsson 2023). In his 1934 book on popular culture in nineteenth-century Iceland, Pastor and ethnologist Jónas Jónasson from Hrafnagil examined the access to writing paper. He noted that writing paper was typically acquired through merchants, yet it was frequently expensive and difficult to obtain. This observation reflects the challenges faced by individuals seeking writing materials during that period (Jónasson 2010).

Sighvatur's diary from his Klúka years holds few direct references to acquisition of writing paper. The first of these is from 21 July 1869 when Sighvatur notes at he had received 12 sheets (Icel., *arkir*) of paper along with various other goods from a trading station in exchange for wool. From such a sheet a booklet or quire of four pages in folio could be produced, eight pages in 4to, or 16 pages in 8vo. Twelve sheets would make 96 pages in 4to, the most common format of handwritten books at the time. In an entry from July 1872, we learn that Sighvatur purchased what he terms "mail-paper" (Icel. *Póstpappír*) and a pencil (Icel., *blýantur*). Later that same month, he notes that he had written a message regarding paper to a local merchant regarding paper (Lbs 2374 I 4to:

22 July 1872 and 28 July 1872). It is then soon after Sighvatur moved from Klúka to Höfði in Dýrafjörður in the summer of 1873 that we see examples of his more extensive paper purchases directly from merchants. On 25 October that year, he took out 3 books of paper from a merchant in Þingeyri. Ten days later he wrote in his diaries that he had bought 27 sheets of paper on his trip to the village of Þingeyri (Lbs 2374 I 4to: 25 October 1873 and 5 November 1873).

Sighvatur's writings can be roughly divided into two sections, writings he did for himself and commissioned tasks. We know from the diaries and other writings done by Sighvatur Grímsson that he was particularly active as a scribe while located at Klúka and that these practices were done in a scribal community with people who wrote, read, bought, sold, presented, and received handwritten material. Sources that have derived from the scribe and lay scholar Sighvatur Grímsson strongly suggest that the more affluent farmers in Kaldrananeshreppur had good access to writing paper in the period studied here, the late 1860s and early 1870s. While he had the skill and the motivation to transcribe bulks of texts and produce hefty books by hand, and an access to and knowledge of rare texts, those who facilitated the practices and supplied the necessary materials were also essential to the circuit of scribal publication (Ólafsson 2023, 351–56).

The source of Sighvatur's paper during his earlier years at Klúka is more uncertain, but it is safe to assume that in most cases it was the client who provided the paper, too, at least for the more substantial writings. In March 1871 Sighvatur recorded that he had delivered a huge handwritten volume of 799 pp. folio and in very small and dense writing" (Lbs 2374 I 4to: 2 March 1871). The text in question here was an Icelandic translation of the ancient chronicles attributed to Flavius Josephus (Josephus 1897). Sections of Josephus's writings had been known to Icelandic learned readers at least since the thirteenth century but this translation of the complete works of Josephus was, however, only available in Icelandic by Gísli Konráðsson in the first half of nineteenth century and has never been published in print. Some months earlier Sighvatur recorded that he had been commissioned by Jón Guðmundsson at Hella to copy *Gyðingasaga*, from three quarto books into one large folio volume (Lbs 2374 I 4to: 11 November 1870.) Never during the writing period of nearly four months does Sighvatur mention how he acquired paper for this task, suggesting that it was simply supplied by Jón Guðmundsson at Hella. It is only at the end of his work that Sighvatur notes that he had received 24 sheets of paper, or around 200 pages in 4to, presumably a part of his reimbursement. We find the most direct information in Sighvatur's diaries after he moved to Höfði in Dýrafjörður. On 25 January 1876, Sighvatur notes that when heading home from the farm Mýrar, he brought with him 10 sheets of paper allocated for a transcript of *Parmesar saga* for a neighbouring farmer (Lbs 2374 I. 25 January 1876). Some days later, the transcript was finalized and delivered on 64 pages in 4to (4 sheets). To make the most of the paper available, Sighvatur began to copy another short saga for the same client that he delivered a few weeks later, noting that the two texts combined were 194 pages.

4.3 *Ink*

Ink used in nineteenth-century Iceland was however both imported and home-made. The overwhelming majority of Sighvatur's manuscripts are written in black ink, but various other manuscripts from the nineteenth century indicate that scribes could obtain colored ink, which was then used for decoration and illustrations. There are only few direct references to ink in Sighvatur's diary but and we see that he made his own ink at least as late as in 1870. In March that year he notes: "I brewed ink in a bottle" (Lbs 2374 I. 8 March 1870). A mid-eighteenth-century travel book detailing a scientific exploration around Iceland describes the traditional method that has been popular for centuries for making ink:

> The bulk of the ink used in Iceland is made of willow, as follows: Broth from bearberry [Icel., *sortulyng*], is taken and mixed with dark clay (Icel., *sorti*) and welded together as if for dying. However, some people use only the blackest of this brew. The twigs of raw willow are added to the liquid, and they are left there for a while. Then the mixture is boiled down, until it becomes thick and sometimes a little adhesive. When it is so thick that a round drop sits still, if it is allowed to drip on a man's nail, the ink is finished. (Ólafsson & Pálsson 1978 (1752–1757), 101)[8]

We have seen above that Sighvatur borrowed an ink jar from the farm Sandnes in the summer of 1872. While there is no indication whether it was a home brew or imported and sold by a merchant, the relatively strong economic status of the Sandnes household would suggest the latter. (Lbs 2374 I. 9 July 1872). Just like with paper it is probable that when Sighvatur was commissioned for larger transcripts, the patron would also provide ink as well.

---

8 "Meginið af bleki því, sem notað er á Íslandi, er búið til úr víði, sem hér segir: Tekið er sortulyngsseyði og því blandað við sortu og soðið saman eins og til litunar. Ýmsir nota þó aðeins hið svartasta af legi þessum. Í löginn eru lagðir spænir af hráum víði, og eru þeir látnir liggja þar um hríð. Síðan er lögurinn seyddur, þangað til hann verður þykkur og stundum dálítið límkenndur. Þegar hann er orðinn svo þykkur, að hnöttóttur dropi situr kyrr, ef hann er látinn drjúpa á nögl manns, er blekið fullgert."

5 Interior Conditions for Literacy Practices

In addition to what we can call the three necessities for scribal production of texts – pens, ink, and paper – we can add another level of objects that facilitate such practices; seats to sit, tables to write on and shelfs and chests to store books and manuscripts, and various means of illumination.

5.1 *Lamps and Light*

In his account Nansen mentions the dimness of the tunnel where Sighvatur stored much of his books and manuscripts, before they entered a brighter common room. In contrast, the upper-level features, according to Nansen's account, a window, allowing natural light to filter in. This window was almost certainly featured glass panes. It became a common feature in households during the latter part of the nineteenth century, replacing earlier use of foetal membrane from cows. In Sighvatur's diary we can see some evidence of this development. On 26 August 1870 he notes that he had received two panes of glass at the farm Svanshóll and two weeks later he bought four panes at Hella. Then on 12 September he registers that he had installed the glass panes in his window (Lbs 2374 I. 26 August and 12 September 1870).

Nansen does not make any mention of alternative light sources, such as candles or lamps. This omission can likely be attributed to his visit in late July and early August, a time when daylight in the northern hemisphere is nearly continuous. It is nevertheless evident that the bulk of Sighvatur's writings was done during the winter months, while outdoor work took priority over the shorter summer period. Similarly, traditional indoor tasks, such as wool work and other household industries, were carried out during the extended and dark winter months, necessitating adequate illumination.

In the beginning of September 1870, Sighvatur notes that the light(s) were lit in the living room for the first time that fall (Lbs 2374 I. 2 September 1870). This testimony aligns with the tradition of the so-called "light time", meaning that lights were allowed to burn in living rooms from September until May (Ólafsson 1987, 350). It is clear that when Nansen called upon Höfði that Sighvatur was using lamps designed to burn kerosene. Kerosene lamps were imported to Iceland from the early 1870s and came relatively fast into common use (Jónasson 2010, 4). This lighting device employs kerosene as a fuel and features a wick or mantle as a light source. It is shielded by a glass chimney, enhancing the overall light output. As early as the fall of 1873 we see for the first-time a reference to such where he registers that he had been to the nearby village of Þingeyri and received both oil and lamp (Lbs 2374 I. 13 September 1873).

Prior to the advent of kerosene lamps most Icelanders used extremely simple lamps made of a small vessel filled with liver oil from cod, shark, seal or whale and a wick made of arctic cotton grass. The oldest and plainest form of lamp was commonly referred to as 'kola' in Icelandic. It was a small container that could be made from various materials, but most commonly from stone. Lamps made of metal, but also fuelled by liver oil became commonplace from the seventeenth century (Icel., *lýsislampar*) (Ólafsson 1987, 354). Typically, households possessed a single lighting device, offering only limited brightness and controlled by the head of the household (Gunnlaugsson 1991).

Jónas Jónasson writes in his 1934 ethnological account that these lamps provided remarkably good brightness if the oil was good, but if cod-liver oil was used, and of poor quality, as it often happened, the illumination was dim, and the houses would become filled with the smoke of the lamp. As this single light had no glass chimney it would have been rather dim as the distance from the lamp increased. Those who were deeply involved in writing sagas and poems would, according to Jónas, sometimes hang a glass globe filled with water near the lamp and let the beam of light from it fall onto the pages (Jónasson 2010, 4). The use of candles – for common households, these candles were usually made from tallow – was at the time primarily linked to Christmas rather than being a part of everyday lighting. This is evident in Sighvatur's diaries, where references to candles only occur during the Christmas season, as he brought home various types of food supposedly received as barter in return for his commissioned writings for the Hella household (Lbs 2374 I. 23 December 1869 and 22 December 1870).

Relentless reading and writing in inadequate lighting from a young age might have had adverse effects on his eyesight. We learn from Sighvatur's diary that he had already acquired glasses as early as 1870 when he took them to a nearby farm for repairs (Lbs 2374 I. 31 July 1870; 21 January 1871). To some extent, eyeglasses became accessible to ordinary people in Iceland during the latter half of the eighteenth century. It is documented that the author Eríkur Laxdal (1743–1816) owned glasses at the time of his death (Jónsson 2014, 58). Anna Heiða Baldursdóttir mentions in her recent doctoral thesis three examples of individuals, a labourer and two widows, who owned eyeglasses at the time of their death in the mid-nineteenth century (Baldursdóttir 2022). Highlighting the connections between social classes and manuscripts, the manuscript collection at NULI features an intriguing example. Dating back to around 1800, this manuscript carries a lasting imprint from the spectacles that a reader may have placed between its pages (Lbs 5272 8vo).

FIGURE 9.4 Impressions of glasses visible in a handwritten book from the nineteenth century. National and University of Iceland. Manuscript collection. NULI Lbs 5272 8vo. "Rímur af barndómi Jesú Krists" by Guðmundur Bergþórsson
PHOTOGRAPHER HALLDÓRA KRISTINSDÓTTIR

5.2 Furniture

According to Nansens's description Sighvatur did not own any shelfs for his books and manuscripts were simply stacks along the walls of the tunnel at the lower floor. Nansen does not make any mention of trunks or chests to store books, but according to the diary, Sighvatur had two trunks (Icel., *koffort*) to transport his books and manuscripts when the family moved from Klúka to Höfði in 1873 (Lbs 2374 I. 30 October 1870; 9 April 1873; 12 April 1873; 25 May 1873). In the second-floor common room however, he kept the books and manuscripts he mostly needed for his writing. These books were arranged on a table under a small study window, under the table, and beyond as space allowed, according to a description given by one of Sighvatur's neighbours, Kristinn Guðlaugsson. The small table near the window was often cluttered with books and manuscripts, but Sighvatur typically wrote with a small platform resting on his knees while seated in his bed. "This came as no surprise to me", Kristinn remarks, "since this was how I myself learned to write shortly before 1880"[9]

9 "Þetta kom mér ekki ókunnuglega fyrir, því þannig lærði ég að skrifa laust fyrir 1880."

(Guðlaugsson 1982, 264–65). Nansen's observation is similar, with the exception that according to him Sighvatur he sat on a stool with a board on his lap when writing (Nansen 1916). Based on his son's testimony, Sighvatur utilized a wooden box as a writing desk, resting it on his lap for all his writing endeavors, at least until 1890. This particular box is now counted among the few artifacts associated with literacy practices, curated at the Manuscript department of the National and University Library of Iceland.

6 Conclusion

Almost six decades following Fridtjof Nansen's visit to the modest farmstead at Höfði, Jón Helgason, a professor of Icelandic studies at the University of Copenhagen, presents a depiction highlighting the lasting vitality of manuscript culture in Iceland throughout the early modern and modern periods. Helgason portrays Icelandic scribal practices during these periods as enduring remnants of medieval traditions. Furthermore, he contends that Iceland, in this respect, stands out as distinctive within both European and global contexts. "Enter the sixteenth century", he wrote, "and copying by hand became an obsolete bookmaking method in neighbouring countries" (Helgason 1958, 8–9). Certainly, printing made its way into the country relatively early, but its control fell under the Church's monopoly. Vernacular genres, however, remained unprinted. Into the nineteenth century, Helgason claimed, "there still existed true medieval men who continued to reproduce books with their pen" (Helgason 1958, 8–9).

While this description is accurate in many respects, in my previous works on nineteenth-century manuscript culture, I contested the idea that the scribal practices of the seventeenth, eighteenth, and nineteenth centuries were mere remnants of a bygone era (Ólafsson 2009, 182). Instead, my argument contends that the significant role played by scribal culture in early modern and modern Iceland was not unique within a European or global context, nor was it a relic of medieval traditions. Rather, it stood as a substantial and integral part of living culture, persisting until the nineteenth century.

During his lifetime Sighvatur Grímsson wrote out thousands of pages and millions of words, in what can only be described as bibliomania. And all this took place within the confinements of material conditions that have been explored in this chapter. Albeit he was by no means typical of people of his status, he was not an isolated or fully exceptional case either. He was, throughout his life, firmly set within networks of like-minded people that wrote, read, bought, sold, commissioned, presented, and received handwritten material of various genres. His case illuminates the pathways of the informal educational system in nineteenth-century Iceland, which were closely interconnected with

FIGURE 9.5 Manuscripts written or owned by Sighvatur Grímsson. The National and University Library of Iceland (NULI)
PHOTOGRAPHER HELGI BRAGASON

the pathways of scribal communication. His case is not only a story about manuscripts and the people who wrote and read them but also a tale of paper, pens, and ink, and of the material conditions surrounding the literacy practices.

Bibliography

Baldursdóttir, Anna Heiða. 2022. "Hlutir úr fortíð. Eigur fólks og safnkostur frá 19. öld." PhD thesis. Reykjavík: University of Iceland.

Bennett, Jane. 2010. *Vibrant Matter: A Political Ecology of Things*. Duke University Press.

Chartier, Roger, ed. 1989. *The Culture of Print: Power and the Uses of Print in Early Modern Europe*. Oxford and New York: Blackwell.

Darnton, Robert. 1982. "What is the History of Books?" *Daedalus* 111: 65–83.

Grímsson, Sighvatur. 1965. "Æviágrip Sighvats Grímssonar Borgfirðings fram til 27. des. 1892 eftir sjálfan hann." *Árbók Landsbókasafns Íslands 1964*, 21: 91–99.

Guðlaugsson, Kristinn. 1982. "Sighvatur á Höfða". In *Frá ystu nesjum* vol 3, edited by Gils Guðmundsson, 64–69. Hafnarfjörður: Skuggsjá.

Gunnlaugsson, Gísli Ágúst. 1991. "Ljós, lestur og félagslegt taumhald." *Ný saga* 5: 62–66.

Harman, Graham. 2018. *Object-Oriented Ontology. A New Theory of Everything*. Pelican Books.

Helgason, Jón. 1958. *Handritaspjall*. Reykjavík: Mál og menning.

Jónasson, Jónas. 2010/1934. *Íslenzkir þjóðhættir*. Fourth edition. Reykjavík: Opna.

Jónsson, Már. 2014. "Bækur og brotin gleraugu." In *Saltari: Stilltur og sleginn Svanhildi Óskarsdóttur fimmtugri 13. mars 2014*, edited by Þórunn Sigurðardóttir, Margrét Eggertsdóttir & Guðvarður Már Gunnlaugsson, 58–60. Reykjavík: Menningar- og minningarsjóður Mette Magnussen.

Josephus, Flavius. 1987. *The Works of Josephus Complete and Unabridged*. Peabody, MA: Hendricson.

Latour, Bruno. 1988. *We Have Never Been Modern*. Harvard University Press.

Latour, Bruno. 2005. *Reassembling the Social: An Introduction to Actor-Network Theory*. Oxford University Press.

Law, John & Hassard, John. 1999. *Actor Network Theory and After*. Oxford: Blackwell Publishers.

Love, Harold. 1993. *Scribal Publication in Seventeenth Century England*. Oxford: Oxford University Press.

Magnússon, Sigurður Gylfi & Davíð Ólafsson. 2017. *Minor Knowledge and Microhistory: Manuscript Culture in the Nineteenth Century*. London: Routledge.

Magnússon, Sigurður Gylfi & Davíð Ólafsson. 2023. "Pen, Paper and Peasants: The Rise of Vernacular Literacy Practices in 19th-Century Iceland." In *The Common Writer in Modern History,* edited by Martyn Lyons, 121–39. Manchester: Manchester University Press.

Mímisson, Kristján. 2025. "The Affect of Relating: The Divergent Manifestations of Things." In *Objects in the Archives: Modern Material Culture and Heritage in the North,* edited by Kristján Mímisson & Davíð Ólafsson. London: Routledge.

Nansen, Fridtjof. 1916. *Frilufts-liv. Blade av dagboken*. Kristiania: Jacob Dybwads forlag.

Nichols, Stephen G. 1990. "Philology in a Manuscript Culture." *Speculum* 65: 1–10.

[NN]. 1940. "Sighvatur Grímsson Borgfirðingur. Aldaminning." *Vesturland* 50: 194.

Ólafsson, Davíð. 2009. *Wordmongers: Post-Medieval Scribal Culture and the Case of Sighvatur* Grímsson. PhD thesis. St. Andrews, University of St. Andrews.

Ólafsson, Davíð. 2013. "Scribal Communities in Iceland: The Case of Sighvatur Grímsson." In *White Field, Black Seeds: Nordic Literacy Practices in the Long Nineteenth-Century,* edited by Anna Kuismin & Matthew J. Driscoll, 40–49. Helsinki: Finnish Literature Society.

Ólafsson, Davíð. 2017. "Post-Medieval Manuscript Culture and the Historiography of Texts." In *Mirrors of Virtue: Manuscript and Print in Late Pre-Modern Iceland,* edited by Matthew J. Driscoll & Margrét Eggertsdóttir, 1–30. Bibliotheca Arnamagnæana XLIX. Opuscula XV. Copenhagen: Museum Tusculanums.

Ólafsson, Davíð. 2023. "Many Pages, Much Ink: The Materiality of Writing in 19th-Century Iceland." In *Paper Stories: Paper and Book History in Post-Medieval Europe,* edited by Silvia Hufnagel, Þórunn Sigurðardóttir & Davíð Ólafsson, 343–60. Berlin: Gruyter.

Ólafsson, Davíð. 2025. "Social Circulanity of Books and Manuscripts: Sharing Economy and the Material Culture of Text in 19th Century Iceland." In *Objects in the Archives: Modern Material Culture and Heritage in the North,* edited by Davíð Ólafsson & Kristján Mímisson. London: Routledge.

Ólafsson, Eggert & Pálsson, Bjarni. 1978 (1772). *Ferðabók Eggerts Ólafssonar og Bjarna Pálssonar 1752–1757*. Vol I–II. Reykjavík: Örn og Örlygur.

Ólafsson, Guðmundur. 1987. "Ljósfæri og lýsing." In *Íslensk Þjóðmenning I. Uppruni og umhverfi,* edited by Frosti F. Jóhannesson, 346–69. Reykjavík: Þjóðsaga.

Ólafsson, Jóhann Gunnar. 1973. "Frá fjölskyldu Bergs hvalveiðimanns á Framnesi í Dýrafirði." *Ísfirðingur* 26–35: 6–10.

Skýrslur um landshagi á Íslandi 4. 1870. Copenhagen; Hið íslenska bókmenntafélag.

Skýrslur um landshagi á Íslandi 5. 1875. Copenhagen; Hið íslenska bókmenntafélag.

Archival Material and Manuscripts

National and University Library of Iceland (NULI).

Lbs 2374–2377 4to. Sighvatur Grímsson. "Diaries 1863–1930."

Lbs 3623 8vo. "Sighvatur Grímsson Borgfirðingur: Æviágrip, skráð 27. des 1892."

Lbs 2358–2373 4to. "Presta æfir á Íslandi, safnað hefur Sighvatur Grímsson Borgfirðingur."

Lbs 5272 8vo. "Rímur af barndómi Jesú Krists. Guðmundur Erlendsson."

Lbs 2325 I–III 8vo. "Syrpa."

Gísli Sighvatsson. Letter to the archive 14 March 1930.

FIGURE 10.1 Thomasson family on display in Hamburg, 1892. Stiftelsen Jamtlis Samlingar "JLM NTh10842". https://jlm.kulturhotell.se/objects/c60-98598/ CC BY-NC-ND 4.

CHAPTER 10

Afterimages: the Afterlife of Photographs of Swedish Sámi from the "Colonial Archive"

JoAnn Conrad

Abstract

What is the role of photographs within the ethnographic museum collection? How can the colonial past and the epistemic legacy that undergirded ethnographic collecting pivot to the demands of culturally diverse histories/memories that may be contained within such photos? What is the relationship between museum collections of archived photos and the diffuse proliferation of the photographic record? How have the inherent instability of the photograph and its excess of meaning, the contingencies of the photographic encounter, the randomness of collection, and the arbitrariness of cataloging affected how histories are made, remade, retrieved, and retold using archived photographs? This article focuses on photographs of Sámi individuals from across Swedish Sápmi from the late 1800s and early 1900s, a time when settler colonialism, anthropology, theories of scientific racism, and photographic technologies combined to form a potent and often toxic alliance. Such 'ethnographic photographs' are often categorized according to the homogenizing and reductionist logic of the "colonial archive". Revisiting the photographs of five photographers in distinct contexts and with dissimilar motivations and dis-aggregating their distinct bodies of work, I argue that although the specter of colonial power persists throughout, it cannot contain the excess of meaning in the collected photographs.

Keywords

Sámi – ethnographic photography – colonial archive – John Bauer – postcards

 | DOI:10.1163/9789004749665_011

In 1892, 12-year-old Nils Thomasson and his family were part of a 'Lappish'[1] traveling troupe in Hamburg – one of the many "living ethnological exhibits" that were popular throughout Europe at the end of the nineteenth century (Baglo 2014). The Thomasson troupe staged "traditional Sámi life" to European spectators – nomadic reindeer herders outfitted in "Lappish Equipage" (Mathisen 2017), ostensibly "performing themselves" (Kirshenblatt–Gimblett 1998). The Thomasson family members were not, however, herders; their performative touring bears witness to the financial precarity of their situation *without* reindeer. They were not "signs of themselves" (Kirshenblatt–Gimblett 1998, 18) but actors playing roles already imprinted in the European imaginary landscape by the flourishing trade in photographs and photographic postcards.

The troupe was photographed with a blank background (Figure 10.1), which served as an ersatz winter scene, erasing the immediate background so that an imaginary ethnographic scene could be effected. The photo was then sold to postcard companies, sometimes colorized or otherwise manipulated, and given the title – "Parti från Lappland", which further generalized the scene and relocated it to an imaginary Lappland. This image of the Thomasson family on tour circulated widely throughout Europe as an image of Lapps in their native 'habitat' in the winter, and yet the photo is simulacrum. The proliferation and distribution of the postcard image, however, created its own type of knowledge of the Sámi.

The photo is from the Jamtli collection of historical photographs from Jämtland.[2] A copy of the image, made into a postcard, is also in the postcard collection (tslp10812) of the Arctic University Museum of Norway in Tromsø. The image has been analyzed in post-colonial critiques (Mathisen 2017; Baglo 2019) as a transparent inscription of colonial power relations – as an item in the "colonial archive". In each collection, however, the systems of classification and organization, as well as their epistemological bases, are different, suggesting a proliferation of meaning that includes, but also goes beyond the colonizing imprint.

The "colonial archive" (Stoler 2002) is a potent, retrospective shorthand used in the critique of the collection, appropriation, reorganization, and representation of material gathered during the ethnographic encounters of the past centuries. The term presupposes a colonial imperative in the collection process and the subsequent (re)-telling of history. Although the "colonial archive" suggests closure and uniformity, and posits an explanatory narrative, it cannot contain the photograph's inherent instability and polysemy and unravels in its generality. The focus on the colonial imprint also detracts attention from

1 The terms 'Lapp'. 'Lappland', and 'Lappish' are anachronistic and pejorative, now replaced with the ethnonym Sámi who inhabit Sápmi, which spans four countries.

2 See https://jlm.kulturhotell.se/item/fotografi/.

photography's role as a "central element [...] in the new commodity economy [...] of consumption and circulation" (Crary 1992, 13), facilitated by the popular press. Photography, one of the major tools of nineteenth and twentieth-century anthropology, once embraced for its apparent objective view, has become a privileged focus of this critique of the colonial archive. It represents the visible sign of the imprint of colonial power onto both the surface of the photograph and that of the body photographed.

But what is the role of photographs within the ethnographic museum collection?[3] How can the colonial past and the epistemic legacy that undergirded ethnographic collecting pivot to the demands of culturally diverse histories/memories that may be contained within such photos? What is the relationship between museum collections of archived photos and the diffuse proliferation of the photographic record in other forms, such as personal photos, photos for tourism, and other popular media? How have the inherent instability of the photograph and its excess of meaning, the contingencies of the photographic encounter, the randomness of museum collection, and the arbitrariness of cataloguing affected how histories are made, remade, retrieved, and retold using archived photographs?

This article focuses on photographs of Sámi individuals from across Swedish Sápmi from the late 1800s and early 1900s, a time when settler colonialism, anthropology, theories of scientific racism, and photographic technologies combined to form a potent and often toxic alliance. I revisit the photographs of five photographers in distinct contexts and with dissimilar motivations, to dis-aggregate their work, and to argue that although the specter of colonial power persists throughout, it cannot contain the excess of meaning in their photographs. I also consider the particular museum context of the photos, and the stories that they seek to tell through their collections. Revisiting photos in this way reveals their ambiguous position and meaning in the colonial archive, and the social lives of photos that contain little stories that disturb the narrative of colonialism.

1 Sámi Photographs in the Museum Context – Case Studies

1.1 *Carolina Charlotta (Lotten) Mariana von Düben (1828–1915)* – Nordiska Museet

Lotten von Düben is remembered today as the photographer for her husband Gustaf von Düben, Doctor of Medicine and Professor of Pathological Anatomy

3 Edwards & Lien (2014) and Edwards (2015) ask this question and conclude that photographs have a "just there" status in museums – "an unquestioned absorption into the design strategies of museums" (2014, 7).

at the *Karolinska Institutet,* in Stockholm, who was one of "the first generation of Nordic men practicing Lappology, a semi-scientific, vaguely well-intentioned, colonialist-inspired, and largely racist bundle of ethnographic and linguistic studies, theories, and prejudices" (Sjoholm 2023, 100). In line with the racial theories both in Sweden as well as Anthropology generally, Gustaf espoused a hierarchical theory of races, which his craniometric research was intended to support. This led to the collection of Sámi skulls, their measurement, and ultimately the measurement of the heads of Sámi individuals in the field along with supplemental photographs provided by Lotten during their expeditions to the South Sámi areas of Sweden in 1868 and 1871.

Neither a professional photographer nor an ethnographer, Lotten is often credited as being a pioneer in ethnographic photography (Modin 2021), but it is important to recognize the extent to which her work was shaped by and limited to the male world of her husband and his pursuit: her photographs were in support of her husband's cranial studies. That is, her photographs were always part of a typological system – first as part of the Karolinska's craniometric studies, and later, in 1939, as part of Nordiska Museet's "Lappish Department".

Von Düben's ethnographic photography stands in contrast to the aesthetics, conventions, and motivations of the studio portrait photography popular among the Swedish bourgeoisie at the time, and the contrast encapsulates and gives evidence to Swedish racial hierarchies. Lotten's "head shots" – the *en face* and profile formula of anthropometric portraiture – reflected contemporary anthropological photographic convention (Bank 2001) and criminology and psychiatry. It was a convention reserved for "primitive cultures" and "deviants" (Sekula 1989, 345), which was appropriated by ethnologists "to link optical empiricism with abstract, statistical truth, to move from the specificity of the body to abstract, mathematical laws of human nature" (Sekula 1981, 16). In keeping with the empirical conceit, her photos included the name, age, and location of her subjects – all necessary for racial classification, which traces ancestry and regional specificity.

Lotten von Düben's photos (Figure 10.2) were thus primarily intended to corroborate and add to the limited data obtained on Sámi crania in order to substantiate her husband's race-based typology. They "provided visual imaginary 'evidence' of racial specificities where the scientific practices otherwise failed" (Lien & Nielssen 2018, 1). Her photos were not, however, strictly limited to the head and facial features, but were three-quarter views, showing the person, usually sitting with their hands in their laps, as well as details of clothing and identifying hats. This kind of "photographic excess"

FIGURE 10.2 Lotten von Düben: Frontal and side view of Inga Kajsa Granström, 22 years old, Tuorpons sameby. Nordiska Museet – NMA.0033118 2.jpg. Original housed in Uppsala University Library

(Nielssen 2018; Edwards 2015) not only destabilizes claims of scientific purity, but also provides cultural clues to re-interpret these photos in subsequent generations.

To complement his cranial research, Gustaf von Düben also produced a volume on Sámi culture and life – *Om Lappland och Lapparna* (1873) – a seminal text that informed later anthropological studies of the Sámi. In *Om Lappland,* Lotten's field photographs were reproduced as grayscale wood-block prints, with all extraneous background imagery removed (Figure 10.3). The names, ages, and locations were also not included alongside the images but relegated to a cumbersome appendix. This shift from individual to type reflects the shift from craniometry to ethnography.

Lotten von Düben's photos explicitly reinforced the racial ideology pervading late nineteenth-century Swedish discourse. And yet, as with their modification in *Om Lappland,* the photos' eventual incorporation into Hazelius's utopian project – Nordiska Museet, were selectively retrieved to de-emphasizes Lotten's contribution to craniometric studies, thereby disavowing a painful colonial legacy and rewriting history to effect a kind of historical aphasia (Stoler 2011).

FIGURE 10.3 Print from Lotten von Düben's photo of now unnamed Inga Kajsa Granström in *Om Lappland och Lapparna*, plate 3, photo 4

1.2 *Emilie Demant Hatt (1873–1958)* – Nordiska Museet

Emilie Demant Hatt is best known for her collaboration with Sámi author Johan Turi on *Muitalus sámiid birra* (Turi 1910), Turi's classic account of the life of Sámi reindeer herders. From a chance meeting with Turi during Emilie's first trip to Lapland in 1904, she and Turi would come to an agreement wherein she would help him realize his book, and in return, he would facilitate her desire to live and migrate with the Sámi on their yearly migrations. Thus, from 1907–1908, Demant joined two different reindeer-herding families in their seasonal migrations (Sjoholm 2017; Demant Hatt & Sjoholm 2013). Her *Med lapperne i højfjeldet* (1913) is a description of that year.

Med lapperne i højfjeldet was not a traditional ethnography for its time, especially when compared to von Düben's *Om Lappland och Lapparna*. Demant lived and experienced life on migration, as did those with whom she traveled; her experience was intimate and proximate. Her text is conversational and describes daily interactions and events along their route. Yet, detailed ethnographic information is packed into her narrative. She also took photographs, although not the standardized, scientific photos with the background hidden by a sheet and people stiffly posed. Demant Hatt's photos are more akin to family snapshots and reveal an ease and friendliness between Emilie and her hosts. The intimacy displayed in the text and photos would today be recognized as both feminist and collaborative ethnographic fieldwork (Kuutma 2003); a shared relationship with her friends and hosts in the *siida* that was able to resist some of "the primitivist image making of her day" (Sjoholm 2014, 357).

Whereas Lotten von Düben's photos can be seen in the context of unequal colonial power relations underwritten by racist pseudoscience (Eglinger 2023), Demant Hatt's are not so easily explained. The bubble of privilege that Lotten von Düben operated in while in the field because of her husband was absent in Emilie Demant's case. Although structurally benefitting from her status as a Dane, Emilie was not only operating within a highly gendered social system at home, but also conducted her fieldwork as an unmarried woman, which problematizes the blanket critiques concerning the privilege afforded white women travelers at the intersection of gender and (male) colonial power. Barbara Sjoholm, who has translated much of Demant Hatt's work into English, disagrees with the notion that Demant Hatt was an agent of colonialism, observing that "the great value of her text is her non-colonial perspective. Instead of that, hers is [one of] participant observation" which helped deconstruct colonialism (Sjoholm 2011).

At a minimum, "Demant Hatt's female perspective provides an alternative, concurrent voice that might decenter and reframe existing imperial representations" (Reepleog 2019, 183). This perspective obtained from her position within the female sphere, illustrated in an encounter during the last and most treacherous day of the spring migration of 1908, when the group she was traveling with, which consisted of women and children, stopped for the first rest of the day:

> One young woman had two reindeer: one carried the baggage, and the other reindeer, completely white, carried the cradle with her youngest child [...] Before we went down into the valley the young woman stopped and nursed her little one in the cradle on the reindeer. (Demant Hatt 2013, 138–39)

The moment was captured in Demant's photo (Figure 10.4), the fate of which exemplifies the complicated nature of collecting, curating, and narrating photographs. In the original Swedish version of *Med lapperne i højfjeldet* the photo accompanying the above passage is entitled *Barnet ammes* [the child is being nursed] (Demant Hatt 1913, 146). In contrast, the photo has been omitted in the English translation from 2013. The photo is also part of Nordiska Museet's digitized collection of Demant Hatt's photos. There, the title is: "A woman, a man and a reindeer in a snowy mountain landscape during the spring migration from Närvä. The caption indicates that the child is being breastfed" (Emilie Demant Hatts arkiv, L.A. 900, vol. Fotografier 1907–1916, K1:1). That the woman is nursing is not evident from the photo, as a reindeer blocks a straight view. It is only Demant Hatt's supplemental text that makes the invisible visible. But the caption to the archived and digitized image also mistakes the young boy for a man, effectively overwriting the visual evidence, effecting an alteration of the group dynamics, which now is rendered a nuclear family.

Demant Hatt continued her fieldwork in Sápmi through 1916, sometimes along with her husband, Gudmund. From this fieldwork, she amassed some 350 photographs (Sjoholm 2013, XI) which, along with notebooks, sketches, notes, and paintings, she gave to Ernst Manker in 1939 when he created the "Lappish Department" in Nordiska Museet. Neither fully in Art nor Ethnography, Demant Hatt's photos did not command the authority of von Düben's, and with very little circulation in popular cultural forms, her work has received little attention. The entirety of her collection at Nordiska Museet is in storage and underutilized. Unlike Lotten von Düben, Demant Hatt does not have a museum web page, instead a small selection of 49 photos and paintings have been digitized and lumped together without curation on DigitaltMuseum.[4]

4 See https://digitaltmuseum.se/021095502305/emilie-demant-hatt.

FIGURE 10.4 *Barnet ammes* – the child is being nursed. Demant Hatt (1913, 146)

The complicated and partial process of 'repatriation' of photos brings up questions of 'ownership', 'preservation', and the excavation of memory. Emilie Demant Hatt's photos and negatives are still in the possession and control of Nordiska Museet [NM] where they are organized as part of her collection – "EDM". Repatriation of these images, however partial, would have the effect of diminishing the importance of EDMs role and instead emphasize the local ties to land and blood. To date, NM has provided digital copies of EDM's photos on the contact sheets, digital copies of the contact sheets with EDH's captions, and NM's meta-data on the contact sheets to the Tromsø municipality who then shared them with the Norwegian Sámi Parliament.[5]

1.3 *John Bauer (1882–1918)* – Jönköpings läns museum

Photography is haunted, per Allan Sekula, by the two chattering ghosts of bourgeois science and bourgeois art: "[t]he first goes on about the truth of appearances, about the world reduced to a positive ensemble of facts, to a constellation of knowable and possessable objects. The second specter has the

5 Personal correspondence (email) from Hanna Horsberg Hansen, Arctic University of Norway and Academy of Fine Art, Academy of Arts at UiT, 9 February 2024 and 15 May 2025.

historical mission of apologizing for and redeeming the atrocities committed by the subservient – and more than spectral – hand of science" (Sekula 1981, 15). If Lotten von Düben's photographs are haunted by the first, the second specter haunts the photographs of John Bauer's trip to Lapland in 1904.

Before his name was forever linked to his images of trolls, gnomes and princesses in fairy tales, Swedish artist and illustrator John Bauer was commissioned by the publisher CAV Lundholm to provide color plates for the large-format *Lappland: Det Stora Svenska framtidslandet* [Lappland: The Great Swedish Land of the Future] – "A description in words and pictures of the nature and people of Lappland" (1908), edited by Olaf Bergqvist and Fredrik Svenonius. *Lappland* is a catalog of the north's resource potential, clearly defined as Sweden's future, reflecting a belief in progress and the benefits of modernization and industrialization. Embedded in this ideology was a paternalistic and ultimately destructive attitude towards the local Sámi – the "Lapp ska vara lapp" policy – which eschewed assimilation in favor of preserving the traditional nomadic Sámi lifestyle in a kind of permanent stasis in the past.

Lappland is illustrated with 32 full-page color plates, mostly of beautiful landscapes. Eleven plates are by John Bauer, who was tasked with providing ethnographic vignettes of Sámi life. Bauer's images are not backed by the scientific authority of von Düben's physical anthropology: they are not based on rigorous measurement, identification, and documentation, but are, rather, generalized, and stereotypic impressions, long since established. In the context of *Lappland*, however, they take on a scientific aspect.

In 1904, when tapped by Lundholm, Bauer was a third-year student at the Art Academy, with little experience. In addition, Bauer was from the south of Sweden, from Jönköping, and had no knowledge of the north other than the depictions of the Sámi and snow-covered 'wilderness' circulating in mass-produced images that were part of the ever-expanding visual discursive terrain. The saturation of the visual surround with these images effected a generalized sense of Lapland as cold, remote, mysterious, and inhabited by reindeer herders locked in a primordial existence (Johansson 2019).

Because of his inexperience, Lundholm set several conditions for Bauer: a one-month trip to Lapland in the summer of 1904, preceded by a visit to the Sámi exhibit at Hazelius's open-air ethnographic museum, Skansen (Bauer May, 1904), where visiting Sámi families demonstrated 'authentic' Sámi life – caring for reindeer, cooking on an open fire, making Sámi handicrafts – much like the touring troupes such as the Thomasson family. Bauer also studied Gustaf von Düben's *Om Lappland och Lapparna*, which he carried with him on his trip (Bauer June 1904). Thus "unhampered by experience" (Facos 1998, 89), Bauer's work was influenced by already circulating representations of the Sámi.

Bauer traveled to Lapland aboard the "Lappland Express", which ran from Stockholm to Narvik, Norway thanks to the newly completed iron-ore rail line that ran from Kiruna to the Norwegian port. Thus, in addition to pre-existing stereotypical images of "postcard-Sámi" (Lehtola 2018, citing Berit Åse Johnsen from the Sámi museum in Karasjok), Bauer's views of the North were further standardized and determined by the linear logic of the rail line. His destination was the Swedish Tourist Association's (STF) station in Abisko, a railway station along the iron-ore line. Thus, Bauer shared an itinerary with hundreds of other tourists, trekkers, and visitors to the north who also shared the specific sites/ sights of that area.

From the STF station in Abisko, Bauer began the preparation for his eventual images – mostly aimlessly wandering, taking photos, painting a few rough watercolors, and writing in his diary. Bauer was not a professional, or even a minimally accomplished photographer. His relatively few photos are blurred and random, and the Sámi in his photos are unnamed and without context. These photos constitute the "data" that Bauer manipulated, once back in Stockholm, to construct his paintings for Lundholm – selectively using his photos and aligning them with already-in-place conventions and expectations. The finished versions that appeared in *Lappland*, rather than "ethnographic" as later historians claim, are fully manipulated to create a fictional landscape – a fantasy landscape that masqueraded as ethnography (Conrad 2020; 2023).

Nordiska Museet has, in its archive, reprints of five of the eleven paintings Bauer produced for *Lappland*. Their location within the museum suggests that these are ethnographic representations of the Sámi. Two of the images, however, are explicitly fictions: "Rast under flyttningen" [rest during (reindeer) migration] [NMA.0103220] and "Lappar i snöstorm. Jukkasjärvi" [NMA.0103241] are winter snow scenes, impossible for Bauer to have observed during his one-month summer excursion. *Rast under flyttningen* moreover features South Sámi, in areas Bauer never visited. Others in the series appear to be ethnographic, but by comparing the final images to Bauer's photographs from the field, it is clear that they are artistic renderings, not ethnographic drawings.

"Pojkar från ett Lappläger vid Ripainen, Jukkasjärvi" [boys from a Lapp camp near Ripainen, Jukkasjärvi] [NMA.0097966] is an example of how Bauer manipulated scenes from his field photos (Conrad 2020). Similarly, "Interiör af Lappkåta" (Figure 10.5)[6] suggests a family scene, and yet Bauer recombined items from at least four different photos to fashion this image while creatively adding other elements (Figure 10.6): his 'family' is thus a reconstructed fiction.

6 The caption in the digital museum's collection is more explicit: "Familj i sin kåta, kvinnan väver" [A family in its *kåta*, the woman is weaving]. https://digitaltmuseum.se/021011757896/fargtryck-efter-malning-av-john-bauer-familj-i-sin-kata-kvinnan-vaver-motivet.

FIGURE 10.5 "Interiör af Lappkåta," in *Lappland* 1908, plate XXII. NMA.0103242

This kind of creative license relegates these images to the world of art, and yet they are housed in an ethnographic museum, where they *become* ethnographic.

Bauer's photographs from his Lapland trip belong to the Jönköpings läns museum. Bauer, a native son, is a central attraction for the regional museum, with the focus on his fairy-tale paintings from the children's Christmas annual *Bland Tomtar och Troll*. These images have come to occupy a special place in the Swedish cultural landscape; beloved for their depiction of a Nature in which the supernatural still exists, "a world where the physical reality and the mythical are present at the same time" (*Nationalencyklopedin*, "John Bauer"), which appealed to the belief that there exists a particularly Swedish biomystical connection to Nature (Facos 2001), a critical aspect of modern Swedish identity formation. This kind of conflation was enhanced by Bauer's conscious self-mythologizing, as he frequently staged himself engaged with his trolls. Bauer's imaginary Lapland bolsters his own narrative for which his photographs provide 'evidence' that authenticates and substantiates Bauer's mythical persona.

Bauer's photographs were thus the basis for two interrelated fictions: as the data from which he produced the "ethnographic" plates in *Lappland*, they created a vision of Sámi life locked in the ethnographic present; and as the documentary evidence of Bauer's experience of the magical North, they allow for the fiction of Bauer as the man who knew Trolls. To return to Sekula's spectral analysis, Bauer's use of his photos to create the ethnographic images in *Lappland* – a

FIGURE 10.6 John Bauer's photos from Lapland 1904. Top left: JM.1988–17–492 "The Sámi Thuri sits in a *kåta* with household goods and skins." Top right: JM.1988–17–494. Bottom left: JM. 1988–17–487. Bottom right: JM.1988–17–483. Jönköpings läns museum

book which celebrates Progress – serves as an apology for the atrocities committed by that progress by creating a picture of the Sámi that will never change. In contrast, the use of Bauer's photographs to authenticate his experience in the mystical North to effect a nostalgic return to a time in which the natural and supernatural co-existed redeems and reconfigures those atrocities as fantasy. Never intended for museum collection, they have been collected after the fact because of Bauer's subsequent fame as a fairy-tale illustrator. Reworked and recontextualized as the basis of troll and other fairy-tale imagery, Bauer's images mask (and normalize) the imperatives of the colonial project.

1.4 *Nils Thomasson (1880–1975)* – Jamtli (*Regional Museum of Jämtland and Härjedalen, a Branch of the Nationalmuseum in Stockholm*)

Nils Thomasson (the same Nils as in the *Parti från Lappland* postcard) was a professional photographer from Jämtland whose career spanned seven decades. His legacy of over 30,000 photos (and glass plate negatives), held in the Nationalmuseum Jamtli, Östersund, includes the documentation of changes taking place in Jämtland during the twentieth century in the life experiences of the local Sámi reindeer herders, Swedish farmers, forest industry workers, and tourists who came to recreate in Nature and to have a 'real' experience with the exotic Sámi. Thomasson apprenticed with Olof Östling in Gävle, afterwhich he returned to Svedje in 1898 to open his portrait business. He photographed locals – Sámi, Swedes, individuals and families, who posed in front of his studio backdrop.

Thomasson's portraits of local Sámi, dressed in their best *gapta* and willingly posing for portraits, is a reversal of the staged photographs described by Baglo (2019) in which Swedes and Norwegians dressed up in South Sámi-like clothing in front of painted winter scenes and also from the postcard imagery of the Thomasson reindeer troupe from 1892, posed in an imagined *Lapplandesque*. Thomasson's portrait of Maria Stengel's portrait (Figure 10.7), neither a tourist

FIGURE 10.7
Maria Stengel from Norway, before 1916
PHOTO BY NILS THOMASSON (INC). STIFTELSEN JAMTLIS SAMLINGAR "JML NTH10490". CC BY-NC-ND 4.0

FIGURE 10.8
Fjällström family photo. Father Jon (1876–1941), Jonas, Paulus, baby John Teodor (b. 2/1906), and mother Brita (1874–1954). Stiftelsen Jamtlis Samlingar "NTH14187" (InC).

nor an ethnographic photo, shows Stengel posing in a version of Appadurai's "colonial backdrop" (Appadurai 1997), as Stengel participates in a ritual of the emergent bourgeoisie.

Thomasson also took photographs in the field, many of them family and group portraits in some specific and often work-related context – at school, with the reindeer herds, at the marketplace, at the farm – some featuring his own family members. These often are misinterpreted as being of a kind with the usual ethnographic portrayal of the Sámi, but most were solicited photos – commemorating a life event, or the marking of time's passing. One of the families Thomasson photographed was that of Jon and Brita Fjällström (b. 1876 and 1874) (Figure 10.8); their photos chronicle their lives across several generations and provide an intimate view of a family transitioning into modernity.

During the same outing (~late 1906) in which the Fjällströms posed for their family portrait, Thomasson took several photos of their children, including their twin sons, Jonas and Paulus (JLMNTh8616 (InC)). Around 1910, Thomasson printed this photo of the boys as a postcard to sell to tourists in Åre (Jonsson 2008, 114), the twins becoming "Lappbarn från Jämtland" on the postcard

FIGURE 10.9
Två Jämtländsk Lappgossar as they appear in Herman Lundborg's *Svenska folktyper* (1919, 20)

(Jonsson 2008, 118). In 1919, a cropped version of the 3-year-old twins, by now 16, appeared in Herman Lundborg's *Svenska folktyper* as "Två Jämtländsks lappgossar" (Figure 10.9). The boys changed from "Lapp children from Jämtland (1910) to "Two Jämtlandish Lapp boys" through the manipulation of the image and captioning, which renders them examples of a specific racial sub-type.

Thomasson's photos in *Svenska folktyper* – a book illustrating, categorizing, and attempting to 'prove' the predominant racial biology theory of the day – does not suggest, however, that Thomasson was complicit or in agreement with its explicit racist ideology as was von Düben. Thomasson, an entrepreneur, was solicited by Lundborg when he was assembling his typology, and responded by sending several existing portraits, including that of Paulus and Jonas, along with a note (27/11/1918, in Jonsson 2008) expressing regret that he could not create negatives per Lundborg's instruction but hoped that the few photos were of some use.

Historical reconstruction of individuals and families is possible in the case of Nils Thomasson's photos because of his fastidious and prescient organization and documentation of his work. His donation of his entire body of work to Jamtli provides a record for the future of a specific past, and Jamtli's online collection is faithful to his originals and provides substantial additional contextual information. But this is offset by the free flow of photos in magazines, books, postcards, etc., that circulate in a promiscuous manner, decoupled from any context and origin. Baglo's 2019 analysis of Thomasson's photos comes from postcards in the UiT Museum's postcard collection in Tromsø, concluding that his photos often conformed to ethnographic conventions, a conclusion belied by the entirety of his work, which includes all manner of people, occupations, recreational forms and landscapes in Jämtland. Thomasson selected specific photos from which he produced postcards for the nascent tourist industry, and he can be credited with creating, rather than following the "postcard aesthetic". To analyze the end product is to erase the process, confusing consumption with production and running the risk of investing his photos with meanings never intended.

1.5 *Borg Mesch (1869–1956)* – Kiruna kommun bildsamling

Borg Mesch, like Thomasson, was a professional photographer in the North. Unlike Thomasson, he was not Sámi, nor was he originally from the North. Mesch arrived in the North and opened a satellite portrait studio in 1899 in what would become, in 1900, the city of Kiruna. From the beginning, thus, Mesch tied his fortunes to those of the city and its mine, as Sweden looked generally to Lapland as its *Framtidslandet* – its "land of the future". Mesch's fame is inextricably linked to the heroic, pioneering, masculinist story he tells in his photographs as they document the story of progress over the wilderness – photographs of Kiruna's founding fathers and the edifices of power they created; photos of the miners; of the navvies who built the railway; and of the Sámi, whose lives would be so dramatically and terminally affected by the introduction of such 'progress' and who represented its antithesis.

Mesch had the "right stuff" for the narrative of the frontier: handsome, photogenic, and rugged, he embodied the imaginary male conquest of the North (Figure 10.10), which he captured in his self-portraits. He was the first to summit the north peak of Kebnekaise (4/7/1907), Sweden's highest mountain, on the sixth of his twelve ascents. He was a member of the exclusive, male-only *De lappländska fjällkarlarnas klubb*, whose members, limited to 20, were required to have undertaken at least two winter tours in the mountains of Lapland and to have been the first to climb a mountain of at least 1200 meters (*Fjelds of*

FIGURE 10.10 Borg Mesch in the mountains, Kiruna Bildsamling BM 10000000; Borg Mesch in Studio, Nordiska Museet NMA.0042490; Borg Mesch in front of his studio, 1900 Kiruna Bildsamling 10000313

Sweden), and he photographically documented his personal, heroic acts of conquests, metonymically linking his body to the body politic, those photos also entering into circulation as postcards.[7] He also belonged to the *Svenska Fjällklubben* and was a tourist outfitter and wilderness guide. Mesch and his male climbing associates were conquering the Northern landscape at a personal, intimate level in parallel with the operations of the mine, the railroad, and the building of Kiruna on Sámi land. Mesch also produced promotional and touristic postcards, which would be consumed by audiences far beyond those tourists looking for a Nature experience. Mesch's photos emphasized an unspoiled Northern wilderness with romantic images of the Sámi as a part of nature.

Mesch's relationship with two of the most important agents of modernization (and of the state) in Kiruna – Hjalmar Lundbohm and Fredrik Svenonius – would further inscribe the narrative of Lapland as Sweden's land of the future – an unapologetic embrace of modernization that included a paternalistic attitude towards the Sámi, ultimately condensed into the "Lapp ska vara Lapp" policy. Both men were situated in overlapping spheres of influence and were networked with other nation-builders in Sweden during this critical moment. Mesch, arriving in Kiruna at the moment of its inception, with his impressive skill set, was useful to these power brokers, and he, in turn, benefitted from their patronage.

7 See for example the postcard of Borg Mesch on top of Kebnekaise, 1906. Kiruna Bildersamling 10001084.

Svenonius, a geologist and glaciologist for *Sveriges Geologiska Undersökning* [SGU – The Geologic Survey of Sweden], was deeply invested in the economic and industrial development of the Swedish north. He was also an avid trekker and mountaineer and was one of the founders of *Svenska Turistföreningen* [STF – the Swedish Tourist Association] in 1885, which, from its inception had a northern emphasis. Borg Mesch was a frequent contributor to the STF yearbook, supplying evocative landscapes of the North for southern would-be tourists seeking experiences in Nature. Svenonius' involvement in and commitment to both northern industrial development and northern wilderness tourism came together in *Lappland: Det Stora Svenska framtidslandet*, which linked descriptions of the north's resources, fledgling industries, geology, and geography, to popularize northern industrial and social expansion. Mesch provided many of the book's black and white photos.

Hjalmar Lundbohm was the first general manager of LKAB (*Luossavaara-Kirunavaara Atiebolag*), the iron ore mining complex in Kiruna. He oversaw the development of the mine and was the founder of the city. His belief that traditional reindeer herding Sámi culture should be kept separate from modern Swedish society motivated his interest in the documentation of traditional Sámi culture and resulted in the creation of his ethnographic series *Lapparne och deras land. Skildringar och studier* [The Lapps and their land. Descriptions and studies]. In roughly twelve years, Lundbohm published nine such Sámi studies, which he financed and over which he exercised considerable control. The first in the series, Turi's *Muittalus Samid birra* and the second, Demant Hatt's *Med Lapperne i Højfjeldet* both contained photographs by Borg Mesch at Lundbohm's suggestion (as do other volumes: IV. Karl-Erik Forsslund's, *Som gäst hos fjällfolket* (1914), and V. Ester Blenda Nordström's, *Kåtornas folk* (1916)).

This mix of contingency and privilege; luck and arrogance; capitalism and colonialism provided the context in which Mesch's fame and influence flourished. Like Thomasson, Mesch left behind a huge body of work. That which pertained to local history – the building of the city, the railroad, the expansion of industry, the "high society" photos, and in general Kiruna life, is not well known outside of Kiruna. Mesch's iconic photos of Sámi life, and the images of the wide expanses of nature, immediately escaped the restraints of the local and recirculated without reference to any specific context or origin. These images were everywhere – in textbooks, magazines, postcards, tourist paraphernalia, academic works, and *National Geographic* (1939).

Borg Mesch's photos of Sámis in both the field and the studio feature the visual tropes enumerated by Baglo (2019). In contrast to Thomasson's studio portraits with his clients in front of a 'western-coded' backdrop, many of Mesch's most famous studio portraits are exaggerated fantasies of an imagined Sámi way of life, such as the iconic studio image of Johan Turi (Nordiska Museet NMA.0034015), in which he is posing surrounded by incongruous props – a coffee pot lying on the ground, a set of antlers, a reindeer pelt tossed into a sledge attached to nothing, all in a faux 'winter' background. With its overt theatricality and absence of subtlety, this photo signifies Lapp to the consuming public in the metropole.

Biographies of Nils Thomasson and Borg Mesch often conflate the inclusion of their photos of the Sámi in Herman Lundborg's *Svenska Folktyper* as a sign of their active involvement in the production of a racist ideology masquerading as science. But such a broad generalization of the colonial legacy requires selective retrieval and omission. Most of Mesch's photos in *Svenska Folktyper*, as with Thomasson's, were portraits taken long before Lundborg's typology and were repurposed for use in it. Their photos also comprise a small portion of the entire collection, the majority of which were taken by Lundborg himself. Of Mesch's 18 photos, only a few display the ethnographic/anthropometric convention.

The Media Gallery of the Kiruna Kommun Bildsamling[8] has a comprehensive collection that contains over 3000 photos by Borg Mesch. This re-aggregation of his work reinserts his photos of the Sámi and Sámi life into the contemporary social context, disallowing their sequestration into the ethnographic present. For the Sámi, the site also provides important personal, family, and regional information, and allows for the re-individuation of people formerly presented as types. In this way, a silence is partially filled, and an alternate history written.

2 Reclamation – Archival Photographs as Resource

Multi-media Sámi artist Nils-Aslak Valkeapää (1943–2001), the unofficial poet laureate of the Sámi Movement, initiated a call to revisit the "colonial archive" with his 1988 *Beavi, Áhčážan* [*The Sun, My Father*], a massive collection of vintage photographs accompanied by Valkeapää's poetry, which served as a strident critique of the violent colonial past.

In the decades since *Beavi, Áhčážan*, online photo collections have expanded access to the archive and facilitated new ways of reconstructing a more granular picture of the past. Contemporary scholars and artists are grappling with the photographic record in attempts not only to connect with their erased pasts but also to assemble an alternative narrative from those individual histories. Veli-Pekka Lehtola, professor of North Sámi cultural history contends that the Sámi approach to archival photographs reveals "'small stories' of *us* and *our* ancestors rather than 'big histories' of colonial circumstances and

8 https://kiruna-mediagallery.imagevault.se.

FIGURE 10.11
Maria Magdalena Matsdotter from *Om Lappland och Lapparna* (1873, plate 6, 13, 14)

contradictions with outsiders' societies" (2018, 1). A recent website investigation into an image in *Om Lappland* points, however, to the impossibility of disengaging the personal from the colonial, and to the complicated reality in which gender and race intersect in the writing of history. Plate 6 in *Om Lappland* presents a headshot of a young woman with braids and a floral shawl (Figure 10.11) – Maria Magdalena Matsdotter, 31 years old, from Åsele. The image, according to the "South Saami History" website, is based on a photo by Lotten von Düben.

The image is a starting point in the recovery of a long-erased history. Maria Magdalena was, in her short life, a crusader for Sámi schools and school reform. She traveled to Stockholm where she met with King Karl XV and received funding for an independent school in Vilhelmina congregation (Heith 2018) and visited feminist Fredrika Bremer (Howitt 1867). Through revisiting her image, her story, long forgotten, has been revivified (Maria Magdalena), but this is not a local history, nor is it a "small story". Rather, Maria Magdalena Matsdotter's photo stands witness to the complex inter-ethnic relations and negotiations within structural inequality.

Lotten von Düben's photographs have also been used by Åsa Johansson (2019) to recover her own family's suppressed history. In her master's thesis, Johansson recounts how her family chose to assimilate, and not acknowledge their Sámi heritage. In some of von Düben's most well-recognized photos – those of Inga Kajsa Granström (Figure 10.2), Eva Brita Mulka, née Granström, and three others – Johansson rediscovered distant relatives and retrieved a portion of her lost heritage.

Post-colonial theorists have similarly opened the possibility of "excavating" Demant Hatt's photographs for "hidden content" (Lehtola 2018). Feminist

FIGURE 10.12 Lena Stenberg, *Anne-Marja-Vad ser du?* (1991)
REPRODUCED WITH PERMISSION FROM THE ARTIST

scholar Hanna Horsberg Hansen sees in Demant Hatt's photos "equality and reciprocity rather than the inequality and objectification" and sees her subjects as people with agency "who themselves choose to be present through their work, their play, their clothes and their landscapes – and perhaps basically tell about their lives as people in the world" (Horsberg Hansen 2022, 136). Elin Anna Labba re-collects and recombines ethnographic and touristic photos by Demant Hatt and others to reconstruct the story of her ancestors that was destroyed by the closing of the Norwegian-Swedish border (1919), which stopped reindeer migration and dislocated the *sidda* (Labba 2020). These contemporary feminist scholars, optimistically and nostalgically return to Demant Hatt's photographs to recreate lost family histories, to excavate stolen past lives, and to see the unseen in the photos when looked at from a different perspective. Their desire to consume past photographs is not tied to consumerism but is restorative and recuperative.

Sámi multi-media artist Lena Stenberg (b. 1961) incorporates the legacy of colonialism while working at an intensely personal level: "Central to Stenberg's work is the notion of historical and political reflection, as she analyses the myriad ways in which the Sami people have been oppressed by the dominant cultures and political systems that have governed their lives [...] [Her] work is a testament to the enduring power of art to bear witness to injustice, to inspire change, and to give voice to the marginalized and oppressed" ("Lena Stenberg – Borders"). In her piece *Anne-Marja-Vad ser du?* [Anne Marja, what do you see?; Ánne Márjaá maid donoainnát?] (Figure 10.12), Stenberg uses a family photo of her mother as a young girl to thematize her own family background in what Jon Helt Haarder (2007) has called "performative biographism" (Grini 2017, 313). In an interview, Stenberg recounts how using "photographs from photo archives, I have traced my family since the end of the nineteenth century: they had to leave their reindeer grazing lands in Norway and Finland ... [and suffered] abuse that still affects the Sámi contemporary day" ("Lena Stenberg" *Sámi Dáiddárráđđi*). Stenberg's *Anne-Marja-Vad ser du?* foregrounds the serial nature of the particular image that circulated in reproductions of the 'Lapps' from early twentieth-century photographs, while reversing Anne Marja's gaze and redirecting it towards the viewer.

Finnish urban Sámi artist and photographer Marja Helander (b. 1965) negotiates her own (mixed) identity as disconnected from the Sámi past. In her work "Katja-Nilla" (1997), in which she not only retrieves a photograph of her Sámi ancestor, but also double-exposes the image onto Helander's own landscape photos of the areas Katja-Nilla traversed as a reindeer herder. The disembodied head haunts the landscapes, intensifying the loss in the aftermath of colonialist and assimilationist policies. Helander's "aftermath-confrontations with the environmental destructions of Sámi culture and landscapes may be understood as a Nordic indigenous version of postcolonial melancholy" (Lien 2020, 44) a kind of "mournful reindividuation" (Sekula 2014, 31) that speaks to a loss that is ultimately irretrievable.

Early critiques of the photographic archive of the Sámi colonial past, such as Valkeapää's, used decontextualized and randomly selected and organized photos from that archive – the photos becoming reified and reduced as uncomplicated signs of colonial domination. Filled with moral certainty, such critiques failed to complexify the colonial, the archive, and the photographic. Photographs in this context were not treated as complex and unstable but were limited to a singular interpretive frame. That is, the critique offered a singular history in the same manner as was attempted by the colonial archive itself. More recent excavations of photos, particularly in their explicitly unstable digitized form, reveal the intersections of multiple histories that intersect in the colonial photos, opening them up to multiple interpretations and histories.

Bibliography

Appadurai, Arjun. 1997. "The Colonial Backdrop." *Afterimage* 24 (5 Mar/Apr): 4–10.

Bank, Andrew. 2001. "Anthropology and Portrait Photography: Gustav Fritsch's 'Natives of South Africa', 1863–1872." *Kronos* 27: 43–76. http://www.jstor.org/stable/41056669.

Baglo, Cathrine. 2014. "Rethinking Sami Agency during Living Exhibitions. From the Age of Empire to the Postwar World." In *Performing Indigeneity. Global Histories and Contemporary Experiences*, edited by Laura R. Graham & H. Glenn Penny, 136–68. Lincoln, London: University of Nebraska Press.

Baglo, Cathrine. 2019. "Puncturing Parts of History's Blindness: South Saami and South Saami Culture in Early Picture Postcards." In *The Indigenous Identity of the South Saami: Historical and Political Perspectives on a Minority Within a Minority,* edited by Håkon Hermanstrand et al., 93–120. Cham Switzerland: Springer Open.

Bauer, John. 1904. Letter to parents dated 7 May 1904. Brev Samling JMA.00233. Jönköpings läns museum. https://digitaltmuseum.se/021029506233/brev.

Bauer, John. 1904. Letter to parents dated June 1904. Brev samling JMA.00235. https://digitaltmuseum.org/021029509882/brev.

Bergqvist, O. & Fredrik Svenonius. 1908. *Lappland Det Stora Svenska Framtidslandet En Skildring I Ord Och Bild af Dess Natur Och Folk*. Stockholm: C.A.V. Lundholm.

Conrad, JoAnn. 2020. "John Bauer På Spåren." In *Artiska spår: Nature och Kultur i Rörelse,* edited by Lotten Gustafsson Reinius, 116–19. Stockholm: Nordiska Museets Förlag.

Conrad, JoAnn. 2023. "Illustrating the Future: Representations of Swedish 'Lappland' in Early 20th Century Publications." *Lychnos*: 111–34. DOI: 10.48202/25047.

Crary, Jonathan. 1992. *Techniques of the Observer: On Vision and Modernity in the Nineteenth Century.* Cambridge, MA: MIT Press.

Demant Hatt, Emilie. 1913. *Med lapperne i højfjeldet.* Stockholm: Nordiska bokh.

Demant Hatt, Emilie & Barbara Sjoholm. 2013. *With the Lapps in the High Mountains: A Woman Among the Sami 1907–1908.* Madison: University of Wisconsin Press.

Edwards, Elizabeth & Sigrid Lien (eds.) 2014. *Uncertain Images: Museums and the Work of Photographs.* Ashgate.

Edwards, Elizabeth. 2015. "Anthropology and Photography: A Long History of Knowledge and Affect." *Photographies* 8 (3): 235–52.

Eglinger, Hanna. 2023. "Traces, Absences, and Distortions in Emilie Demant Hatt's Representations of Sámi Life." *Interventions* 25 (7). DOI: 10.1080/1369801X.2023.2169624.65.

Emilie Demant Hatts arkiv, L.A. 900, vol. Fotografier 1907–1916, K1:1. Nordiska Museet NMA.0107226 (https://digitaltmuseum.se/0210114160729/en-kvinna-en-man-och-en-ren-i-ett-snoigt-fjallandskap-under-varflyttningen).

Facos, Michelle. 1998. *Nationalism and the Nordic Imagination: Swedish Art of the 1890s.* Berkeley: University of California Press.

Facos, Michelle. 2001. "Primitivism in Sweden: Dormant Desire or Fictional Identity?" In *Antimodernism and Artistic Experience*, edited by Lynda Jessup. Toronto: University of Toronto Press.

Fjelds of Sweden. https://www.fjeldsofsweden.com/associations.

FOTOGRAFIHISTORIA – *Fotografer och ateljéer i Hälsingland och Gästrikland. Borg Mesch (Gävle).* 22 January 2022, Lars Eriksson https://www-fotografihistoria-se.translate.goog/gavle/borg-mesch/?_x_tr_sl=sv&_x_tr_tl=en&_x_tr_hl=en&_x_tr_pto=wapp.

Grini, Monica. 2017. "Contemporary Sámi Art in the Making of Sámi Art History: The Work of Geir Tore Holm, Outi Pieski and Lena Stenberg." In *Sámi Art and Aesthetics: Contemporary Perspectives,* 297–323. Aarhus University Press.

Haarder, J.H. 2007. "Ingen fiction. Bara reduktion: Performativ biografism som konstnärlig strömning kring millenieskiftet." *Tidsskrift för litteraturvetenskap* 4: 77–92.

Heith, Anne. 2012. "Valkeapää's Use of Photographs in Beaivi Áhčážan: Indigenous Counter-History versus Documentation in the Age of Photography." *Acta Borealia* 31 (1): 41–58. doi:10.1080/08003831.2014.904996.

Heith, Anne. 2018. "Maria Magdalena Mathsdotter." *Svenskt kvinnobiografiskt lexikon* www.skbl.se/sv/artikel/MariaMagdalenaMathsdotter. Accessed 3 February 2024.

Horsberg Hansen, Hanna. 2022. "Fotografier i Emilie Demant Hatts bok *Med lapperne i højfjeldet*: Fotografiske møter, biografiske inskripsjoner og 'våre' historier." *Kunst og Kultur* 15 (September): 129–41. https://doi.org/10.18261/kk.105.2.6.

Howitt, Margaret. 1867. *Ett år hos Fredrika Bremer.* Stockholm: Flodin.

Johansson, Åsa. 2019. "Photographing in the Safari of Lapland." MA Thesis, Linköping University. https://www.diva-portal.org/smash/get/diva2:1381919/FULLTEXT01.pdf.

Jonsson, Johan. 1997. "Samer i bilder–bilder av samer. Ett fotografioch tre berättelser." *In För Sápmi i tiden. Nordiskamuseets och Skansens årsbok*, edited by Christina Westergren & Eva Silvén, 107–20. Stockholm: Nordiska museet.

Kirshenblatt–Gimblett, Barbara. 1998. *Destination Culture: Tourism Museums and Heritage*. Berkeley, California: University of California Press.

Kuutma, Kristin. 2003. "Collaborative Ethnography Before Its Time: Johan Turi and Emilie Demant Hatt." *Scandinavian studies* 75 (2): 165–80.

Labba, Elin Anna. 2020. *Herrarna satte oss hit: Om tvångsförflyttningarna i Sverige*. Stockholm: Nordsteds.

Lehtola, Veli-Pekka. 2018. "Our Histories in the Photographs of the Others: Sámi Approaches to Visual Materials in Archives." *Journal of Aesthetics & Culture* 10 (4): https://doi.org/10.1080/20004214.2018.1510647.

"Lena Stenberg." Sámi Dáiddárráđđi – Sámi artists council website, https://Sámidaiddar.no/en/Sámi-artist-catalogue/lena-stenberg/.

"Lena Stenberg – Borders" Sámi Dáiddaguovddáš/Sami Center for Contemporary Art. https://samidaiddaguovddas.no/en/lena-stenberg-borders/.

Lien, Sigrid. 2018. "Performing Academic Masculinity in the Arctic: Sophus Tromholt and Roland Bonaparte's Photographic Accounts of Sámi Peoples and Northern Landscapes." *Journal of Aesthetics & Culture* 10: 1–17. https://doi.org/10.1080/20004214.2018.1498677.

Lien, Sigrid. 2020. "Sámi Silence Visualized: Indigenous Loss Negotiated in Contemporary Art." *Negotiating Memory from the Romans to the Twenty-First Century: Damnatio Memoriae*, edited by Øivind Fuglerud, Kjersti Larsen & Marina Prusac-Lindhagen, 44–61. Routledge.

Lien, Sigrid & Hilde Nielssen. 2012. "Absence and Presence: The Work of Photographs in the Sámi Museum, RiddoDuottarMuseat-Sámiid Vuorká-Dávvirat (RDM-SVD) in Karasjok, Norway." *Photography and Culture* 5 (3): 295–310. DOI: 10.2752/175145212X13415789392965.

Lien, Sigrid & Hilde Nielssen. 2018. "Negotiating the Past: Addressing Sámi Photography." *Journal of Aesthetics & Culture* 10 (1): 1–2.

Lundborg, Herman. 1919. *Svenska folktyper: bildgalleri, ordnat efter rasbiologiska principer och försett med en orienterande översikt*. Stockholm: Tullberg.

Maria Magdalena. https://southsaamihistory.wordpress.com/2018/01/19/maria-magd-matsdotter-31-ar-fran-asele-lappmark-i-dito-dito-lotten-von-duben-ser-ut-att-ha-fotograferat-maria-magdalena-matsdotter-litografi-efter-fotogram-av-schlachter-seedorff/.

Mathisen, Silje Opdahl. 2017. "Still Standing: On the Use of Dioramas and Mannequins in Sámi Exhibitions." *Nordisk Museologi* 1: 58–72.

Mathisen, Stein R. 2004. "Hegemonic Representations of Sami Culture. From Narratives of Noble Savages to Discourses on Ecological Sami." In *Creating Diversities: Folklore, Religion and the Politics of Heritage*, edited by Anna-Leena Siikala, Barbro Klein & Stein R. Mathisen, 17–30. Helsinki: Finnish Literary Society.

Modin, Elsa. 2021. “Carolina Charlotta (Lotten) Mariana von Düben.” *Svenskt kvinnobiografiskt lexicon*. www.skbl.se/sv/artikel/LottenvonDuben. Accessed 28 January 2024.

Nationalencyklopedin, John Bauer. http://www.ne.se/uppslagsverk/encyklopedi/lång/john-bauer.

Nielssen, Hilde Wallem. 2018. “Visualising Invisibility: Photography and Physical Anthropology in Norway.” *Journal of Aesthetics & Culture* 10: https://doi.org/10.1080/20004214.2018.1498672.

“Nils Thomasson – Sámi photographer” https://jlm.kulturhotell.se/page/2/?s=nils+thomasson.

Reeploeg, Silke. 2019. “Women in the Arctic: Gendering Coloniality in Travel Narratives from the Far North, 1907–1930.” *Scandinavian Studies* 91 (1–2): 182–204.

Schiller, Harald. 1935. *John Bauer, sagotecknaren*. Stockholm: Sveriges allmänna konstförening.

Sekula, Allan. 1981. “The Traffic in Photographs.” *Art Journal*, special issue: Photography and the Scholar/Critic 41 (1): 15–25.

Sekula, Allan. 1989. “The Body and the Archive.” In *The Contest of Meaning: Critical Histories of Photography*, edited by Richard Bolton. 343–89. Massachusetts: MIT Press.

Sekula, Allan. 2014. “Photography and the Limits of National Identity.” *Grey Room* 55 (Spring): 28–33.

Sjoholm, Barbara. 2011. “Participant Observation of the Human Landscape.” On the Reading Rest. https://floasche.wordpress.com/2011/12/26/participant-observation-of-the-human-landscape/.

Sjoholm, Barbara. 2014. “The Art of Recalling: Lapland and the Sami in the Art of Emilie Demant Hatt and Johan Turi.” *Feminist Studies*, special issue: Food and Ecology 40 (2): 356–93.

Sjoholm, Barbara. 2017. *Black Fox: A Life of Emilie Demant Hatt Artist and Ethnographer*. Madison Wisconsin: University of Wisconsin Press.

Sjoholm, Barbara. 2023. *From Lapland to Sápmi: Collecting and Returning Sámi Craft and Culture*. Minneapolis: University of Minnesota Press.

Stoler, Ann Laura. 2002. “Colonial Archives and the Arts of Governance.” *Archival Science* 2: 87–109.

Stoler, Ann Laura. 2011. “Colonial Aphasia: Race and Disabled Histories in France.” *Public Culture* 23: 121–56.

Turi, Johan Olafsson & Emilie Demant. 1910. *Muittalus Samid Birra: En bog om Lappernes liv*. Stockholm: Nordiska bokhandeln.

Utsi, John E. 1997. *Samefotografen Nils Thomasson*. Östersund: Jamtli.

Valkeapää, Nils-Aslak. 1991. *Beaivi, áhčážán*. Kautokeino: DAT.

von Düben, Gustaf. 1873. *Om Lappland och Lapparne Företrädesvis De Svenske*. Stockholm: P.A. Norstedt & söners förlag.

OBJECT: Photograph

FIGURE 11.1 A lost object?
PHOTOGRAPH OF GERMAN TOURISTS DRESSED IN SÁMI CLOTHES

CHAPTER 11

Staging the North: Narratives of a "Sámi Photograph"

Ulrike Spring

Abstract

This chapter takes as its starting point a lost object, which was then found during the writing process: a photo of the renowned Vienna Burgtheater actor Hugo Thimig in Sámi clothes. Thimig had together with actor friends travelled in summer 1883 to Norway, and his photo was one year later discovered by another tourist in a souvenir shop in Tromsø in the North of Norway, this time presented as a photograph of a Sámi. This chapter explores the multifaceted role of this photograph, in its capacity as a historical source, as a representation of specific nineteenth-century discourses and as a relational material object that both shapes and encapsulates various meanings of the North. It examines the photograph's implications within the growing tourism industry and colonial discourses of the late nineteenth century, and from the perspective of Thimig's professional identity as an actor. In addition, the chapter is a reflection on the process of conducting research, where each new piece of information contributes to collecting new items that lead to new narratives of the North and beyond.

Keywords

tourist photograph – Norway – tourism history – Sámi culture – acting – cross-dressing

1 A Photograph

A few years ago, I came across a passage in a travelogue by Habsburg-Austrian historian August Fournier (1850–1920). He wrote about his visit in summer 1884 to a store in the North Norwegian town of Tromsø to look at Sámi photographs that were displayed in the window. On closer inspection, he realised that one, apparently depicting a woman in Sámi clothing, was actually a man whose features were oddly familiar:

 | DOI:10.1163/9789004749665_012

> kurz, es war unser liebenswürdiger Freund Thimig vom Burgtheater, der sich im verflossenen Jahre in dieser Verkleidung photographiren liess und der nun gewiss schon in so und so vielen Exemplaren als ethnographischer Typus die Albums englischer Ladies und Gentlemen ziert. So blüht deutscher Humor selbst am Ende der Welt.
>
> in short, it was our amiable friend Thimig from the Burgtheater who had himself photographed in this disguise last year and who now certainly adorns the albums of English ladies and gentlemen in so and so many copies as an ethnographic type. German humour thrives even at the end of the world[1]. (Fournier 1884b, 65)

Hugo Thimig (1854–1944), originally from Dresden, had been working as an actor at the renowned Vienna *k. k. Hofburgtheater*, known as the *Burgtheater*, since the mid-1870s. In the summers of 1883 and 1884 he spent several weeks in Norway as a tourist, venturing as far as the North Cape. As I will show in this chapter, Fournier's quote provides access to several narratives and practices that help to shed light on the early stages of mass tourism. The backdrop was late nineteenth-century European nationalism, imperialism and a burgeoning tourism industry, the popularity of the novel medium of photography and of souvenir cultures and various travel practices such as the collecting of destinations and memories, cross-dressing and performance.

In my approach, I have drawn inspiration from both representation and network theories, not wishing to give one more importance than the other. I will consider Thimig (a human actor) and the photograph (a non-human actor) as moving nodes in a network, where the shifting relationships between the nodes generate new meanings (Latour 1992; Sayes 2014). My focus will be on the photograph – in its capacity as a historical source, as a representation of specific nineteenth-century discourses and as a relational material object that both shapes and encapsulates various meanings of the North. The photograph has its own life stories or (auto-)biographies. Some are independent of its creator, the photographer, of Thimig, its main generator and protagonist and of his fellow traveller Demnitz, who was photographed with him. Once it was taken, it was reproduced in multiple copies and entered economic circulation, engendering new meanings. However, for the sake of analysis, I will view the photograph as a singular entity with personalised biographies rather than as

1 All translations by the author, with help from GPT UiO.

a commodity that exists in many different copies, each possessing its own distinct biography or life story. Drawing on Igor Kopytoff, I am interested in "the various singularizations" of the photograph, its place(s) "in an uncertain world of categories whose importance shifts with every minor change in context" (1986, 90).

I have analysed the photograph as a singular entity both linked to Thimig and independent of him. This makes it necessary to critically reflect on my position as a researcher and how I approach the past and its protagonists. I discovered Fournier's travelogue in the library of the Theatermuseum in Vienna where a large part of Hugo Thimig's estate is preserved. Fournier published his notes as a book, with a publisher in Prague (1884b), and in instalments in a shortened version in the feuilleton section of the prestigious Vienna-based newspaper *Neue Freie Presse* (1884a). Fournier dedicated the copy I looked at to his friend Thimig, and it was probably Thimig himself who left the only reader mark in the book: a thin pencil-line marking the paragraph quoted above (Sign. 624.249-B). Various sources suggest that Thimig had a distinctive sense of humour (e.g. Ponstingl 2024), but, as with most biographical data, much must remain as speculation. Unlike researchers working with living people, a historian's subjects are usually long gone and have no opportunity to correct their biographer. Ethical considerations have, for long, been a part of a historian's methodology. However, the rapid digitization of archives in recent decades and the resulting access to 'forgotten' persons and stories (Crossen-White 2015) have presented new challenges. Even though Thimig was a well-known person during his lifetime and remains so among culturally interested people in the German-speaking countries, his thoughts are largely 'forgotten' by the public. They are mostly available to us through commemorative records and biographical notes, his own publications and through unpublished letters and notes in archives. According to his son, Hermann Thimig, a prominent actor himself, Hugo Thimig decided to keep his diaries for posteriority, noting: "Es ist nicht ausgeschlossen, daß sie [Tagebuchnotizen, US] einmal in späteren Tagen einem Historien-Schnüffler Interesse erwecken" ("It is not impossible that they [diary notes, US] will arouse the interest of a history snooper in later days") (Thimig 1962, 7). Although Thimig granted access to his diaries, I will try to avoid being too much of a "Historien-Schnüffler" and will respect the private nature of his two journeys to the North, concurring with my own interest in the photograph itself and its various functions, rather than in Thimig's biography.

However, Thimig and Demnitz are not the only protagonists here. Sámi people are the dominant absent presence in this photograph. Their role was mostly to function as a background for others such as the photographer, the colourizer and the tourists.

The medium of the photograph is a powerful tool for perpetuating cultural stereotypes. While this is typical of tourist photography and applies to all destination cultures, it assumes more weight when applied to a marginalized group that has long been suppressed and discriminated against by majority cultures. Consequently, Norwegian scholars who study Sámi photographs must address the challenge of seeing their work through a colonial gaze, while recognizing the entanglement of histories that hinders the neat division of different cultural groups and their respective pasts (see Lien & Nielssen 2021). My decision to leave my leading object empty, is an attempt to acknowledge some of the challenges researchers encounter when engaging with representations of the past: The blank picture frame reflects my research itinerary, which I started with a lost historical source. Moreover, I wish to make visible the photograph's entanglement with nineteenth-century imperialist, colonialist and racist discourses which may not be apparent if one examines the picture in isolation. I also wish to demonstrate that the tourism industry often views destinations as canvases to be filled with imaginings and desires, potentially leading to the compartmentalization and simplification of destination cultures. Lastly, I wish to acknowledge that neither Demnitz nor Thimig can voice a response to this photograph or to my interpretation of it.

In my approach to the photograph, to Thimig and to the Sámi absence, I have been inspired by a posthumanist approach, mostly by its anti-hierarchical and non-dualistic stances (Ferrando 2019). I will zoom in and out, using the photograph as a basis for reflections on various entanglements: the growing tourism industry in the late nineteenth century, with particular attention to ethnic and cultural imaginings; the increasing relevance of the medium of photography in shaping our gaze; and various performative practices.

2 A Missing Object

There may exist many copies of the photograph, wedged in private albums, as suggested by Fournier. However, when I started my research, I began with a missing object. I was made aware of the photograph's existence by Fournier's quote, but I did not know what it looked like. Consequently, a significant portion of my early research included numerous hours spent scouring photo collections at museums and archives and revolved around this absence – attempting to fill the void with information about the tourists, about their travels to the North and about general tourist practices. As a historian, dealing with these gaps in knowledge is a familiar experience. It is often necessary to accept that one will never be able to fill the gaps with facts or physical manifestations. In this case,

FIGURE 11.2
Demnitz (left) and Hugo Thimig (right), posing as a Sámi couple. Tromsø 1883
PHOTOGRAPHER: SCHJETNE. COPYRIGHT: KHM-MUSEUMSVERBAND, THEATERMUSEUM

I was lucky. A photo archivist at the Theatermuseum came across two copies of the photograph in Thimig's collection, both in cabinet card format, the typical format used in the 1880s. One was in black and white with the photographer's logo and a note to his parents on the back (Figure 11.2). The other was a colourized version (Figure 11.3). The latter one has since been published in a book about Thimig's profound interest in photography (Ponstingl 2024, 96).

I had previously learned from Thimig's diary, also held in the museum archive, that the photograph not only depicted him, but also included his German friend and fellow traveller Demnitz, who posed as the male part of the Sámi couple. While I will predominantly follow the links to Thimig, my choice is influenced not only by the scarcity of material relating to Demnitz, but also by the fact that Thimig's contribution, like the photograph, has become part of public space through Fournier's travelogue and subsequent publications (e.g. Hadamowsky 1962).

With my focus shifting from the absent to the material and physical, new nodes can be added, creating links to the photograph: Sámi clothing and the men's posture; Dresden, the home of Thimig's parents; and the destination of this particular photograph.

3 (Re-)Making Culture

As a historian writing in 2024, there is something uncomfortable about the idea of dressing up in Sámi clothes. This feeling is emphasised in light of the largely derogatory and condescending treatment to which Sámi people were subjected in the late nineteenth century and in subsequent political and cultural discourse. While current discussions about cultural appropriation have heightened our sensitivity in this regard, we need to remember that dressing up in the costumes of visited cultural groups was once a common practice. In fact, it continues to be a popular practice among tourists often condoned and even encouraged by the destination itself. In Norwegian photo collections, images portraying Norwegian bourgeois women, anthropologists and travellers dressing up in Sámi clothing are witness to the widespread popularity of this practice during the late nineteenth and early twentieth centuries (Lien 2018, 12; Baglo 2019, 97–8). Thimig substantiates this when writing to his parents on the back of the black and white photograph (Figure 11.2): "Ich halte es für meine Schuldigkeit, Euch ein Abbild der üblichen Touristenkleidung in Lappland zu übersenden" ("I consider it my duty to send you a picture of the usual tourist clothing in Lapland").

At the time of Thimig's travels, there was a significant rise in the number of tourists, particularly from the affluent middle and upper classes of Europe, embarking on week-long trips to Norway. Thimig, Demnitz and Fournier thus followed in the footsteps of numerous tourists. From his diary, we learn that Thimig departed from Hamburg for Norway on 6 July 1883 and returned via Frederikshavn on the North coast of Denmark on 2 August 1883. On his trip, he travelled by steamship from Trondheim to the North Cape via Tromsø and Hammerfest. He repeated this part of the tour the following summer.[2]

The tourism industry is largely based upon the rapid and superficial consumption of both places and people, often leading to the creation of simplified representations and stereotypes that fail to capture the complexity of the locations visited. John Urry has famously coined the term "the tourist gaze" to describe the increasingly pre-defined and standardized perspective of places and people resulting from the surging mobility and fast visual consumption

2 Hugo Thimig, Tagebuch 1883, HS_VM 2928 Th. Theatermuseum Vienna, Nachlass Hugo Thimig.

that started in the 1840s (Urry & Larsen 2011, 13). During the 1880s, tourists en route to the North Cape would usually only spend a few hours in Tromsø, attempting to gather as many experiences as possible. According to Thimig's diary and assuming that the recorded chronology is accurate, he arrived at Tromsø on 22 July 1883, at 10 o'clock in the morning, took the photograph with Demnitz, and visited the Sámi camp at half past four, before departing at midnight for the next leg of the journey.[3]

In Tromsdalen, just outside Tromsø, the Swedish Könkämävuoma Sámi used their summer camp to sell homemade objects. From the 1870s onwards, the summer camp became an integral part of the itinerary of the steamer tours along the coast (Baglo 2015a, 59–60). Sámi artefacts ranked among the most sought-after souvenirs from Norway. The visit to the camp, located approximately half an hour's walk from the coast, allowed just enough time to take photographs and purchase some souvenirs. Thimig acquired a shirt during his visit to the camp in 1883.[4] The display of Sámi culture, at the camp, at other locations in Tromsø and particularly in the town's shops, served as a kind of tokenism. It condensed the complexity of everyday life into a version tailored for tourists, and reinforced preconceived notions of the Sámi people in Europe. Prior to their journeys, many travellers to Norway would already have specific expectations of what to look for. Albums, books, and other representations of the North shaped what Anne Friedberg has described as "a mobilized 'virtual' gaze", connecting the act of travel with visual representation, mostly photographs (1993, 2–3) and establishing a "visual economy" of the North and the Sámi (Lien 2018, 6).[5]

In the German-language context, Theodor Mügge's novel *Afraja,* partly set in the area around Tromsø (initially published in 1854 and quickly translated into several languages), had a strong impact on perceptions of the Sámi (see also Theodorsen 2008). Fournier explicitly mentions *Afraja* in his travelogue, noting that many of his fellow travellers harboured only vague ideas regarding the North and the Sámi, most of which were derived from their reading of the novel (1884, 49). As Kaasa and Sjelmo point out, the Sámi geography in the novel was fictitious and largely imaginary, bearing little resemblance to real Sámi life (2022, 131–32). It is possible that Thimig and his companions had

3 Hugo Thimig, Tagebuch 1883, HS_VM 2928 Th, 31–32. Theatermuseum Vienna, Nachlass Hugo Thimig.

4 Hugo Thimig, Tagebuch 1883, HS_VM 2928 Th, 31. Theatermuseum Vienna, Nachlass Hugo Thimig.

5 Lien refers to scientific exploration, but her concept can productively be applied to tourism as well.

already encountered Sámi people prior to embarking on their journeys to the North. In the 1870s, Sámi had visited Central and Eastern Europe, participating in international exhibitions and ethnic shows (Baglo 2015, 53).

Late nineteenth-century travelogues include descriptions that often exoticized and drew upon the romanticised image of the 'noble savage' and upon racist notions of so-called uncivilized Indigenous people. The concepts of modernity and progress were at least as important as those of coloniality and imperialism, and they were closely intertwined with nationalism. Nation-building positioned Europeans at the apex of societal evolution on a global scale. As Indigenous, nomadic people (although in fact many were settled), Sámi were frequently used in travel literature as a means to showcase the alleged superiority of Norwegians and other Europeans, while offering a rare opportunity to learn about nomadic communities in Europe. When viewed from this perspective, allowing oneself to be photographed in Sámi clothing, even more so in women's clothing, can be seen as an expression of the patronizing approach adopted by many nineteenth-century travellers towards Indigenous people.

4 The Photograph as Evidence

Siân Jones notes that "experience and negotiation of authenticity also relate to networks of relationships between objects, people and places" (2010, 183). Authenticity is relational and contextual. In heritage and tourism studies it is a contested term (e.g. Salazar 2012), and I use it here to refer to the tourists' perception of the genuineness of the photograph and other Northern objects. After reading numerous similar descriptions of the Sámi in German-language travelogues, my impression is that, for most tourists, Sámi clothing and handicraft served as collectible souvenirs, rendering memories of their journeys to the North. Their interpretation of 'authenticity' did not necessarily relate to the way in which Sámi themselves used these objects in their everyday lives, but rather to the extent to which they were perceived as representations of Northern culture and the Sámi people. As for the photograph, I would like to argue that the authenticity Fournier believed it conveyed, was possibly due to several factors: firstly, the photograph was situated at the junction of two Northern locations – Tromsø with its Sámi population and the shop where other photographs of Sámi individuals were on display; secondly, it featured Sámi clothing as authentic visual markers; thirdly, it used the evidence-based medium of photography; lastly, it was part of the established visual economy of the North and the Sámi (see Lien 2018, 6). It is likely that photographs taken and sold in the North were perceived as more genuine representations of Sámi culture than those taken and bought in Vienna, for example.

By the 1880s, photographs had become a powerful instrument for cultural representation. Halftone technology had just been invented and facilitated the rapid and cost-effective reproduction of photographs (Bjorli 2008, 30). Photographs were however still collectible items. They held documentary value and were often perceived as more impartial and trustworthy than other common visual representations such as drawings, etchings and paintings. Employing photography as a medium at that time was a statement that played upon discourses on authentic representation and evidence.

In the decades around 1900, hundreds of photographers were active in Norway (Larsen and Lien 2007, 49), several of them in Tromsø (Bonge 1980, 471–72). They actively promoted their services and studios through advertisements in local and regional newspapers. At the studios, one had the opportunity to have one's own portrait taken and also to purchase photographs of Tromsø itself. Photographs of Sámi were high on the list of demands. In fact, in its first edition covering Norway and Sweden (1879, 350), the popular German Baedeker guidebook recommended the photographer Vickström, in Tromsø, as a place to obtain portraits of the Sámi. In 1887, a few years after Thimig's trips, tourist Ferdinand Krauß wrote:

> Vor Allem sind es aber die Photographien von Lappen, welche die Aufmerksamkeit auf sich ziehen, und es sind diese Bilder aus dem Leben der letzten Nomaden Europas so interessant, dass nicht nur jeder Nordlandsfahrer, sondern auch jeder andere Freund ethnographischer Studien diese merkwürdigen Typenbilder acquiriren wird. Namentlich die Photographen *Schjetne, J. H. Wonneberg* und *Jörgen Wickström* in Tromsø, haben immer eine reiche Auswahl von derlei Lappenbildern, die gewiss zu den merkwürdigsten Erinnerungen jedes Nordlandsfahrers gehören, am Lager.

> Above all, it is the photographs of Lapps that attract attention, and these pictures from the life of the last nomads in Europe are so interesting that not only every northern traveller, but also every other friend of ethnographic studies will acquire these notable pictures of Sámi types. In particular, the photographers *Schjetne, J. H. Wonneberg* and *Jörgen Wickström* in Tromsø always have a rich selection of such Lapp pictures

> in stock, which are certainly among the most notable memories of any northern traveller[6]. (Krauß 1888, 777; emphasis in original)

The name of the photographer, Lars Andreas Andersen Schjetne, is printed on the back of the black and white photograph of Thimig and Demnitz. Schjetne placed a small advert in 1883 ("L. Schjetnes" 1883, 3) in the local newspaper, *Tromsø Stiftstidende,* and a longer one in 1884, in which he pointed out that he had the latest available equipment ("L. Schjetnes" 1884, 3). According to the 1884 advert, he also had his own retoucher. From the census in 1885, we know that the retoucher he employed was Paul Andreas Mecelsen (also Mikkelsen/Michelsen),[7] who was, himself, later to become a photographer in Tromsø.[8] Given that Mecelsen had worked for Schjetne in 1883, it is plausible that he was the one responsible for colourising the second photograph that is deposited in the Theatermuseum archives (Figure 11.3).

Colourizing clothing on postcards was a common practice in the late nineteenth century. However, unlike many studio portraits from this time, there are no props or painted backgrounds evident in this photograph to convey a specific message. The two men are posing against a dark background and standing on a plain floor. Indeed, the clothing itself seems to serve as a prop, emphasizing the artificiality of the image, further reinforced by the staged posture of the men. In her master's thesis, Schøning analysed numerous postcards from the late nineteenth century depicting Sámi individuals. Her research shows that gákti, the traditional Sámi clothing, was by far the most popular signifier, while pelts were less common but still popular (Schøning 2018, 30, 49).

In the 1880s, ethnographic research viewed portraits as genuine representations that could be studied in depth to better understand racial differences and similarities. They were a means of making the imaginary visible and transforming it into scientific evidence thus placing the focus on the collective rather than on the individual (see Baglo 2012, 48). This scrutinizing approach did not apply to tourist photographs, which aimed to capture memories and ambiance rather than scientific "truth". However, images like that of Thimig and Demnitz capitalized on this scientific methodology by becoming a part of the documentation about Sámi people. Sámi, like Norwegians and member of

6 The term "Lapp" was commonly used to refer to the Indigenous Sámi during that period. Today, this term is considered derogatory and is not in use anymore. I only resort to it when directly quoting from historical sources.

7 Digitalarkivet: Folketelling 1885 for 1902 Tromsø kjøpstad, https://www.digitalarkivet.no/census/person/pf01053306005759, accessed May 23, 2024.

8 Historisk befolkningsregister, https://histreg.no/index.php/person/pf01037546000780, accessed May 23, 2024.

FIGURE 11.3
Demnitz (left) and Hugo Thimig (right), posing as a Sámi couple. Tromsø 1883
COLOURIZED PHOTOGRAPH BY L. SCHJETNE. COPYRIGHT: KHM-MUSEUMSVERBAND, THEATERMUSEUM

other cultural groups, were typically depicted in a generic manner in tourist portraits. Such photographs repeated cultural or national types, rarely considering the individual stories behind them. As Magdalena Naum (2023) discusses in her analysis of 'Indian' souvenirs in nineteenth-century Sweden, souvenirs initially served as metaphors but gradually became metonymic symbols, no longer representing just a single family or an individual, but representing the entire nation. Moreover, domestic discourses intersected with imperialistic ones, accentuating the otherness of the cultural artefacts. Photographs do not just represent the signified, or in tourism discourse the sights, they also produce them. As Lien and Nielssen in their discussion of ethnographic photography emphasize, photography was a powerful tool in shaping the way that Sámi were perceived. The photograph of Thimig and Demnitz then "forms part of a Nordic-European colonial visual culture and perceptual regime" (2018, 1).

In addition to perpetuating stereotypes, classifying and othering people and cultures, the photograph also exposes the mundane aspects of anthropological and souvenir portraits. In the late nineteenth century, one of the aims of anthropological Sámi portraits was to reveal the moral and psychological characteristics of the Sámi people (Baglo 2012, 48). This is also reflected in tourist descriptions of the Sámi. The portrait of Thimig and Demnitz shows the contingency, and we may add, the absurdity, of this approach. Their features could serve as examples of the typical Sámi, caricaturing the scientific belief in the direct relationship between anthropometric data and culture.

Up to this point, my analysis of the photograph has centred predominantly on the perspective of the tourists. However, Sámi were also actively involved in shaping their own image, particularly those Sámi participating in the burgeoning tourism activities in Tromsø. The Sámi in the camp in Tromsdalen and in town charged money for allowing tourists to take their pictures or enter their *lavvos* (Ytreberg 1962, 254; Baglo 2015b, 33). Leveraging this exploitation by tourists (and often their disrespect) and turning it against the visitors by utilizing their economic resources is a common strategy employed by destination cultures. By doing so, the exchange between the visitor and the visited becomes somewhat more balanced and gives agency to both parties.

However, there is still some uncertainty as to the degree to which the Sámi had agency regarding photographs such as that of Thimig and Demnitz. During his second journey in 1884, Thimig returned to the camp, and in his diary on 23 July he noted: "Im Lappenlager werden die Fotografien von Demnitz u. mir als Lappetypen verkauft" ("In the Lapp camp, the photographs of Demnitz and me are sold as Lapp types").[9] On a postcard from Tromsø sent to his mother on the same day, he specified that the photograph was sold by "unternehmende Tromsöer" ("enterprising Tromsø people").[10] Both Thimig and Fournier suggest that the photograph was marketed as authentic. Thimig reported to his mother that "das Halloh war groß" ("there was a great fuss") from the other tourists when they learnt that the photograph was fake.[11] We can only speculate about the extent of Sámi involvement in this. However, we can certainly assume that both the sellers and the Sámi were well aware of the fabricated nature of the photograph.

5 Cross-Dressing and Performance

As with any object, the photograph can have different meanings depending upon its connections to other actors. For tourists who purchased the

9 Hugo Thimig, Tagebuch 1883, HS_VM 2928 Th, 79–80. Theatermuseum Vienna, Nachlass Hugo Thimig.

10 Mappe HT and PT 21, AM 75.060 Th. Theatermuseum Vienna, Nachlass Hugo Thimig.

11 Mappe HT and PT 21, AM 75.060 Th. Theatermuseum Vienna, Nachlass Hugo Thimig.

photograph of Thimig and Demnitz it most likely served as part of a collection of Sámi portraits and a memory of their trip to the North. For the local people who sold the photograph, it represented an economic asset. For them and for the Sámi it probably also served as an ironic commentary on the tourists' desire to collect various elements of destination cultures. For Thimig and Demnitz themselves, the photograph was most likely a symbol of friendship and enjoyment, functioning as a memento of a shared trip.

Tourists view their surroundings and experience their journeys in different ways, depending on their individual life stories. In the case of Thimig, we know from various biographical sources that he had a great sense of humour, which he expressed through his performances in both tragic and comedic roles on the stage. Consequently, two narratives relating to the photograph have to do with the themes of humour and acting.

Acting, re-enacting and the creation of a stage are central aspects of tourist travel and memory formation. The Sámi staged their culture for the growing number of tourists. Their culture was also integrated into museum displays, such as those of Tromsø Museum and the university collections in Christiania (now Oslo). Furthermore, it was visually reproduced in numerous travelogues, tourist albums and postcards. Tony Bennett (1995, ch. 2) points out that, in the latter half of the nineteenth century, the spectator also became a participant. For some tourists, dressing in Sámi clothes while in Sápmi might have been part of an immersive experience. Their encounter with Sámi culture occurred within the context of an emerging leisure industry where, in contrast to the scrutinizing anthropological gaze, entertainment was central (see also Baglo 2015a, 64). In his classic work on "staged authenticity", Dean MacCannell, drawing on Erving Goffman, discusses the tensions between three 'front regions', and three 'back regions'. In the 'front regions' the destination culture is staged to appear authentic for tourists and the 'back regions' represent the everyday lives of the local population, which many tourists desired to enter (1973, 598). The tourism industry typically strives to create an illusion that the front region is, in fact, a back region, granting tourists a glimpse into the 'real world'. Tourists visiting the Sámi camp could believe they were experiencing the everyday life of the Sámi, while in reality, they only saw what the Sámi inhabitants allowed them to see. They misinterpreted the front stage presented to them as if it were backstage reality. Looking into the Sámi tents gave at most a limited impression of their daily life. Similarly, the photograph could be misconstrued as representing the backstage, the real Sámi people. Any culture showcased in tourism is after all a simplified and sanitized version of the complex and messy reality of everyday culture.

Thimig's profession as an actor was probably an incentive for him when donning the Sámi clothes. After all, his everyday occupation was to assume various personas from different historical periods and cultures, attempting to make them appear believable. Thus, dressing in Sámi clothing and portraying a Sámi woman may have been an extension of his regular routine. In theatre culture, it has been a longstanding tradition for men to play women and we know that Thimig dressed as a woman during his career also in other contexts, for instance in 1878 to celebrate his friend Ludwig Gabillon's jubilee (Hadamowsky 1962, 44). Recent research has also demonstrated that Thimig utilized the medium of photography extensively, both professionally, as part of his acting career, and privately among his family and friends. Essentially, he staged personal photographs, making them an expression of his acting and life rather than a memento of specific family gatherings (Ponstingl 2024, 11).

In other words, Thimig enjoyed staging otherness in his personal life, his travels and in his profession. Part of being an actor involves identifying with the concept of otherness and striving to turn it into sameness in order to bridge the gap between the actor and the characters that they portray. The photograph not only tells of a story of performing other cultures, but also of a professional interest in cross-dressing. Given Thimig's background in theatre, assuming a Sámi female identity can be viewed as a way of adopting a new role or experimenting with a different identity. There certainly seem to be deliberate comedic elements in the way both men present themselves in the photograph. The slouched shoulders of Thimig, and the concealment of Demnitz' hands in the long sleeves of his fur, give the impression of intentional mimicry, perhaps even parodying the idea of tourists trying to 'go native'. It is reminiscent of a comedy on stage rather than an attempt to genuinely embody a real Sámi couple. In line with what we know about Thimig's fondness for staging photographs (Ponstingl 2024), it can be interpreted as a conscious performance rather than an act of what we would now call cultural appropriation. This was also a time where theories about the close relationship between facial expression and character abounded. As Ponstingl (2024, 77) points out, Thimig had a series of photographs made in 1888 in which he showed a variety of different facial expressions, calling them "Physiognomische Studien". Thimig's comic posing as well as his disguise as a Sámi woman then implies another narrative: that of alleged childishness and femininity ascribed to Indigenous and colonized peoples in a time of biological racism and sexism. This served as a negative counterpart to the colonialist's alleged hegemonic (and positively connected) masculinity.

Returning to Thimig's biography, the narrative of the photograph creating a stage, and consequently connecting it to Vienna and comedic theatre performance, is strengthened when examining the rest of his voyage through Norway. Indeed, he appears to have regarded both Sámi culture and Norway in general as his theatrical arena, where the blurring of borders between the real and the fictitious added a dash of fun. Hadamowsky does not distinguish between the two journeys. He relates parts of Thimig's tour to the North in 1884, focusing on Thimig's efforts, together with his actor friend and fellow traveller Hermann Schöne, to alter reality by hanging up photographs of their fellow traveller, actor Ludwig Gabillon, at various inns in Norway. Gabillon, who allegedely often lamented the lack of appreciation shown to him as an actor, came to believe that he had gained popularity even in Norway (Hadamowsky 1962, 41–2). In a similar way to the 1883 photograph, which could be interpreted as genuine, Thimig made the portraits of Gabillon appear authentic by framing them in Vienna "nach norwegischem Geschmack" ("according to Norwegian taste"). In a letter from Christiania penned to his mother on 30 July 1884, Thimig recounts the successful pranks he played on Gabillon. He even managed to cajole the captain of the ship transporting Gabillon to and from the North Cape into framing and displaying a photograph of Gabillon in the ship's smoking room.[12] I was unable to locate any of these photographs and, as with the fake Sámi photograph, they may have become part of private or public collections portraying (typical) Norwegian men.

The portraits of Gabillon were certainly hung up in a spirit of light-heartedness and whimsy.[13] These portraits served as a means of transforming the journey into a theatre play, where the travellers effectively created their own stage. In this narrative, the Sámi clothing and the framed pictures assumed the roles of props and were used to toy with the boundaries between reality, fiction and performance.

6 Narratives and Silences

Commodities are embedded within "different systems of values" (Kopytoff 1986, 80) and this includes souvenirs and other things that enter economic circulation. The photograph had a different value for Thimig than for the tourists looking at it in a shop in Tromsø or purchasing it as a souvenir at the Sámi camp. In the latter cases, the photograph became a representation of Sámi culture, detached from the personal biographies of Demnitz and Thimig. It entered a different system of values and economic circulation. The photograph

12 Mappe HT and PT 21, AM 75.062 Th. Theatermuseum Vienna, Nachlass Hugo Thimig.

13 However, Gabillon ([1884]) himself did not mention this anecdote in his published travelogue.

was both a singular collectible and a reproducible commodity (Kopytoff 1986), conveying differing values depending on the relationships it established.

Maybe more than anything, it showcases the often ostentatious nature of cultural representation in tourism and provides a cynical commentary on the superficial nature of tourist encounters with destination cultures. A few signifiers, such as Sámi-style garments, the logo imprint of a Tromsø photographer, and its sale in a Northern context, were sufficient to identify the photograph as that of a Sámi couple. The history of the photograph underscores the power of representation wielded by the tourism industry. It emphasises how tourist interpretations leave traces that contribute to the process of stereotyping cultures. In her discussion of early postcards featuring South-Sámi people, Baglo points out that analysing such picture postcards helps "to reclaim stories, imagery and knowledge that have remained hidden or lost and consequently open for discussions of colonization and decolonization" (2019, 116). As my discussion has demonstrated, the photograph was indeed part of the colonial system of knowledge production, yet at the same time subverted it. After all, the theatrical and carnivalesque harbour the potential for subversion.

Using a, so-called, ethnographic photograph as a historical source invites rather provocative questions. To what extent do we analyse and define material things as representative or indicative of a particular narrative or context, oblivious to the fact that they may entail multiple narratives? How many photographs of individuals labelled as 'Norwegians', 'Austrians', 'Sámi', and so forth exist within museum and private collections, as well as in albums, where, in reality, they do not align with these ethnic labels? By exclusively examining such photographs from the perspective of a compartmentalized ethnic identity, do we risk perpetuating discourses from the nineteenth century? This is a particularly topical question for historians. For a long time (and to some extent even now) they have had the tendency to regard material culture and cultural objects as a means of illustrating or representing a specific argument, assigning them to a singular narrative.

As shown in this chapter, by establishing links with other actors, the photograph gave rise to numerous narratives: that of asymmetrical power relationships between tourists and their destination culture, amplified by many condescending and exoticizing descriptions of the Sámi in contemporary travelogues; tourists' light-hearted sense of humour and their yearning for entertainment; the desire to enact a different persona, in this particular case motivated by one's profession as an actor. Lien (2018) contends that photographs of Sámi taken by the scholar Sophus Tromholt in 1883 and those taken during the expedition led by Roland Bonaparte in 1884, both of whom strongly believed in anthropometric measurement, must be viewed within a multifaceted context involving various actors and intentions. By juxtaposing the photographs with the men's own self-portraits taken during their expeditions, they conveyed particular images of themselves to the scholarly community. On the other hand, the photograph of Thimig and Demnitz took on a life of its own immediately after being taken, independent of their intentions and of their biographies. Unlike many nineteenth-century expeditions, such as those of Tromholt and Bonaparte, where the leaders performed their masculinity through photographs of themselves taken in the field (Lien 2018, 5), Thimig did the opposite, by posing as a woman.

At the very beginning of this project, all I had were a few lines in a travelogue published in 1884 and, for many months, I considered my main subject – the photograph – as a lost object. With each new piece of information that I unearthed, I was able to collect items that led to new narratives. However, I am aware that there are still many items and narratives that remain undiscovered. After all, collecting is a never-ending process. It would be fascinating to discover additional copies of the photograph and to track their biographies across space and time, and I trust that more will come to light over the next few years. Moreover, as with all historical writing, but increasingly so owing to the ongoing digitization of sources, this text represents a work in progress – as well as a collection in progress. I hope to have shown that the narratives I have explored serve to highlight the significance of material culture, and, more specifically, the significance of material objects, when seeking to understand historical processes. Furthermore, object-centred research within historical studies has the potential to reveal a myriad of stories of the North, some of which may intersect with other discourses across Europe. The concept of the North is not static. It is constantly engaged in dynamic relationships with its surroundings. This chapter contributes to an ongoing discussion on how images and imaginings of the North came into existence within a complex network, shaped by the burgeoning Northern tourism industry, characterized by a codification of visual signs and by the utilization of new media. Objects are representations and are relational. They are historical sources. They produce evidence. They are processes. They also evoke emotions (see Nordby 2021). Examining the photograph from various perspectives and recognizing it as a thing of inbetweenness (Basu 2017) with the potential to constantly create new narratives and collections serves as a reminder that individual travelling experiences are highly diverse and vary depending on the traveller and the specific travel situation. These experiences have the potential to both undermine and confirm the grand narratives put forth by the tourism industry, often simultaneously. In the case of this photograph, it affirms the narrative of the Sámi as exotic representatives of the North, but it also represents an individual and unique tourist experience staged in the form of a comedy intended for a private audience – albeit in the broader context of the nineteenth-century colonial and asymmetrical discourse prevalent in tourism.

Acknowledgements

This chapter was written as part of the UiO:Nordic Collecting Norden project group (University of Oslo). I would like to thank Mag. Ilse Eichberger, Dr. Lydia Gröbl and Dr. Christiane Mühlegger-Henhapel at the Theatermuseum in Vienna for their help with the Thimig estate. A special thanks goes to Mag. Ilse Eichberger for uncovering the photographs. Many thanks to my co-editors for their constructive comments on earlier versions of my chapter, and to the numerous participants at various seminars and conferences for their questions and reflections about the relationship between tourism and cultural stereotyping in the nineteenth century.

Bibliography

Baedeker, Karl. 1879. *Schweden und Norwegen nebst den wichtigsten Reiserouten durch Dänemark. Handbuch für Reisende*. Leipzig: Karl Baedeker.

Baglo, Cathrine. 2012. "Fotografiers materialitet: bilders betydning for forestillinger om samisk kulturell fremmedhet." *Tidsskrift for kulturforskning TfK* 11 (3): 45–59.

Baglo, Cathrine. 2015a. "Reconstruction as Trope of Cultural Display: Rethinking the Role of 'Living Exhibitions.'" *Nordisk Museologi* 2: 49–68. https://doi.org/10.5617/nm.3047.

Baglo, Cathrine. 2015b. "Sameleire i Tromsø som turistmål, fotomotiv og møteplass." *Ottar* 306 (3): 30–37.

Baglo, Cathrine. 2019. "Puncturing Parts of History's Blindness: South Saami and South Saami Culture in Early Picture Postcards." In *The Indigenous Identity of the South Saami: Historical and Political Perspectives on a Minority within a Minority*, edited by Håkon Hermanstrand, Asbjørn Kolberg, Trond Risto Nilssen & Leiv Sem, 93–119. Cham: Springer International Publishing. https://doi.org/10.1007/978-3-030-05029-0_6.

Basu, Paul. 2017. "The Inbetweenness of Things." In *The Inbetweenness of Things: Materializing Mediation and Movement between Worlds*, edited by Paul Basu, 1–20. London: Bloomsbury.

Bennett, Tony. 1995. *The Birth of the Museum: History, Theory, Politics*. London & New York: Routledge.

Bjorli, Trond. 2008. "Fotografiet mellom modernisering og modernisme." In *80 millioner bilder: norsk kulturhistorisk fotografi 1855–2005*, edited by Jonas Ekeberg & Harald Østgaard Lund, 28–36. Oslo: Forlaget Press.

Bonge, Susanne. 1980. *Eldre norske fotografer: fotografer og amatørfotografer i Norge frem til 1920. Utarbeidet på grunnlag av Ragna Sollied: Eldre norske fotografer. Bergen 1972*. Bergen: Universitetsbiblioteket i Bergen.

Crossen-White, Holly L. 2015. "Using Digital Archives in Historical Research: What Are the Ethical Concerns for a 'Forgotten' Individual?" *Research Ethics* 11 (2): 108–19.

Ferrando, Francesca. 2019. *Philosophical Posthumanism*. London: Bloomsbury Academic.

Fournier, August. 1884a. "Eine Nordfahrt. III. In Finnmarken." *Neue Freie Presse, Morgenblatt*, October 11, 1884. ANNO/Österreichische Nationalbibliothek.

Fournier, August. 1884b. *Eine Fahrt zum Nordcap*. Prag: F. Tempsky.

Friedberg, Anne. 1993. *Window Shopping: Cinema and the Postmodern*. Reprint 2019. Berkeley: University of California Press.

Gabillon, Ludwig. [1884]. *Ferien im Norden*. Wien: Selbstverlag.

Hadamowsky, Franz. 1962. *Hugo Thimig erzählt von seinem Leben und dem Theater seiner Zeit. Briefe und Tagebuchnotizen ausgewählt und eingeleitet von Franz Hadamowsky*. Graz: Hermann Böhlau Nachf.

Jones, Siân. 2010. "Negotiating Authentic Objects and Authentic Selves: Beyond the Deconstruction of Authenticity." *Journal of Material Culture* 15 (2): 181–203. https://doi.org/10.1177/1359183510364074.

Kaasa, Janicke S., & Marit Sjelmo. 2022. "Verdenslitteraturens geografier: Norge på det litterære verdenskartet." *Agora* 40 (2–3): 114–34. https://doi.org/10.18261/agora.40.2-3.7.

Kopytoff, Igor. 1986. "The Cultural Biography of Things: Commoditization as Process." In *The Social Life of Things: Commodities in Cultural Perspective*, edited by Arjun Appadurai, 64–91. Cambridge: Cambridge University Press. https://doi.org/10.1017/CBO9780511819582.004.

Krauß, Ferdinand. 1888. *Von der Ostsee bis zum Nordcap. Eine Wanderung durch Dänemark, Norwegen und Schweden, mit besonderer Rücksicht auf Kunst- und Culturgeschichte, Sage und Dichtung*. Neutitschein: Rainer Hosch.

"L. Schjetnes". 1883. *Tromsø Stiftstidende*. 8 April 1883. Nettbiblioteket, Nasjonalbiblioteket.

"L. Schjetnes". 1884. *Tromsø Stiftstidende*. 27 July 1884. Nettbiblioteket, Nasjonalbiblioteket.

Larsen, Peter & Sigrid Lien. 2007. *Norsk fotohistorie frå daguerreotypi til digitalisering*. Oslo: Det Norske Samlaget.

Latour, Bruno. 1992. "Where are the Missing Masses? The Sociology of a Few Mundane Artifacts." In *Shaping Technology/Building Society: Studies in Sociotechnical Change*, edited by Wiebe E. Bijker & John Law, 225–58. Cambridge, MA: The MIT Press.

Lien, Sigrid. 2018. "Performing Academic Masculinity in the Arctic: Sophus Tromholt and Roland Bonaparte's Photographic Accounts of Sámi Peoples and Northern Landscapes." *Journal of Aesthetics & Culture* 10 (4): 1–17. https://doi.org/10.1080/20004214.2018.1498677.

Lien, Sigrid & Hilde Wallem Nielssen. 2021. "Negotiating Research: Studying Sámi Photographs as Norwegian Outsiders." In *Sámi Research in Transition: Knowledge, Politics and Social Change*, edited by Laura Junka-Aikio, Jukka Nyyssönen & Veli-Pekka Lehtola, 169–85. London: Routledge. DOI: 10.4324/9781003090830-10.

Lien, Sigrid & Hilde Wallem Nielssen. 2018. "Negotiating the Past: Addressing Sámi Photography. Special Issue Editorial – Sámi Representation." *Journal of Aesthetics & Culture, Special issue: Sámi Representation* 10 (4): 1–2. https://doi.org/10.1080/20004214.2018.1510646.

MacCannell, Dean. 1973. "Staged Authenticity: Arrangements of Social Space in Tourist Settings." *American Journal of Sociology* 79 (3): 589–603.

Naum, Magdalena. 2023. "Souvenirs from North America: Understanding and Representing 'Indianness' in Nineteenth-Century Sweden." *History and Anthropology* 34 (3): 473–96. https://doi.org/10.1080/02757206.2021.1885401.

Nordby, Guro. 2021. "På sporet av en jernbane-uniform: Ulike analytiske tilnærminger til en museumsgjenstand." *Tidsskrift for kulturforskning* 20 (2): 27–42.

Ponstingl, Michael. 2024. *Theater für die Kamera. Fotografische Passionen des Hofschauspielers Hugo Thimig*, mit einer biografischen Skizze von Elisabeth Großegger. Salzburg: Fotohof *edition*.

Salazar, Noel B. 2012. "Tourism Imaginaries: a Conceptual Approach." *Annals of Tourism Research*, 39 (2): 863–82.

Sayes, Edwin. 2014. "Actor-Network Theory and Methodology: Just What Does it Mean to Say Nonhumans Have Agency?" *Social Studies of Science* 44 (1): 134–49.

Schøning, Eli-Anita Øivand. 2018. "From 'Lapp' to 'Margrete:' Representations of Sámi People in Photographic Postcards from Norwegian Sápmi." MA-thesis, UiT The Arctic University of Norway.

Theodorsen, Cathrine. 2008. "Political Realism and the Fantastic Romantic German Liberal Discourse and the Sámi in Theodor Mügge's Novel *Afraja* (1854)." *Nordlit* 23 (February): 355–70. https://doi.org/10.7557/13.1350.

Thimig, Hermann. 1962. "Geleitwort." In *Hugo Thimig erzählt von seinem Leben und dem Theater seiner Zeit. Briefe und Tagebuchnotizen ausgewählt und eingeleitet von Franz Hadamowsky*, 7–8. Graz: Hermann Böhlau Nachf.

Urry, John & Jonas Larsen. 2011. *The Tourist Gaze 3.0*. 3rd edition. London: SAGE Publications. https://doi.org/10.4135/9781446251904.

Ytreberg, Nils A. 1962. *Tromsø bys historie* vol 2. Tromsø: Peder Norbye.

Archives

Folketelling, Digitalarkivet, Norway
Historisk befolkningsregister, Norway
Theatermuseum, Vienna, Austria

OBJECT: Polar bear

FIGURE 12.1 A family polar bear hide has taken part in a birthday celebration. The bear was shot on Spitsbergen between 1910 and 1920.
PHOTO BY THE AUTHOR

CHAPTER 12

Collecting Polar Bears

Brita Brenna

Abstract

This chapter delves into the cultural history of collecting polar bears and especially their offspring, polar bear cubs, during the late nineteenth and into the twentieth century. The chapter starts with memories of polar bears in the family lore and contextualizes these within a broader historical narrative of polar bear exploitation, examining how the animals were hunted, collected, and transported to European collections, zoos, and museums. Using Norwegian newspapers, journals, and literature, the chapter investigates how polar bears were integral to national narratives of exploration and heroism, especially in the context of Norway's ambitions in the Arctic regions during the early twentieth century. This specific collecting culture was underpinned by a complex interplay of economic, affective, and moral economies, reflecting human-animal relations and the transformation in the perception of wild animals.

Keywords

polar bear cubs – arctic exploration – collecting cultures – animal protection

•••

> The live polar bear is a magnificent animal that due to its untamed nature has provided the crew with much entertainment; thus, it devoured its own mother, who lay dead on deck.
>
> Trondhjems borgerlige Realskoles alene-priviligerede Adressecontoirs-Efterretninger 29 September 1860[1]

∴

1 All translations from Norwegian by the author.

 | DOI:10.1163/9789004749665_013

Growing up in the countryside, a visit to my grandmother's tiny apartment in the centre of Oslo was an adventure; there were noises from the trams and the cars outside, the large clock on the wall would tick through the night. And then there was the furry polar bear hide. In my grandmother's small living room, the polar bear, with head and a couple of claws still intact, filled the space between the sofa, the television set, and the window. The hide was a place for children to rest and play.

I cherish the memory of the polar bear on the floor. In my memory there is a golden warm light playing in the fur, and the feel of it still lingers in my fingers. The black snout was cold, the teeth were smooth, and the mouth was scary, but in a safe way, there in front of the television set. Today the polar bear hide continues its afterlife as a child-friendly indoor embellishment: adorned with a pink birthday crown, it rests on the mezzanine in my cousin's cabin.

The bear was killed by my grandfather, who had spent around fifteen summers and a couple of winters in the Arctic as a topographer and a technician in the 1910s and 1920s. The hide was a token of the North, a souvenir perhaps, not from a tourist trip, but a reminiscence of a life in the Arctic. The bear was probably shot while spending the winter on Bear Island in 1917, or maybe later. The specifics are lost in the mist of memory. So is the story of a polar bear cub who died on my great grandparents' farm in Hadeland, which according to the locals is the Tuscany of Norway – not a place for polar bears. My mother remembers believing that the dead polar bear – the hide – was the mother of the cub she had heard so much about. The cub was, as far as my mother recollects, put in the pen with the dogs. One day it got loose, climbed to the second floor of the main building and into the bedroom of the children of the house. No one was hurt, except the cub, who was shot. According to my mother, my grandfather remarked: "Why did they shoot it, they could have sold it to Hagenbeck!"

The polar bear hide and the cub are part of family lore; they were kept, the one as dead as a polar bear hide can be, the other alive. The dead one is still here one hundred years later, the other one alive only in the ever more fragile family memory. Both were once actors in a culture of collecting that has left polar bears dead and endangered. Polar bears have been collected to be put on our shelves of imagination or in our living rooms, in zoological gardens and museums, or in trophy rooms of stately homes. Polar bears were one of these objects of the North that were killed and collected or caught alive and collected throughout the time when the North was a geography of collecting. Walrus tusks, narwhal teeth and polar bear hides became valuable items in European collections as Europeans expanded into the Arctic from the sixteenth century, on a quest for resources and routes to new markets in the East. The objects were valuable because they were rare and curious, and because they came from what Europeans perceived as unknown lands – distant geographies.

This chapter will use the sensation of the hide and the story about the cub as an occasion to enquire into the trade of animals from the North in the period between late nineteenth and the first half of the twentieth century. What kind of collecting was this? What desires have polar bears invoked that rendered them objects of collecting? By researching Norwegian newspapers, journals, and books for polar bear stories, I will attempt to answer these questions by relating them to histories of collecting. The specific story of each of the two animals, or the men that brought them to Norway, is not what interests me. Neither the animals nor the men in question would regard themselves as involved in collection. Rather, by framing their movements within a collecting practice, I believe I can shed light both on the practice of collecting and on the fate of so many polar bears. Collecting animals has its own moral, affective, and monetary economy. These economies have undergone historic changes – as have human relationships to polar bears.

For me, the story of the hide that I once cherished and the polar cub in Hadeland opened a view to a form of collecting that implicated violence, torture, and death for the animals. Researching this chapter, I was deeply affected by the stories of the polar bears that I read about in newspapers and books. However, my perception of polar bears is squared with a cultural appreciation of polar bears and wild animals more broadly that has been transformed since these animals were hunted and collected. As I will come back to, collecting and killing wild animals met little resistance at the time the polar bears in my family were caught in the 1910s and 1920s. On the contrary, as I will show, fighting with and killing polar bears were integral to heroic tales of Norwegian national appropriation of the Arctic. Further, the catching and hunting of polar bears (and other Arctic animals) provided a livelihood to inhabitants and contributed to the growth of cities in the North. Territorial expansion towards the North was a Norwegian aspiration from around 1900, aptly termed Arctic Imperialism by the historian Einar Arne Drivenes (for an English explication, see Drivenes et al. 2006). Indigenous relations to polar bears are not relevant here, since the polar bear area of the Arctic belonging to Norway since the Svalbard Treaty of 1920 was unpopulated. In this article, I will tie the hunting and capturing of polar bears to the European collecting culture. I will claim that an important reason why polar bear hunting could provide income for many Norwegians was that they were part of an international collecting market. This was a market for wild animals that were dead or alive, as luxuries, rarities, or natural history objects. The craze for killing – and caring – for polar bears was driven by demand, by the search for economic gain, and by a cultural acceptance for the collecting and killing of wild animals. Lastly, the availability of bears and their presence in areas of interest for Norway as a state and the possibility to hunt freely in areas with no previous population was important. However, the animals were also hunted and caught, transported, and sold, by people who related to the bears in different affectual ways. As I will show in this chapter, there were increasingly protests being raised against the capture and transport of live bears, and later against their killing. Some protests appeared early in the 1900s whilst the demand for better regulations was voiced most strongly after World War II.

I am interested in the affectual relations between polar bears and their predators – the humans – who were increasingly plying the North. During the period from roughly the mid-nineteenth century to 1973 when polar bears received legal protection, the relations between people and polar bears were complex and changing – and this article will provide a glimpse into how polar bears were part of moral, affectual, and monetary economies of collecting in this period.

The polar bear has now become an icon of climate change and a melting Arctic. Even before becoming a climate change icon, the role of the polar bear in popular imagination and the sensibilities of people towards it had undergone huge transformations. What is important in this context is the way human-animal relations can change significantly in short time spans, where technology, science, mediation, and economy all play roles in how we humans perceived the animals. Historian Kristoffer Archibald has interrogated the development of polar bear imagery in periodicals and public media and traces a radical shift between 1950 and 1980. He claims that "[w]hereas the polar bear had been conceptualized as frightening in early-modern expeditionary literature and visual art, in the second half of the twentieth century the bear was depicted as vulnerable due to overhunting and in need of a science-based management plan" (Archibald 2015, 266). He traces the role of technologies in these changes. The large toll that trophy hunting took on polar bears through hunting expeditions with aeroplanes and advanced weapons in the aftermath of World War II provoked a public campaign to save the polar bear from extinction. Scientists came in with their new technologies of aircraft, tranquillizers and tagging, and they were depicted as men in control of the polar bear. "In lieu of the powerful bear that had long terrified the Arctic's audiences, observers now saw a vulnerable beast whose survival depended upon scientists equipped with modern technology" (Archibald 2015, 278). The change that Archibald describes is also a transition in natural history collecting practices. Where once many polar bears had been hunted and killed as natural history objects, now the scientists no longer collected the bears themselves, instead collecting information that could be used to manage their population. However, as we shall see, the image of the polar bear had started to change earlier. Today the colonial legacy of natural history museums is increasingly questioned and interrogated. Trophy hunting is deemed problematic for so many reasons, and not least there is international legal enforcement of the trade in "exotic" animals. That said, polar bears are still traded and highly visible as objects of attraction, but the regulations, practices and relations to wild animals have undergone large transformations.

1 Counting Polar Bears

The polar bears in my family history play an insignificant role in the history of polar bears, as did my grandfather who brought them home. Polar bears are the largest carnivores on Earth, and they belong to what have been perceived as inhospitable and unknown regions of the North. As dead trophies they could represent the courage of those who shot them, and the wealth of those who could afford them. Alive they represented the ultimate wild animal, the captured king of the North. In the second half of the nineteenth century, the traffic in wild animals intensified and accelerated as part of colonial expansion and imperial exploration. Historian Nigel Rothfels has studied the capture of live animals for zoological gardens and private collectors and describes those taking part in the traffic as naturalists, colonial administrators, "great white hunters", commercial hunters and increasingly professional catchers (Rothfels 2002, 59ff). Polar bears made up a tiny minority of the animals that were caught or killed; however, they were, together with a few other species like musk ox and walruses, the most important animals from the North.

In 1973 collecting and killing polar bears became regulated by the International Agreement on the Conservation of Polar Bears and Their Habitat, signed by Canada, Denmark, Norway, the United States, and the Soviet Union. It allowed the killing of polar bears only for scientific reasons, in self-defence, or in Canada, by Indigenous Peoples hunting with traditional methods (Engelhard 2017, 5; Larsen & Stirling 2009). The agreement also forbade the capture of live polar bears for circuses and zoos. Measures for protection had been taken earlier, most notably by the Soviet Union, where hunting had been banned in 1956. The possibility of extinction seemed a real threat, and from what we know of the numbers of killings and the extent to which the bears were hunted, it most likely was one. Between 1871 and 1973, 30,294 polar bears were reported as caught to Norwegian officials; of these 1,683 were caught alive, according to the statistics on bears brought to Norway from the Arctic (Søbye 2000). Of the 28,611 that were brought back in the form of hides, more than 20,000 were caught by sealers, around 6,000 by land-based hunters, approximately 1,500 by meteorologists at the weather stations on Svalbard, and 758 by trophy hunters. The fur on my grandmother's floor was certainly one dead bear out of many.

The polar bear cub who went to Hadeland was also part of an even larger sleuth than accounted for in the numbers above. If we compare the capture numbers to the registers of foreign export, the number of polar bear cubs exported was even higher than the 1,683 mentioned above, increasing the number of cubs brought to mainland Norway to approximately 2,000.[2] In the year 1924 alone, 104 living polar bears were exported from Norway: an all-time high. The following year many were still exported, 77, but up to a third of the price for each. According to the registers, the 1920s were responsible for a quarter of the total number of exports of living bears from Norway (Statistisk sentralbyrå 2015). The reason why the 1920s loom so large in the statistics, whether the catches were especially good, the number of other animals so low, the need for income desperate, or whether the demand from zoos and circuses especially large, we cannot know. Historian Espen Søbye who prepared the statistics and wrote an accompanying article says offhandedly that the cubs were sold to zoological gardens, fairgrounds and variety theatres around the world (Søbye 2000). Indeed, they were. 2,000 living bears seem like a lot, and the author Michael Engelhard presents the same number as a total of live polar bear cubs that went to "zoos, carnivals and vaudeville theatres all over the world" (Engelhard 2017, 4). However, the worldwide numbers are much higher, as the numbers above are only based on animals brought to shore or exported from Norway. Polar bears were also taken from Siberia, Greenland, Alaska, and Northern Canada to other nations.

The total number of polar bears killed and caught will never be ascertained. The numbers are enormous, and they require that we try to understand why all these animals were slaughtered or caught alive. One answer is that collecting polar bears could bring money and status and were in high demand in the collecting culture of modern Europe.

2 Capturing and Killing

Many different actors took part in keeping up the trade in polar bears in the years around 1900. What stands out as the largest outlet for live polar bears from

2 Numbers are hardly accurate. The article "Catching Up" is a recent attempt to quantify the catches of large animals in the Svalbard region, but does not use the export data of polar bears from Norway, and thus has a somewhat lower estimate of polar bear catches (Kruse 2017).

the European Arctic was the Hagenbeck firm in Hamburg, the one my grandfather would have sold the cub to. Initially a trader in exotic animals, the founder Carl Hagenbeck started another enterprise in the 1870s consisting of travelling caravans that displayed animals and Indigenous People. The caravans were a huge commercial success and in the second half of the nineteenth century Hagenbeck emerged as the world's leading provider of exotic animals, with a worldwide network of trappers and hunters on one hand, while on the other hand possessing a huge network of zoological gardens, museums, circuses and private menageries as customers (Rothfels 2002). Imperial expansion and massive exploitation of nature and people were integral to this network, which drew together scientific and natural historical research institutions, popular entertainment through the trade of animals, people, and objects. In this enterprise, polar bears were brought to Western countries as objects of naturalists' curiosity, as well as serving as objects of popular entertainment. This was to be an enduring fate for the bears.

Zoological gardens and museums portrayed themselves – and were seen as – institutions that did research as well as teaching lessons about nature to the public (on the history of zoos, see Baratay & Hardouin-Fugier 2004). William Hornaday, a powerful museum man, writer, and director of the New York Zoological Garden (today, the Bronx Zoo), was also an active nature preservationist, a taxidermist, and one of Hagenbeck's customers. Due to his actions and writings, he became an influential voice in the establishment and running of museums and zoological gardens. Hornaday issued a handbook in taxidermy in 1891, a year before my grandfather was born, providing instructions for collection, preparing, and presenting specimens, i.e., killing and preserving animals. As Hornaday stated in the handbook's introduction:

> The rapid and alarming destruction of all forms of wild animal life which is now going on furiously throughout the entire world, renders it imperatively necessary for those who would build up great zoological collections to be up and doing before any more of the leading species are exterminated. (Hornaday & Holland 1894, VII)

A naturalist who wants representatives of all the soon-to-be extinct species must set about it at once, he stated further. What he says, in so many words, is that they must start the hunt and kill before others get to the specimen. A naturalist's task was to collect and preserve the animal for posterity, in the confines of a natural history institution. Far from being institutions that held animal exploitation in check, natural history museums contributed to legitimising and enforcing the hunt of rare and exotic animals (Thorsen, Rader & Dodd 2013). On the one hand, people like Hornaday expressed affection for the animals. Hornaday wrote in the introduction to the aforementioned book that the "sight of a particularly fine animal, either alive or dead, excites within me feelings of admiration that often amount to genuine affection: and the study and preservation of such forms has for sixteen years been my chief delight" (Hornaday & Holland 1894, VII–VIII). The curious use of the term "forms" here might speak to the understanding of animals as expressing the forms of nature – stuffed or otherwise. On the other hand, Hornaday wrote to Hagenbeck during negotiations to procure a rhino for the Bronx Zoo:

> we must be very still about forty large Indian rhinoceroses being killed in capturing the four young ones. If that should get into the newspapers, either here or in London, there would be things published in condemnation of the whole business of capturing wild animals for exhibitions. (Rothfels 2002)

The killing of numbers of animals to get hold of a few that could be exhibited alive was taking place globally. What counted was getting to the right specimens. The legitimation for this was the importance of having and showing the animals for a public. "For my part", Hornaday continued:

> I think that while the loss of the large Indian rhinoceroses is greatly to be deplored, yet, in my opinion, the three young ones that survive will be of more benefit to the world at large than would the forty rhinoceroses running wild in the jungles of Nepal, and seen only at rare intervals by a few ignorant natives. (Hornaday to Hagenbeck, as cited in Rothfels 2002, 67)

Love for nature took many forms, also for naturalists. Most museums were concerned with preserving dead animals, not only for the sake of science, but also to make public displays that could serve to educate about nature. The love of nature that Hornaday and naturalists working in these museums expressed should also be instilled in the public using entertaining and educational displays. The logic of collecting, preservation and public display is eloquently examined in an article by historian of science Donna Haraway with the appropriate title: "Teddy Bear Patriarchy: Taxidermy in the Garden of Eden, New York City, 1909–36." With Theodor Roosevelt as its patron saint, the American Museum of Natural History presented nature to city dwellers in stunning lifelike dioramas: "In the upside down world of Teddy Bear Patriarchy", Haraway writes, "it is in the craft of killing that life is constructed, not in the accident of personal, material birth" (Haraway 1984, 23). Life as it was presented was thus dependent on collecting and killing. However, there was also a porous demarcation between the living and the dead: "In the past the distinctions between zoological gardens, museums and botanical gardens were blurred, and some museums accommodated live bears to attract visitors" (Davis 2019, 107). What was deemed necessary in this culture of collecting was to present the animal in cities, where the masses of people were living, as lifelike – or as alive – as possible.

In the 1890s Hagenbeck set up a permanent zoological garden outside Hamburg, while continuing to perform and travel with menageries (and people). Polar bears were a major attraction and part of the first "open diorama" that Hagenbeck patented in 1896. Creating an open diorama was an attempt at removing boundaries, fences and bars to let the animals be seen in "natural surroundings", with moats serving to protect the enclosures. The first diorama set up is of interest in connection to both polar bears and Norwegians, as the first Arctic diorama featured *Fram,* the ship of the Norwegian polar explorer, Fridtjof Nansen. Nansen had become famous worldwide after returning from his quest for the North Pole. Thus, a connection between polar bears, Arctic heroic explorations and the Norwegian explorer Nansen was duly established by a German animal collector and entrepreneur.

The zoological garden was a huge success, but Hagenbeck also trained animals. One of the most renowned shows was a polar bear performance in which between 70 and 75 animals appeared together. In the journal for the Norwegian Society for Protection of Animals in 1909, Hagenbeck is quoted with dismay:

> Many of the [polar bears] have been taught tricks in a few months. Others I have been working patiently with for years. The polar bear is the most stubborn of all animals. It is evil by nature and can turn on its master without any reason. *Only two of my bears really perform their tricks voluntarily and with pleasure.* (*Dyrenes Ven* 13 (12) 1909)

The author of the article has italicised the last sentence and ends the piece with emphatic words: "Poor animals. How much they must suffer for the sake of humans – to no avail!" The animal welfare organizations in different countries were to small degrees mobilizing against zoological gardens. As they were also seen as educational establishments, they were not a prime target of these organizations which spread from England in the early nineteenth century. In Norway the first association started activities in 1859 (see for example Schilbred 1959). The Association for Animal Protection was first and foremost concerned with the welfare of domestic animals. The quote above concerning the training of wild animals was not a typical article in the journal. But increasingly through the 1900s, arguments against taming and "inhuman" conditions for wild animals came to play a larger role in public discourse in Norway as they did in other European countries (see Dirke 2000; Kean 1998). However, as this discourse developed, there was a parallel intensification of hunting, and catching animals in the far North.

3 "A Ship, a Polar Bear, and Nansen"

Polar bears provided revenue for Norwegians – for sealers and whalers, for those wintering on Spitsbergen or other places in the Arctic Sea. This activity was supported nationally, and the connection between these commercial ventures and the quest for territorial gains and expansion is, along with the thriving market for polar bears, important for understanding why so many bears were killed and captured. Polar bears were to be used extensively as national emblems and icons. One example can be found in a lyrical description of the Norwegian pavilion at the 1900 Paris Exposition. The author took issue with the other nations at the exhibition that showed luxuries and non-necessities. The Norwegian pavilion was different. Fishing nets crowded the corners, ship models were placed on the floor and the gallery, the walls were adorned with halibut, cod, herring, anchovy, and wellington boots. There were paintings showing skiers in real Norwegian valleys as well as a stuffed polar bear and in the middle of the room: a bust of Fridtjof Nansen. "It was clear to me that the Norwegian pavilion is grandiose in its simplicity. This is Norway, this is how the Norwegian people live. […] It is all so simply stated: A ship, a polar bear and Nansen" (*Moss Avis* 28 May 1900). The message was that Norway delivered polar bears to Europe and the world and could proudly display them as part of what Norway was, notwithstanding that polar bears were not to be found on Norwegian *territory* at this point in time. The Spitsbergen Treaty, giving Norway sovereignty over what came to be called the Svalbard archipelago, was signed in 1920. Before this, these areas of the North were *terra nullius*: no man's land. The Arctic, however, was becoming a national aspiration for Norway. Throughout the nineteenth century, Sweden had sent numerous scientific expeditions to the European Arctic, Spitsbergen and the surrounding islands and Greenland. Denmark had Greenland as their colonial possession. Norwegian explorers were just recently starting to explore the polar regions as a new century dawned, and polar bears feature in all the polar hero stories. The *Fram* polar expedition that was headed by Fridtjof Nansen from 1893 to 1896 set out to prove that the North Pole was in the middle of an ocean covered with ice. As the ship that was frozen in the ice did not reach the pole, Nansen and Hjalmar Johansen left the ship to do the trip on skis. They had to turn around and, on their way back, one incident involving a polar bear was especially widely published. Lying on the ground with a polar bear on top of him, Johansen politely asks Nansen (his superior) to hurry up, as if not it would be too late. The bear was shot, and the situation resolved in the best Arctic heroic manner – even when the fierce bear attacked, formalities were kept. This polar bear story, which is still widely retold in Norway, and reached an international audience through Nansen's own books and lectures, has a moral that implied Norwegians were taking part in the civilizing of the Arctic.

The Norwegian papers also afforded extensive coverage to the polar bear cub Marie that explorer Roald Amundsen had taken on board on his trip towards the North Pole in 1920. Pictures of Amundsen with his cub featured in a host of papers, showing images of Amundsen feeding the cub. "An amusing picture", where we see "the polar researcher feeding his cub", one newspaper wrote. (*Nordlandsposten* 26 September 1921). The cub was later shot, and its fame secured. The bear was stuffed and exhibited standing in the staircase of Amundsen's house, which was turned into a museum. The press was important for making polar heroes, but also in securing the polar bear a place in the national story of heroic conquests of the Arctic.

4 Caring Heroes of the Hunt

In the first part of the twentieth century, polar bear hunting was a heroic feast. The town of Tromsø was home to and the starting point for much of the Norwegian Arctic Ocean sealing, hunting, and trapping fleet, and home to many of the legendary "polar people" (*polarmennesker*, cf. Giæver 1955). The polar people, as Giæver describes them, were the people that lived their lives and made their living by harnessing the natural resources of the North. Tromsø was also where most of the polar bear hides as well as live polar bears were brought ashore. One of the most notable of the Norwegians spending many winters in the Arctic, Henry Rudi, went under the nickname "The Polar Bear King" (*Isbjørnkongen*). He shot 713 polar bears between 1907 and 1947. The nickname was given respectfully – to be a top polar bear shooter was not an offence. Hardship, dangers, loneliness, and skill were required to earn such a name. But Rudi was also a caretaker of dogs and polar bear cubs. In her discussion of Rudi's relation to the polar bear, folklorist Marit Anne Hauan points out that the bears had both an economical and an emotional use. To capture one or more live polar bear cubs and sell them could salvage the profit from a wintering season. The first cub that Rudi took care of in 1909 was given the name Maja. It was bottlefed, allowed to stay in his small cabin, given toys to play with and after arriving in Tromsø, taught to drink beer (Hauan 2021, 179). "During the time the man and the bear spent together", writes Hauan, "the man had the occasion to implement and practice caregiving skills, had the opportunity for close contact and the possibility of having his needs for showing and receiving affection fulfilled" (Hauan 2021, 180). A heroic hunter did not become less of a hero through caring for cubs, and the care and closeness did not undermine what Hauan terms Rudi's "heroic masculinity". The relation between the hunter and the bear is complex. The polar bear cub is cute and in need of care but it is also an economic asset, and the stories of the relations between hunters and small bears can be read as stories of taking care of what is economically valuable, as well as stories about a relationship of love and intimacy.

Rudi was the Polar Bear King, but as we have seen, thousands of cubs arrived at Norwegian ports. The library and archive of the Norwegian Polar Institute contain many polar photos, sources, photos and books concerning the polar regions, and some of them testify to the closeness between trappers and polar bear cubs. A polar bear cub named Sebben is the subject of one manuscript, written by a trapper called Bjørvik Jacobsen who took care of the cub and brought him to Tromsø to sell him in 1947. The stories of this cub later featured in his book about overwintering in Svalbard (Jacobsen 1983). Polar bear stars are not unusual; one need only consider the more recent example of Knut at the Berlin Zoo, whose story travelled worldwide and who obtained a large crowd of fans (Davis 2019; Engelhard 2017; Flinterud 2013). Sebben's fame was of a more local character in his hometown Tromsø, but even he reached a national audience by way of an interview on the radio, one of the few polar bear interviews given. Sebben is unique in the sense that he would experience Tromsø differently from his fellow travelling cubs. They were mostly left in their small crates at the quay, whilst Sebben was taken on tours through town, getting a lift in a taxi and visiting the liquor store (Jacobsen 1983).

5 "I Can Hear them from My Room!"

As a young girl, the Norwegian author Sara Fabricius (using a different pseudonym to the one she would become best known by: Cora Sandel) wrote to the journal *Dyrenes Ven* (*Friend of the Animals*) in January 1904 (Bro 1904, 1). Sandel described in detail how she could hear the polar bears from her room in Tromsø:

> In the morning, the first thing I hear is their hoarse, drawn-out roars that end in a tortured hiss, and in the evening, it sounds through the darkness,

> desperate, filled with futile anger and defiance – and at times, it rises to a furious scream, followed by wild boys' jubilation. (Bro 1904, 1)

She described her search for the animals at the quay, finding them in the baking sun, helpless in small wooden crates. The article closes by asking about the humanity of the current age. Directing her attention directly to a fictional cub, she writes:

> And when you finally reach your destination, you might get a small pond to splash in. Oh yes, you will, for we live in a humane era. A small pond with lukewarm, murky water and floating bread crumbs from a well-meaning audience. Then we can hope that you have forgotten the sea and the ice and the great stars of the polar night. (Bro 1904, 1, 2)

Her outcry did not reach a large audience, probably due to the fact that her article was printed in the journal of the Norwegian Society for the Protection of Animals, which at this time was more concerned with improving the life of domesticated animals.

Also voicing concern was Mikkjel Fønhus, the author of nature novels featuring different animals as protagonists. Fønhus issued the novel *Under polarlyset: en fortælling fra Spitsbergen* (Under the Polar Light: a story from Spitsbergen) in 1922 which was mainly about a trapper and polar bear, but also contained degrading storytelling about the Sami people (Fønhus 1922). Most of the books by the prolific author were about the Southern Norwegian nature on the mainland that he knew by heart, and the animals that were close to him. Spending time in Spitsbergen in the early 1920s, he was on unfamiliar terrain, where animals, hunters and the landscape were of a completely different nature than his native forests. His admiration for the polar king – the bear – as well as for the hunters that survive the winter in Arctic, seems unconditional. They are both described as belonging to the North, and therefore as strangers to cities and civilization. His depiction of the suffering of the polar bear in a zoological garden was also highly evocative, and it was reprinted in newspapers and journals. What the book makes clear is that the killing and hunting of polar bears was not an issue; on the contrary, to be a hunter was an honourable endeavour. It was the transport of live bears and selling them to zoological gardens that was beyond the "natural" contest between man and animal as it included moving the cubs from their natural habitat. Fønhus described this situation: "each year, such small white captives are taken south from the perpetually cold ice fields by the pole, down to the stifling cages in Europe's zoos" (*Nordland* 17 January 1923). At the time when Fønhus wrote his book, the same time when my grandfather brought the cub back to his family's farm, measures to control the capture of live bears were put in place.

6 Torture

After World War II, much started changing concerning polar bears. There was an emerging debate about killing polar bears because of the fear of extinction. In the Norwegian papers trophy hunting by foreigners was deemed negative, either because they killed too easily, or on the other hand because they were not good enough shooters and would wound the animals and leave them to suffer. It took another twenty years before a total ban was enforced. The living cubs, however, got more immediate attention. What the newspaper and journals published also highlighted the different issues at stake. A newspaper from Northern Norway, *Lofotposten,* asked quite simply in 1949: "Should the capture of polar bear cubs be stopped?" (*Lofotposten* 2 February 1949). New regulations should be put in place, the paper argued, as the cubs suffered under transport. Cubs were often not fed properly, and some were housed in poorly built crates that were much too small. To ban trapping would not be a good idea, as those earning their living by the small scale capture of seals in the Arctic waters were dependent on the extra earnings they got from selling the cubs. The article measures quite earnestly the value of suffering animals against economic gains and argues that it would be unfair to forbid the trade in cubs as long as other animals are held their whole life in captivity – like fox and mink (*Lofotposten* 2 February 1949). Another newspaper argued that the cubs could be installed in a Polar Bear Park in Tromsø, roaming free there before being sent south. The argument was that this would solve the problem of the suffering of the cubs (*Nordlys* 24 October 1955). Again, as one of the national newspapers, VG, reported in 1956, a discussion emerged in the press both in Norway and in other Scandinavian countries about the treatment of polar bear cubs that Norwegian sealers brought in. This had, according to the newspaper, mobilized associations for the protection of animals, and with good reason, the journalist admits:

> When one has heard the polar bear cubs howling in their cramped crates on a scorching summer day after spending weeks either on board the hunting ships or on the docks in Tromsø, one cannot help but understand that the animals have suffered terribly, accustomed as they are to living in extreme cold and with the freedom to move around. (Kleivan 1956)

Public opinion was indeed about to turn also in Norway. As seen above, at this time the Soviet Union had put a total ban on polar bear hunting.

The most striking turn in Norway was taken by the trapper, journalist and author of polar literature, Johan Giæver, who presented his views in 1956 in a large-scale magazine article, the national publication *Magasinet for alle* (Magazine for All). He self-consciously admitted to having caught many cubs himself but argued that age had made him more aware of the pain he inflicted on others. He described in detail how the small bears were caught and not least how they reacted to their treatment, calling this the torture of the "animal-children of the North" (*Magasinet for alle: Arbeidermagasinet og For alle* 29 (6) 1956). The cubs were caught with a rope and dragged towards the boat, according to Giæver. They did everything they could to resist. When onboard, boxes were made from wooden planks, insulated with iron plates on the inside. The cubs should be able to move inside the boxes, but not too much, as this could give them the opportunity to break out. "If the catch was brutal, this was when the real misery begins", Giæver wrote. The cubs scratched and roared constantly. "A consuming hopeless sound" filled the ship for the rest of the trip, according to Giæver. He described how the cubs were so filled with anxiety and longing that they sometimes collapsed and lay with eyes squeezed shut: "[I]t is like witnessing a suffering child surrender to agony". The insistent anthropomorphism in Giæver's article allowed him to render the cubs' suffering understandable and urgent. He ended his article thus: "The children of the wild animals have the same feelings, physically and mentally, as our own children. At least!" (*Magasinet for alle: Arbeidermagasinet og For alle* 29 (6) 1956).

7 Conclusion: to Shoot or Not to Shoot?

Polar bears have attracted affection and been important resources for people who were struggling to make a living. They have secured food for dogs and hunters, and their hides and their cubs could safeguard the family of a winterer in the North. The cubs could also provide entertainment and enable the possibility of caring relations in the lonely, dark winter months in the ice. An important national heroic myth was also woven, with the polar bear as an important symbol of Norwegian civilized expansion into the Arctic. Further, once removed from the North and reinstalled as living entertainment in circuses or bear pits; lying flat on walls or floors; or as upholstered specimens in shops, hotel lobbies, homes of the wealthy or in museums, the polar bears would invoke affections of curiosity and fear (see for documentation of this the amazing book *nanoq: flat out and bluesome*, Snæbjörnsdottir & Wilson 2006). In natural history museums and zoological gardens, polar bears were deemed important for teaching urban populations about distant nature. Increasingly voices were raised about the cruelty these wild animals endured; other voices were afraid for their extinction. What is alarming and makes for a more general question about the responsibility of humans to animals is the realization that everyone understood that the polar bear cubs were suffering. In the town of Tromsø, they could be heard howling, but their cries were deemed less important than the money they brought. I never knew my grandfather, but I do not think he was an evil man. But he would rather have sold the polar bear cub to Hagenbeck than to have it shot. The history of collecting in the North is a history of exploitation of animals and nature, of pain and suffering. But people rarely collect to inflict pain; they kill and collect for many reasons. The issue here is how a culture of collecting led to such an amount of misery and pain. This is a moral lesson that might help us think again when we cherish the memories of polar bear cubs or of rare and beautiful spaces with objects of the North.

Bibliography

Archibald, Kristoffer. 2015. "From Fierce to Adorable: Representations of Polar Bears in the Popular Imagination." *American Review of Canadian Studies* 45 (3): 266–82. https://doi.org/10.1080/02722011.2015.1064979.

Baratay, Eric & Elisabeth Hardouin-Fugier. 2004. *Zoo: A History of Zoological Gardens in the West*. London: Reaktion Books.

Bro, Svend. 1904. *Isbjørnene. Dyrenes Ven* 8. Oslo: Foreningen til dyrenes beskyttelse.

Davis, Peter. 2019. "The Bear in the Museum." In *The Bear: Culture, Nature, Heritage*, edited by Peter Davis, Owen Nevin & Ian Convery, 107–18. Woodbridge: Boydell & Brewer. https://www.jstor.org/stable/j.ctvktrz6r.16.

Dirke, Karin. 2000. *De värnlösas vänner: den svenska djurskyddsrörelsen 1875–1920*. Acta Universitatis Stockholmensis. Stockholm: Almqvist & Wiksell.

Drivenes, Einar-Arne, Harald Dag Jølle, Ketil Zachariassen & Thor B. Arlov. 2006. *Into the Ice: The History of Norway and the Polar Regions*, translated by Bruce Bawer. Oslo: Gyldendal akademisk.

Engelhard, Michael. 2017. *Ice Bear: The Cultural History of an Arctic Icon*. Seattle, Washington, London, England: University of Washington Press.

Flinterud, Guro. 2013. "Polar Bear Knut and His Blog." In *Animals on Display: The Creaturely in Museums, Zoos, and Natural History*, edited by Liv Emma Thorsen, Karen A. Rader & Adam Dodd, 192–213. University Park: Penn State University Press.

Fønhus, Mikkjel. 1922. *Under polarlyset: en fortælling fra Spitsbergen*. Kristiania: Aschehoug.

Haraway, Donna. 1984. "Teddy Bear Patriarchy: Taxidermy in the Garden of Eden, New York City, 1908–1936." *Social Text* 4 (2): 20–64.

Hauan, Marit Anne. 2021. *Polare maskuliniteter: fra oppdagelsen av Svalbard til heltetidas siste time*. Stamsund: Orkana.

Hornaday, William T. & W. J. Holland. 1894. *Taxidermy and zoological collecting; a complete handbook for the amateur taxidermist, collector, osteologist, museum-builder, sportsman, and traveller*. 4th ed. 24 plates and 85 text ill. New York: C. Scribner's sons. https://doi.org/10.5962/bhl.title.2771.

Jacobsen, Bjørvik. 1983. *Halvmåneøya: en fangstberetning fra Svalbard*. Oslo: Tiden.

Kean, Hilda. 1998. *Animal Rights: Political and Social Change in Britain since 1800*. London: Reaktion Books.

Kleivan, Kåre. 1956. "Isbjørn går som i saueflokker i dalene på Kong Karls Land." VG. https://urn.nb.no/URN:NBN:no-nb_digavis_verdensgang_null_null_19560901_12_203_1.

Kruse, Frigga. 2017. "Catching Up: The State and Potential of Historical Catch Data from Svalbard in the European Arctic." *Polar Record* 53 (5): 520–33. https://doi.org/10.1017/S0032247417000481.

Larsen, Thor & Ian Stirling. 2009. *The Agreement on the Conservation of Polar Bears: Its History and Future*. https://brage.npolar.no/npolar-xmlui/handle/11250/173342.

Magasinet for alle: Arbeidermagasinet og For alle 29 (6) 1956. Oslo. https://urn.nb.no/URN:NBN:no-nb_digitidsskrift_2021010783227_001.

Rothfels, Nigel. 2002. *Savages and Beasts: The Birth of the Modern Zoo. Animals, History, Culture*. Baltimore: Johns Hopkins University Press.

Schilbred, C. S. 1959. *Dyrene og vi: utgitt av Foreningen til dyrenes beskyttelse til 100-års jubileet 17. oktober 1959*. Oslo: Foreningen til dyrenes beskyttelse.

Snæbjörnsdóttir, Bryndís & Mark Wilson. 2006. *nanoq: flat out and bluesome, A Cultural Life of Polar Bears*. London: Black Dog Publishing.

Statistisk sentralbyrå. 2015. "Utførsel av levende isbjørn. Antall og verdi. 1871–1978." In *Statistisk årbok 2000*. Oslo: Statistic Norway. https://www.ssb.no/a/histstat/aarbok/tab-2000-09-06-02.html.

Søbye, Espen. 2000. "Isbjørnfangst 1871–1973: Fra fangst til fredning." *Fra forrige årtusen* 2. https://www.ssb.no/a/histstat/artikler/art-2000-09-06-01.html.

Thorsen, Liv Emma, Karen A. Rader & Adam Dodd (eds). 2013. *Animals on Display: The Creaturely in Museums, Zoos, and Natural History*, University Park: Penn State University Press.

OBJECT: Wax phonographic cylinder

FIGURE 13.1 Wax phonographic cylinders of folk songs that Percy Grainger collected. Grainger's self-designed museum contains many examples of folk songs, including those by English, Danish, Rarotongan, and Maori folk singers.
USED WITH PERMISSION FROM THE GRAINGER MUSEUM, UNIVERSITY OF MELBOURNE, AUSTRALIA

CHAPTER 13

Collecting Songs, Forging Identities: Percy Grainger and Pain as Catalyst for Narrating the North

Ryan R. Weber

Abstract

Few tasks occupied the imagination of composer and pianist Percy Grainger (1882–1961) as much as collecting folk songs. Beginning in 1906, he amassed a vast array of phonograph cylinders from across the globe. Yet while many scholars have examined the significance of this reservoir from different perspectives, there are two key aspects that have received relatively little critical analysis. The first entails the degree to which Grainger's early expeditions to record Danish folk songs were formative to his life-long obsession with Nordic culture. The second is the way Grainger's engagement with this repertoire became entangled with circulating discourses around pain. Therefore, in this chapter, I begin by surveying key debates surrounding biomedical models of pain before summarizing the deliberations over its shareability. I then discuss how Grainger's dangerous conceptualization of 'Nordic race' became the foundation for a lifelong fascination with reading meaning out of pain. Next, I apply an adaption of the "peak-and-end-rule" (Kahneman and Riis 2005) to three intersecting domains in Grainger's life and work: witnessing Pain, Social Pain, and Pain as Pleasure and Performance. Collectively, this chapter uncovers the composer's strategies for employing pain as a catalyst for narrating the North.

Keywords

Nordic – pain – suffering – Percy Grainger – folk song – eugenics – social isolation – aesthetics – narrative

∴

The Australian-born composer and pianist Percy Grainger (1882–1961) maintained a voracious appetite for collecting many objects throughout his life.

 | DOI:10.1163/9789004749665_014

Yet perhaps few tasks occupied his time – and his imagination – as much as his penchant for collecting folk songs. Beginning in 1906, he started gathering recordings of tunes using the phonograph. The device was important to Grainger for many reasons. It enabled him to preserve the nuances of a singer's performance and allowed him to capture the "narrative song" qualities that contributed to their public appeal. Thus, in his 1908 article in the *Journal of the Folk-Song Society*, Grainger surveyed the advantages and challenges of so-called "mechanical collecting", proclaiming, "It cannot be made too widely known that the phonograph puts valuable folk-song […] collecting within the reach of all possessed of the needful leisure and enthusiasm" (Grainger 1908, 149). Before long, he amassed a vast array of wax cylinders from across the globe, especially England and Denmark.

Yet while many scholars have examined the significance of this reservoir from different perspectives, including its contribution to the preservation of British culture, there are two significant aspects that have received much less critical attention. The first entails the degree to which Grainger's early expeditions to record Danish folk songs were formative in creating a life-long obsession with Nordic culture – a category that was as malleable as it was pervasive in his thinking. The second is the degree to which folk song collecting generally and Grainger's engagement with this repertoire specifically contributed to circulating discourses around pain. Further afield, some scholars have looked beyond the biological mechanisms of pain to better understand pain's meanings as a way of both palliating its effects and comprehending its role in shaping identity. For instance, Stephen Buetow has demonstrated how pain offers one the opportunity to "obtain a new vantage point to create trajectories of meaning in the person-environment" (Buetow 2021, 59). His novel approach encourages us to reconsider the nature of pain by moving from a paradigm that conceives of it as a sensation centrally located within the physical body to one that regards it as a complex set of relational processes. Accordingly, Joanna Burke has argued that pain is not a "thing" that one "has", but an "event" that is "infinitely shareable" (Bourke 2014, 234). But when it comes to the precise role folklore and folksongs have played in the process of sharing of pain, many questions remain unexamined. For instance, how can the collecting of folk songs serve as what Javier Moscoso has termed "a common space" that unites "physiological pain and its subjective expression" (Moscoso 2012, 8)? In what ways have these narratives been used to mediate social and somatic theories of suffering while privileging cultural hierarchies?

In this chapter, I take up these questions by revealing how pain served as a catalyst for narrating the North. I begin with a summary of key debates surrounding biomedical models of pain before discussing the deliberations concerning its purported shareability. In doing so, I expand my findings from an earlier study on Grainger's use of the environment to create social cohesion wherein I argued, "Despite many groundbreaking achievements in these areas, the shift in authority to biomedicine over the past century has not been accompanied by an equally extensive gain in our understanding of pain's many meanings and their relevance to one's quality of life. Moreover, the use of personal narrative and aesthetics to gain access to this knowledge remains underutilized" (Weber 2023). I then discuss how Grainger's problematic and often dangerous conceptualization of "Nordic race" became the foundation for a lifelong obsession of reading meaning out of pain to create inclusion on one level while ensuring exclusion on another. Next, I apply an adaption of the "peak-and-end-rule" (Kahneman & Riis 2005) to three intersecting domains in Grainger's life and work: witnessing Pain, Social Pain, and Pain as Pleasure and Performance. Collectively, this chapter will reveal the composer's strategies for employing pain as a mediator amidst the liminal spaces created by the interaction of material objects and immaterial narratives, soma and sensation, and self and society.

1 The Problem of Pain: Searching for Meaning in the Space between Two Selves

In her eponymous study on communicating pain, Stephanie de Montalk speaks to the limitations of biomedical approaches to quantifying pain, especially through the creation and circulation of pain scales. In testifying to the challenges in reporting her own chronic pain, de Montalk questioned, "Was my rating lower because I was more familiar, than some, with physical pain? Would it rise as time went on?" (de Montalk 2019, 38). In addition, de Montalk interrogates the supposed universality of these tools, writing, "As someone inclined to push on during illness and pain, I queried whether my behavioral Yellow Five equated with another's Yellow Six" (de Montalk 2009, 38). Her skepticism echoes that of many others who struggled to quantify their experiences of chronic pain, including Eula Biss, who candidly acknowledged, "Pain presents a unique problem in terms of measurement [...] Assigning a value to my own pain has never ceased to feel like a political act [...] This exercise itself is painful" (Biss 2005, 28). Yet one of the problems that looms largest in de Montalk's narrative as well as the debates that occupy disciplines ranging from neuroscience to comparative literature pertains to the perceived shareability of pain. As I noted at the onset, Joanna Bourke has argued for one of the most expansive views. Drawing on the metaphor of pain as a type of event, she argues, "Pain is

[…] inherently social. There is no such thing as a private pain-event […] From the moment of birth, infants are initiated into cultures of pain" (Bourke 2014, 17). By contrast, de Montalk offers a more limited perspective, confessing, "As I have found, it is impossible to share fully the sensory and emotional impact of chronic pain with anyone but a fellow sufferer" (de Montalk 2019, 4).

In this chapter, I do not wish to draw borders around the intractability of pain's most intimate exchanges. Nor do I wish to rupture the boundaries that de Montalk and others have suggested. Rather, I want to dwell in the paradoxes afforded by the presence of both limiting and liberating perspectives, which are united by the mutual conviction of the possibilities that arise from the sharing of pain. The discursive space created by these tensions encourages us to shift our focus away from futile attempts to certify the universal accuracy of pain scales and instead find redemption in the universal applicability of individual and collective narratives. For instance, in their recent article, Paley et al. compare contemporary understandings of pain to medieval attitudes. Based upon their findings, they argue that "sharing personal stories of life is a fundamental human attribute to foster social cohesion, and that nowadays sharing personal stories about pain is difficult during biomedically-focussed [sic] time-constrained clinical consultations" (Paley, Johnson & Paley 2023, 1).

It is in that same spirit of creating transdisciplinary approaches that I will demonstrate how the collection of Nordic objects can be significant to the dual processes of mediating pain and forming social cohesion. In keeping with Grainger's perspective, I employ the concept of 'Nordic' in the broadest terms to reflect his unstable category of identity. Similarly, I use the term 'object' to denote both his fascination with collecting material items, ranging from phonographic cylinders to clothing, and immaterial items, such as narratives that spring from stories and songs. My methodology hinges upon the adaptation and application of the so-called "peak-and-end-rule", or PE-rule, to Grainger's career (Kahneman & Riis 2005). More recently, Strijbosch et al. have described this theory as one "which postulates that how we remember our experiences is determined by the emotions associated with the most intense moment of an experience [the peak] and the emotions associated with the end of an experience [the end]" (Strijbosch et al. 2019, 2). In short, this principle suggests that the most intense and the concluding moments are often granted greater significance than the duration of an experience. In what follows, I argue that it can be a useful model for understanding how the significance of objects becomes encoded in broader narratives due not to the predictive value but the relational value of these experiences. The PE rule also helps to shed light on a dynamic form of affective entanglement that brings together the object(ive) and subject(ive), human and nonhuman, self and Other – thereby rendering the boundary between these dualistic categories fuzzy and porous and, at times, even indistinguishable.

My methodology is similar to that described by Atul Gawande in his book *Being Mortal*. Citing the work of Kahneman, he explains, "People seem to have two different selves – an experiencing self who endures every moment equally and a remembering self who gives almost all the weight of judgement afterward to two single points in time, the worst moment and the last one" (Gawande 2014, 237). These two selves can also yield significant contradictions, Gawande contends, because "the remembering self and the experiencing self can come to radically different opinions about the same experience" (Gawande 2014, 238). He then concludes, "In the end, people don't view their life as merely the average of all its moments [...] For human beings, life is meaningful because it is a story [...] Measurements of people's minute-by-minute levels of pleasure and pain miss the fundamental aspect of human existence" (Gawande 2014, 238). Thus, in what follows, I will illustrate the ways in which Grainger's penchant for collecting Nordic objects opened up a discursive space between his "two different selves", which, in turn, functioned as a catalyst for formulating narratives of the North. As we will see, the act of collecting stretched across many domains of his life and satisfied the call of his "remembering self" while the value he placed on pain catered to his "experiencing self". Together, the interaction of these selves made possible the creation of stories that enhanced the shareability of pain while paradoxically engaging in the process of, in Biss's words, the political act of quantifying pain. For this reason, my analysis is also meant to serve as a cautionary tale of the dangers inherent in seeking out meaning for pain at all costs, particularly by moderating its isolating effects with exclusive hierarchies of belonging.

2 Grainger and Nordic Folklore: Tracing the Path from Interest to Obsession

Percy Grainger was born in Australia in 1882. After showing an early interest in music, he took up formal instruction in music at the age of ten. It was in that same year that another passion began to materialize: his burgeoning fascination with Nordic culture. At this early stage in his life, Grainger envisioned Nordic identity as something shared by an amalgam of countries extending from Greenland to Sweden around which he traced fuzzy borders of belonging. Yet for the young Percy, it was the Icelandic sagas that captured much of his early imagination; so much so that he taught himself Icelandic to read them in their original language. His other ventures into Nordic culture included a tour of Denmark with Danish cellist Herman Sandby in 1904 followed by a meeting with Edvard Grieg in Norway two years later. Their short-lived friendship (Grieg died in 1907) would serve as inspiration for Grainger throughout his life, for he would go on to write articles on Grieg through midcentury (Gillies & Ross 1999, 318–37). Concurrently, beginning in 1905, Grainger started to collect English folksongs. According to Graham Freeman, "Grainger brought a philosophy of music and subjectivity that is sympathetic to late twentieth-century post-structuralist thought to a discourse operating with a methodology derived from antiquarianism and nineteenth-century nationalism, and a conflict between them was therefore inevitable" (Freeman 2011, 435). Despite the ambiguity of his aims, Grainger embraced the tension between these two philosophical impulses as a conduit for expanding his horizons. By the close of the first decade of the twentieth century, he had assembled a repertoire of folksongs from Polynesian and Maori cultures – a process that set the stage for a lifelong interest in procuring folksongs from across the globe. Ironically, it was his relocation to America at the outbreak of war in 1914 that coincided with a return to his self-proclaimed Nordic roots, which he explored alongside a robust composing and performance agenda.

However, by the time the Great War had dissipated, Grainger's interest in collecting Nordic objects merged with his burgeoning interest in racial identity leading him to redraw the boundaries of Nordic identity once again. Consequently, he conflated his love of Nordic culture with the more insidious narratives of eugenics in a series of essays and lectures, including "The Value of Icelandic to an Anglo Saxon" in 1920/1, "Nordic Characteristics in Music" soon after in 1921, and later "Characteristics of Nordic Music" in 1933 (Gillies & Ross 1999, 121–42, 258–66). It is within these documents that one can observe his initial interest spiral into a racist obsession. In these essays, Grainger credits American eugenicists Madison Grant (1865–1937) and Lothrop Stoddard (1883–1950) with helping him to "know the meaning of the term 'Nordic'" (Gillies & Ross 1999, 131). In contrast to his early conception, by the 1920s, his definition transformed into a blood-based paradigm making what was once mutable and inclusive based upon different "characteristics" now fixed and exclusionary based upon an erroneous biological construction. Furthermore, Grainger's radicalization is manifest in the ways he chastised Nordic musicians for their "racial unconsciousness", which he coupled with frequent calls to action:

> Nordic musicians should realise that they have two separate tasks to perform, for their own and the world of music's good. (1) To collect, study, elaborate and preserve and adequately present their own specifically Nordic characteristics in music […] (2) To realise that, in so far as many

> of them are now city-dwellers and civilized folk ... they should take their share in the duties and work and expense of the cosmopolitan world of music, above all supporting consciously and continuously [...] their own musical geniuses. (Gillies & Ross 1999, 138)

But by the close of the decade, his project of merely collecting Nordic folksongs turned even more sinister as his racial interests began to fuel an urge to not just own things from the North but to *become* Nordic. In 1927, his self-proclaimed "Nordic revolt against civilization" had reached such a peak that he stripped Greek and Latin influences from his language and developed a Nordic-English dialect (Ross 1986).

The intersection of cosmopolitanism and race in his writing reveals a split in his own thinking between inclusive, horizontal notions of democracy and exclusive, vertical hierarchies of race. Numerous scholars have investigated this contradiction, so it requires no further unpacking here (Harris 2000; Gillies, Pear & Carroll 2006; Kirby 2017; Weber 2021). Instead, I want to add another layer to the existing analyses by exposing the various functions of pain narratives that reside at the core of this schism. As Sarah Collins and Simon Perry have argued, pain brings his art and identity into closer proximity through a belief in "self-transformation through suffering" because "like other British moderns, Grainger sought improvement through heroic suffering, an uncompromising *askesis* or rupture that was meant to generate radical transformation" (Collins & Perry 2015, 32). For these reasons, I will trace the presence of Grainger's remembering self and his experiencing self in three domains of his life and work. By applying the PE-rule to evidence from his written records, I argue that these selves functioned as the framework over which Grainger hung his complex tapestry of a conflicted Nordic identity. Thus, I turn now to the common point of origin for each of these "lines of vision" (de Montalk 2019, 13), which was also the source of his earliest exposure to Nordic culture: his relationship with his mother.

3 Witnessing Pain: Rose Grainger's Suicide and the Redemptive Powers of Collecting

Grainger frequently proclaimed his mother, Rose, to be "the strongest factor of my artistic life" (Gillies & Pear 1994, 26) and credited her with influencing his love of the North in several ways. First, she imparted an interest in Hans Christian Andersen, Henrik Ibsen, Bjørnstjerne Bjørnson, and other Scandinavian writers. Second, she shared a love of Nordic languages, particularly Danish.

Third, she conveyed an interest in Nordic music, especially folksongs. And fourth, Rose shared a penchant for collecting "old Scandinavian embroideries, 'aaklaer', wood carvings and other relics" (Gillies, Pear & Carroll 2006, 27–32).

Above all, the sharing of pain served as the most enduring source of their bond. Rose fell ill on numerous occasions, leaving Grainger to care for her. In a letter dated 30 April 1922, he wrote: "Let us be full of faith & love & unity, that is the cure for all [...] I feel *sure* I can help you to get well *much much* quicker when I am with you [...] All these problems will be smooth when you are stronger again" (Gillies & Pear 1994, 54). Unfortunately, Rose Grainger never received this letter. She lost her life to suicide before it ever reached her. Grainger's experiencing self was captured in the subsequent letters he sent in the immediate aftermath of her death. For example, in his letter to Balfour Gardiner just a few days later, he recorded:

> My poor darling mother died last Sunday, fell, or threw herself, from an 18 story window in New York City ... She had long been worried about losing her mind, as a result of her long nervous illness ... I find it hard to bear the strain. My heart & head alarm me & I wonder if I can live thru it all. (Gillies & Pear 1994, 55)

It was not long after, however, that the tensions between his experiencing self and his remembering self began to emerge. For Grainger, reframing events in his life was central to constructing a narrative that could be broad enough to be accepted by wider audiences and resilient enough to be preserved for generations. It is in this respect that his narratives of pain and suffering and his conception of Nordic identity frequently intersected. Reflecting on his memories of Rose in 1947, Grainer authored an essay in two languages, stating,

> The parts written in English give an account of those events & points of view that are not likely to give any trouble to the somewhat childish & undeveloped (irresponsible) mentality of English-speaking readers. Passages in Danish require the more grown-up attitude of Scandinavian readers [...] Therefore, the more tragic & puzzling details of my mother's & my together-life are addressed to him, while those details presenting no problems to a timid & evasive mentality are recorded for English-speaking readers. (Gillies, Pear and Carroll 2006, 43)

In this instance, Grainger extends these two selves outward and maps them onto different languages, which are symbolic of his inner conflicts. Thus, in passages like this, we are given a glimpse into the many places in which he demarcates Nordic cultures from other groups based upon their ability to endure suffering. In fact, he often conflated images of his mother with his favorite figure from the sagas, Grettir the Strong (Gillies, Pear & Carroll 2006, 173).

How, then, could he reconcile the sick and vulnerable Rose with the hardy and stoic version? While his letters contain no shortage of attempts to merge these identities, his most extensive effects were concentrated in his collecting activities that began after her death. One branch of these undertakings entailed a return to collecting Nordic folksongs. According to Bruce Clunies Ross, "Between 1922 and 1928 Grainger made three collection expeditions to Jutland with the Danish folklorist Evald Tang Kristensen, where they recorded and notated over 170 Jutish folk songs" (Ross 1986, 1). Uncoincidentally, it was during the same period that he set to work on what would become the central project of the rest of his career: the construction of a museum dedicated to his life and work. But Grainger ironically eschewed any place in the North when it came to his choice of location and instead favored his country of birth.[1] He eventually explained his reasons for this decision in his essay "Music is a universal language", writing,

> While studying music at the Hoch Conservatorium in Frankfurt from 1895 to 1899 I was struck by the fact that the most gifted composition students were all from the English-speaking and Scandinavian countries [...] So I began to collect manuscripts, musical sketches, letters, articles, mementos, portraits, photographs, etc., by and of those English-speaking Scandinavian composers that seemed to me the most gifted and progressive – always with the intention of some day putting this collection on permanent display in Melbourne. (Grainger 1955)

Belinda Nemec has noted, "The suicide in 1922 of his mother, Rose, served as the catalyst to Grainger's museum project" (Nemec 2007). That Grainger would turn to collecting during a period of upheaval and distress was unsurprising for his parents were both collectors and "he was also concerned with preservation of his material and musical heritage at times of crisis such as war" (Nemec 2007).

Her extensive analysis of Grainger's fascination with collecting helps to shed light on the many roles that these activities played in reconciling the layers of his identity (Nemec 2007). In so doing, Nemec provides further evidence

1 The museum is currently housed at the University of Melbourne. For more information, visit https://grainger.unimelb.edu.au/.

to support Moscoso's thesis that "people construct objects as much as objects construct people" (Moscoso 2012, 147). One paradoxical way they do so is by both severing and creating continuity: "Rose's death divided Grainger's life into two parts. This is reflected in his later labelling of certain items of clothing and accessories in his collection 'Mors Tid', meaning 'Mother's time' in Danish. But he tried to establish continuity between these two periods by carrying on his childhood traditions such as creating a birthday gift for her" (Nemec 2007). Grainger internalized the divisions afforded by this pain event in his life as a source of new meaning making, which Nemec summarizes:

> Grainger's perception of continuity between past, present and future, his disinclination to rank old objects over new or new over old, and his enthusiasm for the creative manifestations of many different places and historical periods, are all reflected in his museum collection and the way he documented and displayed it. Grainger treasured many things which happened to be old, but not simply because they were old. He treasured them for their associations and their signification of the past for the present and the future. The emotions they excited in him were immediate and strong. (Nemec 2007)

But Grainger's (re)turn to folksong collecting and his decision to begin work on a personal museum during this period bears deeper resonances than memorialization alone. I argue that the convergence of these events is evidence of Grieg's employment of a highly selective type of memory – one that served as a strategic way for reading meaning out of pain. This significance emerges from the tensions between his experiencing self and his remembering self. Nemec has consequently postulated that "Grainger's sense of obligation towards his mother had long been tinged with guilt, perhaps triggered initially by sympathy for her physical suffering" (Nemec 2007). More generally, Bourke has similarly discussed the effects of witnessing pain, noting that through imagination and identification, observers of people-in-pain "tremble and shudder at the thought of what he feels" (Bourke 2014, 234). Her analysis substantiates how the experience of witnessing pain can function as a prism that refracts new intersubjective vistas for the observer leading to shared sensations, meanings, and identities. But with the passage of time, these identities often undergo a transformation. Nothing could be truer in the case of Grainger, whose remembrances carry with them the weight of an added agenda: Nordic supremacy. In other words, he did not just passively experience "peak" and "end" moments. He actively *created them* by superimposing Nordic culture onto the most intense aspects of his life.

The untimely death of Grainger's mother, combined with his search for redemption in the collecting activities that followed, reveals how, as David

Morris argued in his seminal volume, "pain is always personal and always cultural" (Morris 1991, 25). By etching the "peak" and "end" moments of his autobiographical accounts into the collections he assembled, both materially through the building of his museum and immaterially in the construction of pain narratives, Grainger demonstrated the myriad meanings he was able to co-construct for pain. With each fracture that emerged in his life, collecting provided a valuable source of cohesion that was not limited to his own suffering. Rather, he folded these experiences into his broader vision for society. It is in this respect that he drew closer to his cherished medieval sources than he may have even known. Returning to the observations of Paley et al., they note "In the Medieval Period, pain was a multifaceted shared social experience with several meanings, and people sought to alleviate pain using physical, spiritual, and social interventions. Sharing pain promotes social bonding, cooperative behaviour, camaraderie, and well-being" (Paley, Johnson & Paley 2023, 4). Applied to Grainger, his yearning for social cohesion amid pain transformed collecting into a type of analgesic. De Montalk speaks to this same property in her own illness narrative when she referred to Spider Robinson's *Callahan's Crosstime Saloon* (1977), musing, "Joy shared is joy increased, and pain shared in pain lessened" (de Montalk 2019, 129). By fostering interconnectedness between himself and his mother, between their relationships and Nordic culture, and between Nordic culture and the global community, he hoped to mediate suffering by celebrating the "oneness & harmonious togetherness of all human souls" (Gillies & Ross 1999, 217). Indeed, contemporary scientific research continues to support the notion that the sharing of pain can serve as a type of "social glue" (Bastian, Jetten & Ferris 2014).

However, the sinew between self and society was not always a source of relief for Grainger. On the contrary, one of the most glaring paradoxes of his career emerged precisely in this nexus, or, in Grainger's estimation, its fracture. As I noted earlier, the tension between his quest to earn universal recognition through shared suffering and his agenda of making Nordic culture universally superior was always a source of frustration. It is in this respect that another incongruity transpired: if the sharing of pain can bring people closer together, then it can also potentially drive them apart.

4 Social Pain: Grainger's Rejection and the Plague of Modern Life

Just as the task of measuring pain has provided obstacles to researchers, clinicians, and patients alike, so too has the problem of locating pain. By the dawn of the twentieth century, many researchers formed a consensus that the causes of pain could be found exclusively within the individual. This idea was later ratified in 1979 with the release of the International Association for the Study of Pain's definition (a description that still dominates biomedical approaches): "an unpleasant sensory and emotional experience associated with actual or potential tissue damage, or described in terms of such damage" (Cervero 2012, 16). But what if, instead of emanating from *inside* bodies, the source of pain transpired *between* them?

In many instances, Grainger flips the centuries-old understanding of the causal mechanism of pain inside out when he insinuates that it is not tissue damage but the destruction of social bonds that can lead to some of the greatest forms of pain – a concept that scientific researchers have more recently started to pursue (Eisenberger 2012, 421–22). For him, the pain of isolation afforded by modernity could prove to be one of the most enduring and unbearable forms of suffering. Returning to the analysis of Collins and Perry, they explain: "Grainger repeatedly and explicitly decried 'civilization', and this attitude forms a constant in several aspects of his life-writing" (Collins & Perry 2015, 29). As evidence, they cite his extended essay of 1953 "My Wretched-Tone Life", which was the draft of his planned autobiography. In it, Grainger attributed a host of factors, including race, class, and his compositional procedures, to his perceived lack of public recognition. He accordingly attributed the reasons for the mismatch to the tension between the sophistication of his artistic goals and the inferiority of his audiences. For instance, he asserted, "At any given moment there is always one race (or group of races) that is clever, while others are stupid. It is quite clear that the English-speaking countries & those other countries that share their general outlook on life (Scandinavia, Holland) are the clever ones in our time" (Gillies, Pear & Carroll 2006, 170). Extending from this sociological examination, Grainger then established a direct line to his own racial identity, which, by midcentury, had morphed into a messy hybrid of 'Anglo-Saxon' and 'Nordic'. He thus complained, "A Nordic musician can earn a living only if he becomes a slave to the music of the inferior (the 'stupid') nations [...] A proper Australian musician feels about playing 'enemy' music as an Australian feels about the immigration of the colored races or the inferior European races into Australia" (Gillies, Pear & Carroll 2006, 171–72).

These narratives served as the backdrop for rebranding his pain as heroic suffering. Recalling his early fascination for the Icelandic sagas, Grainger then moved on to note that "Whether it was Hector, or Beowulf, or Grettir-the-Strong, the heroes all were self-sacrificial. I see no good in self-sacrifice today, tho I still see infinite good in heroism. For whatever reason, my music early settled down to being a celebration of heroic self-sacrifice" (Gillies, Pear & Carroll 2006, 173). It is difficult to accept his claim given the fact that the remainder of his essay amounts to an attempt to reframe his career by positioning himself as an artistic martyr. Indeed, it was his willingness to endure *his own* suffering that prompted Grainger to declare, "I think the entire musical world is entirely oblivious of the whole world of bitterness, resentment, iconoclasm & denunciation that lies behind my music. If they were aware of it I am not sure it would make any difference" (Gillies, Pear & Carroll 2006, 174). Then, in a final gasp of resignation, he concluded, "If my music ever makes its mark, it will do so because of the intense tragic feeling behind it, because of my insight into racial qualities, because of my compassionate sympathy with the tragedies of men, & my bitter resentment at the needless cruelties & injustices of life" (Gillies, Pear & Carroll 2006, 173–74). In a single sentence, Grainger conflates the significance of race and pain while blaming the public for their failure to recognize his sacrifices.

Given this context, we can better understand another function of Grainger's fascination with collecting. Faced with uncertainty over his career trajectory, he hoped to craft – and correct – his life narrative. Collecting thereby functioned as a type of coping mechanism for the pain of rejection. As physician Shirley Mueller has noted, "collectors seek pleasure when they collect" (Mueller 2016). And one of the sought-after benefits or psychological reinforcers for some collectors is social recognition. Thus, by (over)compensating for a lack of acknowledgment in his artistic life with a museum project that could potentially bring him the lasting attention he yearned for, Grainger aimed to reshape the surface of pain. In sum, it enabled him to find pleasure amidst his suffering by placing these experiences into counterpoint in the same manner that he juxtaposed items belonging to "English-speaking Scandinavian composers" with objects belonging to so-called "primitive cultures" from around the world in his museum. Yet while he often claimed that music was a universal language, his ultimate goal eschewed equitable participation for all. Rather, the combination of pain/pleasure and Nordic/non-Nordic cultures was often employed as a strategy for accentuating the differences between these categories – all with the purpose of advancing a race-based agenda of self-promotion.

In many ways, his obsession with collecting can be viewed as an external manifestation of the internal tensions he experienced surrounding his career path and his 'racial' identity. Accordingly, his claims of rejection are further evidence of the conflict between his experiencing self and his remembering self. Much of his pain thus differed from the expressions of de Montalk in that it was not the isolating effects of pain sensations, but the pain derived from social isolation that disturbed Grainger the most. This led to one of the central conflicts in his life: the struggle to find a meaning for this disconnection between self and Other that prompted his remembering self to advocate for heroism over weakness and a 'clever' racial identity over other 'stupid' forms.

It is also the remembering self that seemingly went into systemic overdrive by attempting to ascribe meaning to his discomforts whenever and wherever they emerged, often clinging to flawed, illogical, and racist ideologies to offer significance to these pain experiences. In a previous study, I argued that "some of the most deleterious effects of pain result not just from physical sensations alone, but from the experience of *pain without meaning*" (Weber 2023). In this instance, however, I propose that Grainger offers another equally true lesson. It is the paradoxical notion that the search for meaning at all costs carries with it the same deleterious effects as pain without meaning.

The case of Grainger is also a cautionary reminder that, underneath the narratives constructed by the remembering self reside the sensations of the experiencing self. Did Grainger not realize, then, that his penchant for collecting folk music, which he once claimed as a source of cohesion, also became a self-imposed divider between himself and others? We cannot know the precise answer to this, of course. But what can be ascertained is that, given this impasse, Grainger perceived that he was left with only one option: to lean into another form of suffering. It is one that would again bring the somatic and social forms of pain, as well as the categories of pain and pleasure, into closer proximity.

5 Pain as Pleasure & Performance: Grainger's Embodied Aesthetics of Pain

Throughout his career, Grainger exhibited an affinity for writing his autobiography into his musical language. In many ways, then, his eclectic compositional style can be viewed as an analogue for his personal ideologies. Much of this relationship was made possible by the intersection of the late romantic tradition, which tied pain to aesthetic greatness, and burgeoning strands of modernism, which valorized the pain of isolation. The convergence of these styles led to a series of juxtapositions in his works held together by a firm belief in the aesthetic resonance of his pain experiences. Peter Tregear has written extensively on this aspect of Grainger's career, noting his scientific precision in collecting and recording folk song belongs to a category of modernist approaches to folksong that could be a "foil against which an urban elite can define and test itself" (Tregear 2015, 95). Tregear further explains, "the undeniable emotive power of Grainger's music arises because its ironic frame suggests an implicit recognition of the modern self's estrangement from itself, or more particularly from the social context of the folk repertoire he exalted" (Tregear

2015, 103). He later referred to this overriding aspect of Grainger's oeuvre as the "art of agony", or music that "offers a challenging and occasionally compelling perspective on what he recognized as fissures in the modern psyche (his own not least) and what it might take to continue to live an undamaged – as well as undamaging – life in the modern world" (Tregear 2019, 41). Tregear's research has thus offered invaluable insights into how, for Grainger, musical style, folk song collecting, and personal experiences went hand-in-hand in the service of promoting a morally superior view of his personal preferences (Tregear 2022). It also reveals the ways in which his experiencing self played a pivotal role in the development of what has more recently been described as "embodied aesthetics". In their study, Ticini et al. have critically explored the ways in which "aesthetic experiences are grounded in the embodied simulation of the actions, emotions, and corporeal sensations represented in artworks" (Ticini, Urgesi & Calvo-Merino 2014, 103).

Grainger evoked such corporeal sensations on two levels. The first pertained to the demands he placed on his audiences. For instance, in the same essay on his "Wretched Tone Life", he wrote, "The worth of my music will never be guessed, or its value to mankind felt, until the approach to my music is consciously undertaken as a 'pilgrimage to sorrow'" (Gillies, Pear & Carroll 2006, 177). The second level applied to the burdens he placed on himself when it came to enduring forms of self-inflicted pain, or flagellation. He offered a vivid description of his sexual fantasies on various occasions, including his note in 1956, stating, "I attach enormous importance to flagellantism. I feel that flagellantism [...] is a means of turning hostile, harsh & destructive elements in man into harmless channels. Much of civilization consists of turning hostility into playfulness" (Gillies, Pear & Carroll 2006, 161). Several years earlier in his "Notes on Whip-Lust" of 1948, he offered a more comprehensive account in his fabricated Nordic-English, declaring, "I cannot give a true picture of my tone-art & of my art-life if I do not tell of the cruel-joy (sadism) that is one of the main stirs of my being" (Gillies, Pear & Carroll 2006, 166). After contemplating the root of these feelings in his early childhood, Grainger proceeded to formulate a nexus between his corporal sensations of pain, his various life experiences, and the development of his musical style, writing,

> If the whippings I had had were the only root of my cruel-fain-th, why did quite other forms of cruel-hood delight my inner eye? [...] Were not such stirs just an outcome of the vast wildness of my nature [...] the same wildness that fills my Hillsongs & English Dance? (Gillies, Pear & Carroll 2006, 167)

Nemec has also shed light on this relationship between Grainger's music and his fondness for self-inflicted pain, noting,

> Rose's death also triggered in Grainger an autobiographical urge which henceforth ran parallel to his collecting, resulting in a large quantity of memoirs, essays and anecdotes, mostly incomplete, and all unpublished in his lifetime. In these works he was seeking principally to understand his relationship with his mother and her family, to place himself and his musical circle in a position of primacy in the history of twentieth-century composition, and to explain his flagellantic sexual preferences. (Nemec 2007)

Thus, in his development of an embodied aesthetic, Grainger relied on forms of somatic pain to escape the overwhelming suffering caused by the social pain he experienced. It was, perhaps, but another way of trying to gain control over what he perceived to be an out-of-control world. In another nod to the importance of collecting Nordic sagas and folklore, he explained, "There is no doubt that the sad & tragic poems (North-English, Scottish, Faeroe Island, Scandinavian) I worship really affect me personally [...] everything in my art is based on violently sentimental emotionalism & must be received on that basis to get anything out of it" (Gillies, Pear & Carroll 2006, 178). Reading pain out of the sagas and into his music thus became the basis for a complex relationship between joy and suffering, which he credited with instilling as much bravery as "Grettir entering the ghost's grace to get the sword" (Gillies, Pear & Carroll 2006, 179). By placing himself on an altar of self-sacrifice, Grainger hoped to demonstrate that the experience of physical pain could be employed as part of a universalizing strategy to minimize social pain, making flagellation an act of reconciliation between the experiencing self and the remembering self.

6 Conclusion

In her analysis of illness narratives, Anna Leimumäki demonstrates that "Narrating can be understood as identity work" (Leimumäki 2012, 279). Her investigation speaks to the paradoxes of the temporal domain, noting, "Narrativising can create fragmentation and coherence at the same time" (Leimumäki 2012, 280). Similarly, in my survey of three intersecting domains in Grainger's life and work above, I have argued that the tensions between the "past I" and "present I" that Leimumäki outlines can be mapped onto the tensions afforded by what Gawande has referred to as the "remembering self" and "experiencing self". Taken together, their studies support the need to embrace uncertainty and to dwell in paradox if we are to better know pain as a sensation, an event, and a bridge between our own identities. For Grainger, these paradoxes surfaced in the discursive space between his two selves, which made possible an epistemology that expanded the temporal domain embedded in the contemporary PE-rule by emphasizing relationality. Thus, by fostering an awareness that "people construct objects as much as objects construct people" (Moscoso 2012), that narratives can sever and create continuity, and that pain can serve as a catalyst for social cohesion and isolation, we can minimize or even prevent inflicting further pain through the application of "objective" pain scales or by manufacturing meaning at any cost.

Above all, I have argued that Grainger's overlapping and contradictory pain narratives shed light on the role of collecting in expressing, mediating, and even alleviating pain. The process of collecting stories, songs, and objects was always a work in progress for Grainger. From his mother's earliest impressions to his final directive in the construction of his museum, Grainger reveals how collecting can function as a source of meaning making resulting from an affective entanglement, thus transforming it into an analgesic in the midst of pain, a sinew between self and society, and a conduit for connecting our past, present, and future selves. But just as acts of collecting can dismantle existing barriers by creating space for the establishment of new individual and collective identities, the case of Grainger also reveals that they can erect new ones by protecting biased, racist, and hierarchical ideologies.

Finally, peeling back the multivalent layers that resulted from Grainger's propensity to engage in collecting throughout all aspects of his journey through pain sheds light on one further connection: the relationship between the body as experienced and Nordic culture as remembered. Nordic culture was always as central as it was problematic in Grainger's imagination. It therefore represents what Leimumäki has referred to as a type of "focalisation", or "point of view from which the story-world is perceived" (Leimumäki 2012, 281). By adopting a Nordic focalisation, he was able to broaden or narrow the temporal distance between the "past I" and "present I" and between various pain events and personal narratives he developed to mediate his own suffering. This process is similar to that which Moscoso traces in his study of "images adorning the walls of medieval churches" when he claims that "They do not represent the states of things, but states of mind. Their referent is not exactly reality, but rather experience. They do not show us history, but emotions" (Moscoso 2012, 11–12). Thus, in much the same way, Grainger's narratives do not draw fixed boundaries around the shape of suffering any more than pain scales accurately depict the depth and duration of chronic pain. Rather, they represent glimpses into the states of mind that transpire when pain emerges from its elusive chambers and casts shadows on the horizon of possibility.

Bibliography

Bastian, Brock, Jolanda Jetten & Laura J. Ferris. 2014. "Pain as Social Glue: Shared Pain Increases Cooperation." *Psychological Science* 25 (1): 2079–85. https://doi.org/10.1177/0956797614545886.

Biss, Eula. 2005. "The Pain Scale." *Harper's Magazine* (June).

Bourke, Joanna. 2014. *The Story of Pain: From Prayer to Painkillers*. Oxford: Oxford University Press.

Buetow, Stephen. 2021. *Rethinking Pain in Person-Centered Care*. London: Routledge.

Cervero, Fernando. 2012. *Understanding Pain*. Cambridge: The MIT Press.

Collins, Sarah & Simon Perry. 2015. "'The Beauty of Bravery': Alternative Modernisms, De- Historicizing Grainger." In *Grainger the Modernist*, edited by Suzanne Robinson & Kay Dreyfus, 17–32. London: Ashgate.

De Montalk, Stephanie. 2019. *Communicating Pain: Exploring Suffering through Language, Literature and Creative Writing*. New York: Routledge.

Eisenberger, Naomi. 2012. "The Pain of Social Disconnection: Examining the Shared Neural Underpinnings of Physical and Social Pain." In *Nature Reviews Neuroscience* 13: 421–34. https://doi.org/10.1038/nrn3231.

Freeman, Graham. 2011. "'It Wants All the Creases Ironing Out': Percy Grainger, The Folk Song Society, and the Ideology of the Archive." *Music & Letters* 92 (3, August): 410–36.

Gawande, Atul. 2014. *Being Mortal*. New York: Metropolitan Books.

Gillies, Malcolm & David Pear. 1994. *The All-Round Man: Selected Letters of Percy Grainger 1914–1961*. Oxford: Oxford University Press.

Gillies, Malcolm & Bruce Clunies Ross. 1999. *Grainger on Music*. Oxford: Oxford University Press.

Gillies, Malcolm, David Pear & Mark Carroll. 2006. *Self-Portrait of Percy Grainger*. Oxford: Oxford University Press.

Grainger, Percy. 1908. "Collecting with the Phonograph." *Journal of the Folk-Song Society* 3 (12, May): 147–62.

Grainger, Percy. 1955. "Music is a Universal Language." Accessed 23 February 2024. https://grainger.unimelb.edu.au/discover/aims-of-the-grainger-museum.

Harris, Amanda. 2000. "The Nature of Nordicism: Grainger's 'Blue-Gold-Rosy-Race' and his Music." *Musicology Australia* 23 (1): 19–48.

Kahneman, Daniel & Jason Riis. 2005. "Living and Thinking About It: Two Perspectives." In *The Science of Well-Being*, edited by Felicia Huppert, Nick Baylis & Barry Keverne, 285–304. New York: Oxford University Press.

Kirby, Sarah. 2017. "Cosmopolitanism and Race in Percy Grainger's American 'Delius Campaign'." *Current Musicology* 101 (September). https://doi.org/10.7916/cm.v0i101.5353.

Leimumäki, Anna. 2012. "What Does Falling Ill Mean? Illness Narratives as Elucidation of Experience Expertise." In *Knowledge and Pain: Probing the Boundaries*, edited by

Esther Cohen, Leona Toker, Manuela Consonni & Otniel E. Dror, 259–92. Amsterdam: Rodopi.

Morris, David B. 1991. *The Culture of Pain*. Berkely: University of California Press.

Moscoso, Javier. 2012. *Pain: A Cultural History*. New York: Palgrave MacMillan.

Mueller, Shirley. 2016. "The Psychology of Collecting." *Medical Economics* (April). https://www.medicaleconomics.com/view/the-psychology-of-collecting.

Nemec, Belinda. 2007. "'I Am Hungry for Fame-After-Death': Percy Grainger's Quest For Immortality Through His Museum." *reCollections* 2 (2, September). https://recollections.nma.gov.au/issues/vol_2_no2/papers/i_am_hungry_for_fame_after_death#12.

Paley, Emma G., Mark I. Johnson & Carole A. Paley. 2023. "Understanding Pain in Modern Society: Insights from Attitudes to Pain in the Medieval Period." *Frontiers in Pain Research* 4 (May): 1–8. https://doi.org/10.3389/fpain.2023.1162569.

Ross, Bruce Clunies. 1986. "Percy Grainger's Nordic Revolt Against Civilization." *Musicology Australia* 9 (1): 53–65.

Strijbosch, Wim, Ondrej Mitas, Marnix van Gisbergen, Miruna Doicaru, John Gelissen & Marcel Bastiaansen. 2019. "From Experience to Memory: On the Robustness of the Peak-and-End-Rule for Complex, Heterogeneous Experiences." *Frontiers in Psychology* 10 (July): 1–12. https://doi.org/10.3389/fpsyg.2019.01705.

Ticini, Luca F., Cosimo Urgesi & Beatriz Calvo-Merino. 2015. "Embodied Aesthetics: Insight from Cognitive Neuroscience of Performing Arts." In *Aesthetics and the Embodied Mind: Beyond Art Theory and the Cartesian Mind-Body Dichotomy*, edited by Alfonsina Scarinzi, 103–15. Dordrecht: Springer. https://doi.org/10.1007/978-94-017-9379-7_7.

Tregear, Peter. 2015. "Giving Voice to 'the Painfulness of Human Life': Grainger's Folk Song Setting and Musical Irony." In *Grainger the Modernist*, edited by Suzanne Robinson & Kay Dreyfus, 93–105. London: Ashgate.

Tregear, Peter. 2019. "'The Art of Agony': Aspects of Negativity in Grainger's music." In *Nineteenth-Century Music Review* 16 (1): 27–41. https://doi.org/10.1017/S147940981700057X.

Tregear, Peter. 2022. "The Grainger Trap: At the Altar of the Curator's Moral Superiority." *Australian Book Review* (February). https://www.australianbookreview.com.au/arts-update/101-arts-update/8900-the-grainger-trap-at-the-altar-of-the-curator-s-moral-superiority-by-peter-tregear.

Weber, Ryan. 2021. "Dismantling Borders, Assembling Hierarchies: Percy Grainger and the Idea of Democracy." In *Finding Democracy in Music*, edited by Robert Adlington & Esteban Buch, 37–57. New York: Routledge.

Weber, Ryan. 2023. "Aestheticizing Pain: (Re)Claiming Identity through Music, Literature, and Landscape." *Re:visit. Humanities & Medicine in Dialogue* 2. https://journal-revisit.org/jr/issue/view/6.

OBJECT: Whale

FIGURE 14.1 The dolphin specimen in the Natural History Museum of Oslo collection
PHOTO: LENE LIEBE DELSETT. REPRODUCED WITH PERMISSION FROM THE NATURAL HISTORY MUSEUM, UNIVERSITY OF OSLO

CHAPTER 14

The Whales Are Leaving: a Dolphin Skeleton and Its History of Collection, Curation and Research

Lene Liebe Delsett

Abstract

Nordic natural history collections were assembled during the nineteenth century, but are receiving renewed attention today, because their objects form part of our common heritage, and might hold clues for a sustainable future. Visitors to the museums do not see most of these objects, as they are stored in the collections. In the Norwegian museum collections in Oslo and Bergen, whales are a unique set of such objects, because of Norway's whaling history, the evolutionary and ecological role of whales, and because they will not be collected again. After collection, some of these whales have contributed to knowledge and outreach. Whereas many specimens remain in these collections, others have left, and for many, the collection history is lost. This contribution uses one seemingly trivial dolphin specimen in the collection of the Natural History Museum in Oslo to highlight its collection history, and how it has contributed to scientific knowledge and outreach. This specimen also shows how physical objects are valuable for science, and how the story of this object still unfolds today and will continue to do so in the future.

Keywords

natural history museums – collections – whales – specimen-based research – whaling

Natural history museums hold a special place in society as educational, recreational and research institutions and, not least, as keepers of collections: more than one billion specimens held in special facilities (Hilton, Watkins-Colwell & Huber 2021; Johnson & Owens 2023; McLean et al. 2016). Some of these specimens are whales. Approximately eighty species of baleen and toothed whales exist in the world's oceans and in museum collections. It is likely that many of these species will never be collected again, due to the dramatic effect that whaling had on the populations: industrial whaling killed 2.9 million large whales before an international moratorium was negotiated in 1982 (Rocha, Clapham & Ivashchenko 2014). This means that most of the baleen whale species are still significantly affected and cannot be actively collected, and this is also true for the toothed whale species that are threatened with extinction.

Together, the two natural history museums in Bergen and Oslo house more than one thousand whale specimens. These collections are important not only because of the large number of specimens, but also because they are situated in Norway, which used to be a major whaling nation. Previous work has shown that most of these specimens entered the collections prior to 1900. Local whaling, research trends and the interests of scientists were important factors in building the collections. However, the immense scale of Norwegian involvement in industrial whaling in the Southern Ocean is not reflected (Delsett 2024; Kyllingstad & Rørvik 2011). The two collections contain whales mainly from Norwegian coasts and oceans as a result of these drivers and limitations. These objects have the potential to contribute to our understanding of the North, but only if they are taken care of and explored in the right context. This requires knowledge of the history of collection and research.

After their inclusion in collections, most of the animals are only seen by curators and scientists. However, whales have found their way out in several ways. Some are centrepieces in exhibitions, as whales are believed to encourage museum visitors to care about the environment. Other whales leave the basement collections as datasets used in scientific papers, talks or books. Many Norwegian specimens have travelled physically to museums all over the world. Lastly, many whales have left the collections not as celebrated symbols or interesting data points but in garbage bins and foul-smelling fires.

In this chapter, I discuss how whale specimens from the collections in Bergen and Oslo have contributed to the production of knowledge, and other ways that the whales have "left" the collections. I also explore the history of one particular specimen in order to understand why and how it was collected, how it has contributed to public and scientific knowledge and how it relates to specimens physically leaving a collection. The specimen is a small dolphin skeleton in the collection of the Natural History Museum of Oslo that also showcases the importance of recorded collection history and the curation of metadata. Such information is often missing and this hinders insights. The in-depth discussion of this particular specimen is an effort to write the often unknown collection history behind an object, thus highlighting the importance of physical objects for understanding the world around us.

1 Research on Museum Whales: Many Foetuses and a Few Skeletons

Museum specimens are important for whale research. For a long time, they were the only data source available for research. Even with the development of new field-based techniques, collections remain important because whales are notoriously hard to study in the wild. Museum specimens are a direct record of actual nature (Pyenson 2017b) and research on these has traditionally been carried out according to specimen-oriented disciplines such as systematics, taxonomy, morphology, and anatomy (Heyning 2002). In a Norwegian context, zoology professor Robert Collett's book on Norwegian mammals (Collett 1912) is a classic. Its chapters on whales focus largely on museum specimens, including their collection stories and shows that the actual, physical objects are valuable. Collections are thus an important way of getting scientific access to animals that live out of reach.

However, in order to build these collections, whales must be reached. This is possible in cases of stranded dead whales, but otherwise cooperation with the whaling industry is necessary. Which whale species that are available for museum collections have therefore changed over time, often in relation to changes in whaling methods. The most significant shift happened in the 1860s with the development of the explosive harpoon gun, of steam-powered whaleboats and more effective processing on land, attributed to the Norwegian Sven Foyn. In the twentieth century, large ships both hunting and processing the whales gave rise to highly efficient, industrial whaling (Tønnesen & Johnsen 1982) that made all whale species reachable. This influenced not only whale populations, but also science. For almost a century (1860s–1960s), Norway was the world's leading whaling nation, and the relationship between whaling and science has been discussed in previous work (Burnett 2012; Kalland 2014). The large number of whale specimens in the collections and observations made by whalers meant that Norwegian specimens and scientists played a role in cetacean research in the 1800s. A typical example is the study of unborn whales, and species that were previously unreachable (Burnett 2012; Collett 1911–1912; Kalland 2014).

Whales are typically held in collections in various ways:

- as complete or incomplete skeletons (osteological specimens), which can be mounted or kept as separate bones in boxes or on shelves in the dry collection;
- in liquid (formalin or ethanol) – either complete (often foetuses) or body parts such as a piece of skin in the wet collection;
- as dried and/or stuffed soft tissue parts, such as a fin.

A specimen can also be a DNA or tissue sample, and sometimes photographs or other documentation. Museums also often have models, casts and illustrations of whales.

The zoological collections in Bergen are famous for their complete whale skeleton exhibits. However, for science, it is the wet collection with its 295 foetuses that has probably left Norway's largest mark on international whale research. The Danish professor Daniel Frederik Eschricht (1798–1863) claimed that, by collecting whale foetuses, one gets the entire whale, in a miniaturized version (Eschricht 1844; Kalland 2014, 312). This has been an important factor behind this collection. Many such specimens were sent to Copenhagen to be studied by Eschricht, rather than by local scientists – especially specimens of minke whales. Some decades later, Gustav Guldberg, Fridtjof Nansen, and James A. Grieg did research in Bergen on baleen whale foetuses (Collett 1912; Kalland 2014, 316). This research aimed to determine life history parameters, such as when copulation and birth occur. The researchers were able to obtain robust results because of the large number of specimens (C. Kinze, pers. comm.) of mainly minke whales, but also of sei whales (Guldberg 1886).

Foetal development, reproduction and growth were also studied for several toothed whale species – orcas, porpoises and other species – based on specimens held in what is today the University Museum of Bergen (NHM Bergen), as well as some from the Natural History Museum in Oslo (NHM Oslo). This provided new knowledge (Collett 1912, 721; Guldberg 1900a, 1900b; Guldberg & Nansen 1894; Kalland 2014, 317–21). Guldberg and Nansen's paper from 1894 stands out and is still cited because of its documentation of foetal development. Many of the foetuses used for research were obtained from whaling generally, from the local hunt of toothed whales or from whaling around Spitsbergen. In some cases the collection history was published alongside the results (Collett 1911–1912, 667, 681; Grieg 1889; 1912).

However, for scientists concerned about human exploitation of nature, the knowledge provided by specimens from whaling also had a downside. In his encyclopaedia on Norwegian mammals (1912), Robert Collett wrote (my translation):

> In 1864 Foyn invented his grenade harpoon and could then, with luck, target the speedy and valuable fin whales, which had previously been overlooked. Through this invention, our knowledge about whales and their biology has increased more in a few years than over many previous centuries combined. It became apparent that species that were, until now, mostly known only through occasional (mostly stranded) specimens, such as the blue whale (*Balaenoptera musculus*) and the sei whale (*B. borealis*), occurred in the Atlantic Ocean in large groups. However, the invention was disastrous. The Foyn grenade harpoon still brings death through all the world's oceans, with more power than ever. Still, and for some time now, large fortunes pour out from the sea, but the enormous

> toll on whales – a group with a relatively limited reproduction capacity – will undoubtedly reduce populations so much that intensive whaling will stop by itself, and after that circumstances will determine whether these will regain their power to increase. (Collett 1912, 557)

It is worth noting that this was written at a time when Norwegian whaling had reduced the whale populations along the Northern Norwegian and Greenlandic coasts and in the Barents Sea to such an extent that whaling was prohibited in 1904, before the extensive whaling in the Southern Ocean started, which had its largest catches in the years after WWII.

2 A Whale Specimen Enters the Museum

Robert Collett was one of the main drivers behind the expansion of the zoological collection at the Natural History Museum in Oslo. He was working in the period with the largest increase in whale specimens, which, amongst others included most of the foetuses in the wet collection (Delsett 2024). However, the collection efforts at the museum started a long time before Collett's work, and included not only the wet specimens, but also dry ones – the skeletons. One such object was collected in 1842 and serves to highlight many aspects of collecting efforts and research history.

The object is still part of the NHM Oslo collection and is a small whale in the dolphin family (NHMO-DMA-31220/1-0, hereafter called the 'Drøbak dolphin') (Figure 14.1). This is not a wet collection specimen, like the foetuses discussed earlier, but an osteological (skeletal) one, and it is adult. Nor is it a large and spectacular baleen whale that usually lives far away. Instead, this dolphin skeleton serves to highlight that there are other important stories about knowledge production.

A whale is a vertebrate, just like us, which means that it has an inner skeleton. It consists of a skull (jaws, braincase); a vertebral column from the back of the skull and through the tail (vertebrae and ribs); a shoulder (pectoral) girdle connected to the forelimbs (evolutionary modified into flippers); and lastly an innominate, a small bone consisting of the fused pelvic girdle (hip) and long-lost hind fins (Gatesy et al. 2013). The ancestors of whales were land-living animals, and the body of today's whales have gone through heavy modifications to enable them to live fully submerged. The series of fossils from quadrupedal ancestors to fully aquatic (Gatesy et al. 2013; Pyenson 2017a) have made whales a "poster child for macroevolution" (Thewissen & Bajpai 2001). However, there are still many gaps in our knowledge about their biology (Cozzi, Huggenberger & Oelschläger 2017; Gatesy et al. 2013; Kelley & Pyenson 2015; Pyenson 2017a).

This museum specimen discussed here, the Drøbak dolphin, (Figure 14.1) is an almost complete skeleton, split into several parts for curatorial purposes. The skull and one vertebra are housed in a cardboard box. The vertebral column is mounted, articulated as in life, but on a stick and split into three: the frontmost portion with shoulder blades and fore fins attached; the middle part of the vertebral column; and lastly, the skeletal elements from the tail. For scientific purposes, skeletal elements contain valuable information. Hard parts (bones, teeth) are more frequently preserved for a longer time than soft parts (muscles, skin), which means that a skeleton stands a better chance of being preserved and thus collected, introducing a taphonomic bias. Even if the skeleton provides valuable information, it is worth noting that most of this individual is missing, compared to the living, breathing animal it once was.

The specimen has four different labels, all giving at least a species name. The labels provide three different species names for the specimen, which indicates that its taxonomical placement has changed several times. Species are the fundamental units in all biology, and, as such, are central to all possible research on the specimen. Two labels are old and hand-written on two or three separate occasions, providing the names Hvitskjæving and *Lagenorhynchus albirostris*. The third label is machine-printed, naming the specimen as *Delphinus leucopleurus*, whereas the fourth is hand-written with a marker pen, telling us that this is *Lagenorhynchus acutus*. In addition to the labels, the number 2645 is written directly on the skeleton with a marker pen – this is a museum number from the previous numbering system that is no longer in use.

None of this is unusual. Labels are critical information about specimens (Lane 1996). Even if new labels are made, the collections often keep the old ones as they provide valuable metadata and specimen history. As will be clear from this work, assigning a specimen to a species is not a trivial task, and subject to ongoing research, but the task becomes easier if the history of the specimen is preserved. Labels and other notes are also a form of communication between users of the collections. If I find a clearly misidentified specimen when visiting a collection, I usually leave a note of my opinion for the next person.

The word 'TYPE' on the oldest label along with a thick red line, and the word 'SYNTYPE' on the printed label, show that this specimen has a special role. These are internationally acknowledged terms for specific kinds of object: type specimens. Such specimens represent the physical framework of all species-centred biology, regulated through the International Code on Zoological Nomenclature (Ride et al. 1999). A type specimen is one designated collection specimen that represents the species and to which the scientific name is linked. The golden standard is a holotype, the one specimen on which the original scientific description for the species was based. In addition, several other type specimens can be raised if the situation requires, either to show variation within the species, or if the holotype is incomplete or missing. A syntype is a specimen mentioned in the original description, but not specifically stated to be a type. This commonly happens in cases where the description was written before the nomenclature code was adopted. Type specimens are thus important objects, and the dolphin skeleton is one out of 19 type specimens held in this collection (Wiig & Bachmann 2013).

One of the two old labels says "Drøbak 1842", providing geographical and date information. Drøbak is a small city south of Oslo near the Oslo fjord. The printed label says "November 30, 1854" and that the collector is anonymous, but also says "det. Halvor Heyerdahl Rasch", "det." denoting who assigned the specimen to a certain species. Often, collection history is missing from whale specimens, which means that the whale is preserved, but the story is gone, and with it, valuable context (Delsett 2024; Simon-Ekeland & Delsett 2024). However, for this specimen, much of its history is recorded (Collett 1911–1912; Wiig and Bachmann 2013). The main source on how it entered the collection is the original article (in Norwegian) by zoology professor Halvor Heyerdahl Rasch (1805–1883) (Rasch 1845), and the following summary is based on this paper (Figure 14.2).

In 1842, a local fisherman (no name given either in the publication or the museum database) in Drøbak, when scouting for mackerel, observed a group of small toothed whales in the bay outside his house. He and the other fishermen were used to porpoises, but these whales were larger, faster and had different coloration to any whale they had ever encountered (Rasch 1845). The man's reaction was to hunt the animals, as he had previously done with porpoises. Hunting of toothed whales has a long history in Norway. It has never been industrialized, but has rather built on local knowledge and been executed in an opportunistic manner (Kalland 2014). This is also what happened here. The fisherman already had equipment in place for the next mackerel catch, and together with his men, he closed off the bay. They were afraid that the powerful whales would destroy their equipment. Instead, the whales turned and did not seem willing to cross it. The men made the available space in the bay smaller and smaller, thus herding the whales. The men did not dare use the net for dragging the whales onto dry land, and attempts to manually roll them were unsuccessful because of their powerful movements. The men solved this by making a loop on a long rope and putting this around the tails of the animals. The whales were then dragged onto dry land by a horse.

Once on land, most of the whales were killed by putting "tightly packed" moss in their breathing hole (Rasch 1845). Just like us, whales breathe air with lungs, which is why it was necessary to actively block respiration to kill them. A few individuals managed to free themselves, but did not seem willing to abandon the others and so were also caught. The numerous bystanders reacted to the sounds that the animals were making with empathy and sadness (Rasch 1845).

FIGURE 14.2 Illustration from Rasch 1845. The original caption read (translated): The two lithographs ("steentryktavler") that accompany this contribution show [...] the animal from the side, from above and from below

It is interesting that Rasch noted this in the paper, perhaps emphasizing how similar these animals are to humans and how killing them often impacts people.

At first sight of the whales, Rasch was convinced that this was something interesting and a type of whale that he had never previously encountered. Together with professors Rathke and Boeck, he acquired 22 of the 23 captured specimens for the Natural Museum in Christiania (today's Natural History Museum in Oslo) (Rasch 1845). Workers were hired for several weeks to work alongside the technical staff of the museum for the necessary preparation. They collected the skeletons from all specimens, as well as ten skins and ten sets of internal organs, preserving the latter in ethanol for further research. Four specimens were female, one with milk, and 18 were male (Rasch 1845).

In this case, the whales were acquired by the museum and prepared for collections. However, specimens can be lost before they enter the museum, often for economic or logistical reasons (Simon-Ekeland & Delsett 2024). Collett mentions repeatedly that specimens are lost because they are destroyed for industrial purposes (e.g. sent to guano factories) before the museum can acquire them (Collett 1912, 688). Whereas in other cases, specimens were split between research and business. In one case several *Hyperoodon ampullatus* specimens were killed and stranded along the Romsdal coast, and, while most of these went to guano factories, a few specimens became part of the Bergen Museum collection (Collett 1911–1912, 640).

3 Knowledge from the Drøbak Dolphin

The acquisition and spread of knowledge based on collection objects are key activities for a natural history museum. This was the case for the dolphin specimens caught in 1842 and these two activities started early, during the process of transforming the animals into museum objects. The large scale of the preparation process led to substantial interest from the public and press in Christiania. Which species the dolphins belonged to, was discussed in the media, especially because Rasch believed strongly that they were something new to science (Rasch 1845). Due to the rare situation and the possibility of new scientific discoveries, the specimens were already contributing knowledge during the process of becoming part of a collection. Later, one of the specimens from this capture has been part of an exhibit for the public (Collett 1912).

Rasch published a paper (my translation): "Description of a new dolphin species caught in the Christiana fiord"[1] in the journal Nyt Magazin for naturvitenskaberne (Rasch 1845). The description and discussion of the new dolphin were aimed at the scientific community with the purpose of naming it. Rasch started his paper by referring to a talk given by Professor Eschricht in 1840 that had made a strong impression on him. Eschricht had stated that knowledge of whales, particularly of species in the Nordic areas, was substantially incomplete. He went on to urge anyone to speak up, both scientists and the public, if they had observations that could add to the knowledge of the "very dark natural history of these organisms".[2] This is in contrast to the important Norwegian marine biologist, Michael Sars, who stated in 1851 that "the vertebrates in the Norwegian oceans are well known, but that the invertebrates need research attention" (Sars 1851).

In the 1845 paper, Rasch gave a classical anatomical description of the animal, with comparisons to closely related taxa, in order to discuss whether or

1 Orig. title: "Beskrivelse over en i Christianiafjorden fanget nye Delphinart."

2 Orig. text: "Hr. Professor Eschricht [...] udviklede Aarsagen til vor endnu höist ufuldstændige Kundskab om de hvalartede Dyr [...] og opfordrede til samme Tid ikke alene Nordens Zoologer, men ogsaa enhver Anden, som fandt Leilighed til at gjöre Iagttagelser, der kunde bidrage til at oplyse det meget Dunkle i disse Skabningers Naturhistorie, til at meddele disse."

not the dolphins from Drøbak represented a new species. For instance, the cranium was seen as diagnostic to species level, and fifteen of the collected crania were compared to other dolphins. The outer morphology was described, along with measurements of body parts and skeletal elements taken from one male specimen. When describing the outer parts of the reproductive organs and nipples, one male and one female were compared.

In addition, the paper contains a reprint of parts of a manuscript from the late 1700s. According to Rasch, this manuscript was probably written by "Christie", later copied and kept by the zoologist Peter Stuwitz (1806–1842). Rasch found the manuscript when he examined his papers after Stuwitz's death while on duty (Nordvik & Fischer 1985; Rasch 1845). Rasch included this to give an overview of the distribution of the new dolphin species and to establish whether it was previously known to zoologists. The older manuscript discussed this dolphin's distribution along the Western Norwegian coast, where it had been known to the public for a long time. This also provided the traditionally used common name, *hvidskjæving* (still in use today), and stated that one such skeleton had been sent from Bergen to Professor Boeck in Christiania a few years previously.

After the catch in Drøbak, Professor Eschricht in Copenhagen was sent one of the specimens. As a response, he wrote to Rasch that he thought this was identical to the species named *Delphinus dubius* by Cuvier. However, Rasch did not agree and concluded that it was a new species, which he named *Delphinus leucopleurus* (Rasch 1845). In summary, the paper by Rasch was written as a well-founded argument that the dolphins that were caught in Drøbak belonged to a species not previously known to science.

Published descriptions of new species are important, as species are the fundamental units in the biological sciences, which also means that species' names are in fact hypotheses about evolution. Whales have often been given several common names, probably at least in part because there was limited scientific understanding of the morphology and distribution of whale species. This also highlights a major question for whale researchers: which species are present in a given area, for instance along the Norwegian coast, and are they the same as in other countries?

Whales and pinnipeds (seals and walruses) were indeed the topic of the first zoological text known from Norwegian literature, *Kongespeilet* from 1250 (Broch 1954). Kalland (2014) suggests that the first scientific measurements of whales in Norway can be found in an unpublished text from 1844, by W. F. K. Christie, the Bergen Museum director founder. This is a description of the long-finned pilot whale, which he suggested might be a new species (Kalland 2014, 307). Krøyer's discussion (1839) of species delimitations of fin whales might have been the first to use the collection in Bergen for scientific work. In 1883, Collett presented an updated list of Norwegian mammals, concluding that there are six whale species in Norway and seven additional ones that occasionally reproduce (Collett 1883).

Compared to the extensive work on whale foetuses from the NHM Bergen collection, Kalland (2014, 325) concluded that the whale skeletons are used in few research papers, even though the collections of harbour porpoise, Atlantic white-sided dolphin, long-finned pilot whales, Northern bottlenose whale, as well as the baleen whale skeletons, are unique in a European context. There are however some exceptions to this rule, and some are worth mentioning, for instance research that has been done on orcas and the white-beaked dolphin (Grieg 1889). With regard to the orca, there is a long-lasting and still ongoing debate about how many species of orcas exist, and early work on Norwegian collection specimens is summarized by Kalland (2014, 327).

Another example is a lower jaw of what is today called Sowerby's beaked whale (NHMO-DMA-31849/1-O) that was collected in 1860, and that still is in the osteological collection at NHM Oslo. This was the first occurrence of this species along the Norwegian coast. According to the file index card, the synonymy of the species name was intensely discussed and, at the request of the Norwegian Professor Boeck, the specimen was described and illustrated by the famous scientist van Beneden (van Beneden 1883). James A. Grieg later published papers on this species, based on specimens in NHM Bergen (Grieg 1898; 1905a; 1905b; 1908), and a cast was sent from Bergen Museum to NHM in London. There has also been Norwegian research on the fin and humpback whales by Professor Ossian Sars (Collett 1912, 572, 610). Other work on mapping and discovery of new species based on the museum specimens relates to the blue whale, the sei whale, the common bottlenose dolphin, and the sperm whale (Collett 1912, 589, 598, 695; Guldberg 1901).

The focus on anatomy and taxonomy, as well as on collecting whale specimens for the museums, has been reduced over the last hundred years (Delsett 2024), albeit with a few exceptions. One example is the intake of skeletons of the striped dolphin *Stenella coeruleoalba* by NHM Oslo and NHM Bergen during the last decades (Båtvik 1999; H. Meijer, pers. comm.). This species is currently becoming more common along the Norwegian coast due to warming waters and is thus important to complement the collections. Another example is the large number of porpoise specimens acquired by NHM Oslo for a research project in 2000 (Delsett 2024; Ø. Wiig, pers. comm.).

4 Specimens Leaving the Collection

Rasch's research focused on species description and geographical distribution in the Nordic countries, at a time when there was a large potential for discovery of new taxa. While working to understand whether the Drøbak dolphin was, in fact, a new species, he sent a drawing to a naturalist meeting in Stockholm to obtain information. However, drawings were not the only objects to be sent. Whale specimens from Norway, usually baleen whales, were often sent to or exchanged with other museums. Many specimens were sent to Eschricht in Copenhagen, but also to Leiden, to several museums in the Czech Republic and to Pisa in Italy (Braschi, Cagnolaro & Nicolosi 2007; Broekema 1983; Robovsky and Benda 2006). Sometimes this happened through direct contact between the museums, sometimes via dealers. These could also be economic transactions, whereby whale specimens were traded for animal specimens that the museums could not otherwise afford (Kalland 2014; Simon-Ekeland & Delsett 2024).

In the case of the Drøbak dolphins, drawings, but also physical specimens left the collection for scientific use. After preparation, specimens of the new dolphin species were distributed domestically and abroad, to the natural history museums in Copenhagen and St Petersburg amongst others (Collett 1911–1912; Kinze, pers. comm.; Rasch 1845), and some of the specimens can be found in the collections today. Naturalis in Leiden preserves two specimens, one of which is a mounted skeleton (RMNH.MAM.31209) and the other made up of disarticulated skeletal elements and mounted skin (RMNH.MAM.31211a and b) (Naturalis database). The Smithsonian National Museum of Natural History in Washington DC, houses one cranium received from NHM Oslo as an exchange (Flower and Garson 1884) (NMNH database).

In other cases, specimens do not leave for purposes of business, science or outreach, but as garbage. Rasch (Rasch 1845) remarks that the heat and the large number of specimens made the preparation of the dolphin group from Drøbak challenging, not least because of the smell. This is a common problem, as making and maintaining a whale collection has a lot to do with logistics, not just related to the whales' size. If whale specimens are not prepared and stored adequately, they degrade, producing the characteristic smell. Defatted skeletons are far easier and more pleasant to transport, mount, store and research. In summary, some situations make it necessary to discard specimens (Pyke & Ehrlich 2010).

This is probably what happened to the soft tissue samples (skin and inner organs) from the Drøbak dolphin that became part of the NHM Oslo collection. One specimen (NHMO-DMA-25156) is known to have been discarded (in 2018, database). It has not been possible to obtain information regarding the others. This is a common situation, as discarded specimens are often hard to track in museum records. Museum databases, and especially the physical books recording the acquisition of specimens, sometimes provide information on discarded specimens, but do not give a complete list. Not being able to locate a specimen does not necessarily mean it has left. It may also be stored somewhere else in the museum for practical reasons. However, it seems that, generally, soft tissue specimens are more often discarded, which is to be expected based on the greater possibility of decay.

In some cases, specimens referred to in the literature cannot be matched with any of the objects still in a collection, either because of missing labels, or because they might have been discarded. This is the case for several foetuses mentioned by Collett (Collett 1912, 676, 688, 714) and for two sperm whale skeletons. In 1908 and 1909, two complete, large sperm whale skeletons caught by Norwegian whalers in Iceland, were donated to the museum in Oslo. One was a donation by a whaler, Ellefsen, and the second was donated by a lawyer, Barth (Collett 1912, 632). These skeletons cannot be found in the Oslo collection today. They might have been intended for a long-planned whale museum as part of the Natural History Museum in Oslo, but that in the end was never built (Museum of University History 2012).

5 A Specimen Story Still Unfolding

In the case of the small dolphins from Drøbak, it is fortunate that NHM Oslo has not discarded the specimen discussed here, as the mammal diversity of the world is still not fully understood. New species are regularly discovered, and the taxonomies are revised (Reeder, Helgen & Wilson 2007). Rasch himself intended to do more research on the soft tissue and skeletons of these whales (Rasch 1845), but this seems not to have happened. The remaining knowledge gaps and new research questions that are continually being posed highlight the importance of keeping physical animal specimens as a part of collections.

The example of the small dolphin specimen turned out to showcase how the story of a museum object is never over. During my research on this specimen for this chapter, I discovered that it not only has an interesting collection history, but that today, it holds a more important position than previously believed.

The labels belonging to the specimen gave different species names, indicating that the opinion on what it is has changed over time. Rasch described it as a new species, *Delphinus leucopleurus*, for which the specimen in Oslo would be the type specimen (Rasch 1845). Later scientists disagreed that it was a new species, and it has instead been referred to other taxa, most recently to *Lagenorhynchus acutus* (Gray, 1828). The species name given by Rasch, *Delphinus leucopleurus*, is treated as a synonym in a nomenclatural sense (Collett 1911–1912; Wilson & Reeder 2005).

However, Rasch's dolphin name has been interpreted differently by NHM Oslo. A 2013 paper summarizes the type specimens in their zoological collections (Wiig & Bachmann 2013). Here, the specimen is listed as a syntype, but

Rasch is quoted as having withdrawn the new species status and its name, *Delphinus leucopleurus,* in his original paper, because of work by Eschricht. However, my research shows that this interpretation is based on a mistake. Rasch's (1845) paragraphs about the withdrawal of a new species name do not refer to the Drøbak dolphin, but to a discussion about rorquals, a group of baleen whales. This stems from Rasch's reprint of the old manuscript, about whales along the Norwegian coast, within his own paper. This reprint not only concerns the Drøbak dolphin, but also occurrences of baleen whales. In his paper and in somewhat of a detour, Rasch briefly discussed Eschricht's idea that there are two rorqual species: the minke and sei whales. In this discussion he does indeed withdraw one of his name suggestions.

This is relevant, as the name of the dolphin species is still debated, because the genus *Lagenorhynchus* is under scrutiny. New molecular methods used in combination with morphology suggest that the taxonomy of several dolphin species should be organized differently. In recent work on this genus, Vollmer et al. (2019) suggest that the six species currently assigned to the genus *Lagenorhynchus* do not form a monophyletic group. This means that they are probably not each other's closest evolutionary relatives, and that, as a consequence, should not share a genus name. Instead, it is suggested that the former *Lagenorhynchus* genus should be split into three genera: *Lagenorhynchus, Sagmatius* and *Leucopleurus.* This last genus has one species, *Leucopleurus acutus.*

The Drøbak dolphin plays an important role here. The genus name *Leucopleurus* is taken from the original species name invented by Rasch based on the Greek words 'leukos' and 'pleura' meaning white side (Rasch 1845; Vollmer et al. 2019). It means the same as the common names in both English (Atlantic white-sided dolphin) and Norwegian (kvitskjeving). The small dolphin skeleton (NHMO-DMA-31220/1-O) is now a syntype, the species a type species, and the Oslo fjord the type locality for the resurrected genus name, *Leucopleurus* (Fordyce & Perrin 2023; Vollmer et al. 2019). This status, for the name and for the specimen itself, stems from its age, whereby the oldest name and earliest collected specimen have priority (Ride et al. 1999). This recent development, combined with the fact that Rasch never withdrew the species name, will probably mean that NHM Oslo will move the specimen into the type collection and document it more thoroughly (L. E. Johannesen & Ø. Wiig, pers. comm.).

6 Conclusion

The small specimen in the collection of NHM Oslo might once again be given a new name, and it might again rise in status as part of the work to understand the dolphin family tree. The record of its collection history is fundamental to future research. The unique history of the dolphin hunt, in Drøbak and Rasch's work to determine the status of these animals, enhances the value of this specimen and of this whale collection. This may open future opportunities for outreach purposes.

Today, many natural history museum collections are experiencing a renewed interest in the data that they contain. Natural history collections were fashionable and scientifically important in the late 1800s, but later, specimen-based biology was, to a large degree, replaced with other methods. However, because of the acute situation for much of the world's biodiversity today and as a result of the development of new tools to obtain information, museum specimens can be used to deliver much-needed data on geographical and temporal changes in biodiversity (Holmes et al. 2016; Lister 2011). A recent example is the DNA-based research on the hybridization between blue and fin whales over time (Jossey et al. 2024) using museum specimens, including items from NHM Bergen and the whaling museum in Sandefjord. This shows that old specimens are valuable precisely because they are old. This also presents an opportunity for arranging new exhibitions to raise public awareness. Despite their value, many collections are under threat due to inadequate funding over many years, putting specimens at risk of degradation or loss.

Museum specimens are also at the centre of several ethical discussions. One such discussion is around the relationship to threatened species and conservation (Byrne 2023). Another considers the ethical commitment to the best use of specimens from now threatened animals in collections (Lindqvist et al. 2009). Research on whales is a large and active field, relying today on a combination of field studies, molecular methods and museum collections. Knowledge is still missing for many species and filling these gaps may be important for their future conservation. Because of the conservation status of some whale species, they cannot be killed for scientific purposes. However, collection is still possible from stranded specimens through stranding networks, and from the few species that are hunted locally by countries such as Norway and Japan and Indigenous communities. Still, the current situation makes existing museum collections important, and there are many possible research questions that they might answer.

Acknowledgements

The author wants to thank the present and former curators of the mammal collection in Oslo; Lars Erik Johannessen, Øystein Wiig, and Kjetil L. Voje, for access and information about the whale specimens. Hanneke Meijer, Terje

Lislevand, and Anne Karin Hufthammer at the University Museum in Bergen are thanked for help in accessing and understanding the collections. For discussions on Norwegian whaling and collection history, Kirsten Osen, Alexandre Simon-Ekeland, Gard Paulsen, Anne Vaalund, Kirsten Borse Haraldsen, Ulrike Spring, and Janicke S. Kaasa are warmly thanked. Sandra McElroy is thanked for careful copy-editing.

Bibliography

Braschi, Serenella, Cagnolaro. L. & P. Nicolosi. 2007. "Catalogo dei Cetacei attuali del Museo di Storia Naturale e del territorio dell' Università di Pisa, all Certosa di Calci. Note osteometriche e ricerca storica." *Atti della Società Toscana di Scienze Naturali Memorie, Serie B* 114: 1–22. https://www.msn.unipi.it/wp-content/uploads/2016/12/esposizione-cetacei-1.pdf.

Broch, Hjalmar. 1954. *Zoologiens historie i Norge til annen verdenskrig*. Oslo, Norway: Akademisk forlag.

Broekema, Jan W. 1983. "Catalogue of Cetacea in the Collection of the Rijksmuseum van Natuurlike Historie, Leiden." *Zoologische Mededelingen* 57 (7): 67–79. https://repository.naturalis.nl/pub/318666.

Burnett, D. Graham. 2012. *The Sounding of the Whale: Science and Cetaceans in the Twentieth Century*. Chicago, US: The University of Chicago Press.

Byrne, Allison Q. 2023. "Reimagining the Future of Natural History Museums with Compassionate Collection." *PLOS Biology* 21 (5): e3002101. https://doi.org/10.1371/journal.pbio.3002101.

Båtvik, Jan Ingar. 1999. "Stripedelfin *Stenella coeruleoalba* funnet i Kråkstad- fjorden i Råde, fjerde funn i Norge." *Natur i Østfold* 18 (2): 123–29.

Collett, Robert. 1883. "Meddelelser om Norges Pattedyr i aarene 1876–1881." *Nyt Magazin for Naturvidenskaberne* 27 (2–3): 217–60. https://www.biodiversitylibrary.org/item/109438#page/224/mode/2up.

Collett, Robert. 1912. *Norges Pattedyr. Norges Hvirveldyr. Korte meddelelser om arternes udbredelse, levevis og forplantning*. Oslo, Norway: Aschehoug.

Cozzi, Bruno, Stefan Huggenberger & Helmut Oelschläger. 2017. *Anatomy of Dolphins: Insights Into Body Structure and Function*. London, UK: Academic Press.

Delsett, Lene Liebe. 2024. "Collecting Whales: Processes and Biases in Nordic Museum Collections." *PeerJ* 12: e16794. https://doi.org/10.7717/peerj.16794.

Eschricht, Daniel Frederik. 1844. *Undersögelser over hvaldyrene. Förste afhandling. Bemærkninger over cetologiens tidligere og nærværende skjebne*. Copenhagen, Denmark: Bianco Lunds Bogtrykkeri.

Flower, William. H. & John. G. Garson. 1884. *Catalogue of the Specimens Illustrating the Osteology and Dentition of Vertebrated Animals, Recent and Extinct, Contained in the*

Museum of the Royal College of Surgeons of England, Part II. Class Mammalia, Other than Man. London: J. & A. Churchill.

Fordyce, Ewan. & William F. Perrin. 2023. "World Cetacea Database. *Delphinus leucopleurus* Rasch, 1843." Accessed through*: World Register of Marine Species at:* https://www.marinespecies.org/aphia.php?p=taxdetails&id=383869, 6 November 2023.

Gatesy, John, Jonathan H. Geisler, Joseph Chang, Carl Buell, Annalisa Berta, Robert W. Meredith, Mark S. Springer & Michael R. McGowen. 2013. "A Phylogenetic Blueprint for a Modern Whale." *Molecular Phylogenetics and Evolution* 66 (2): 479–506. https://doi.org/10.1016/j.ympev.2012.10.012.

Grieg, James A. 1889. "Hvalstænget ved Bildøen den 15de april 1889." *Bergen Museums Aarsberetning 1888* 3: 3–17.

Grieg, James A. 1898. "*Mesoplodon bidens*, Sow." *Bergen Museums Aarbog 1897* 5: 3–31. https://www.biodiversitylibrary.org/item/130735#page/9/mode/1up.

Grieg, James A. 1905a. "Bidrag til kjendskaben om *Mesoplodon bidens*, Sow." *Bergens Museums Aarbog 1904* 3: 3–39.

Grieg, James A. 1905b. "Spidshvalen." *Naturen* 29: 110–15.

Grieg, James A. 1908. "Spidshvalen." *Norges Fiskeritidende* 27: 264–68.

Guldberg, G. 1900a. "Embryometriske undersøgelser av delfinfostre: *Delphinapterus, Globicephalus* og *Delphinus delphis.*" *Nyt Magazin for Naturvidenskaberne* 37 (4) (5): 361–73.

Guldberg, G. 1900b. "Ueber die Körpertemperatur der Cetaceen." *Nyt Magazin for Naturvidenskaberne* 38: 65–70.

Guldberg, G. 1901. "Cetologische Mittheilungen. I. Bemerkungen über das Auftreten und den Fang von Pottwalen (sL.) an den nordeuropäischen Küsten in letzen Decennium." *Nyt Magazin for Naturvidenskaberne* 39: 343–54.

Guldberg, Gustav. 1886. "Bidrag til Cetaceernes biologi. Om forplantningen og drægtigheten hos nordatlantiske bardehvaler." *Christiana Videnskabs-Selskabs Forhandlinger* 9 (1): 1–55.

Guldberg, Gustav & Fridtjof Nansen. 1894. "On the Development and Structure of the Whale. Part 1. On the Development of the Dolphin." *Bergens Museums skrifter* 5: 70. https://www.biodiversitylibrary.org/item/246929#page/8/mode/1up.

Heyning, John. 2002. "Museums and Collections." In *Encyclopedia of Marine Mammals*, edited by William F. Perrin, Bernd Würsig & J. Thewissen, 778. New York, US: Academic Press.

Hilton, Eric J., Gregory J. Watkins-Colwell & Sarah K. Huber. 2021. "The Expanding Role of Natural History Collections." *Ichthyology & Herpetology* 109 (2): 379–91. https://doi.org/10.1643/t2020018.

Holmes, Michael W., Talisin T. Hammond, Guinevere O. U. Wogan, Rachel E. Walsh, Katie LaBarbera, Elizabeth A. Wommack, Felipe M. Martins, Jeremy C. Crawford, Katya L. Mack, Luke M. Bloch & Michael W. Nachman. 2016. "Natural History Collections as Windows on Evolutionary Processes." *Molecular Ecology* 25 (4): 864–81. https://doi.org/https://doi.org/10.1111/mec.13529.

Johnson, Kirk R., Ian F. P. Owens & the Global Collection Group. 2023. "A Global Approach for Natural History Museum Collections." *Science* 379 (6638): 1192–94. https://doi.org/doi:10.1126/science.adf6434.

Jossey, Sushma, Oliver Haddrath, Livia Loureiro, Jason T. Weir, Burton K. Lim, Jacqueline Miller, Stephen W. Scherer, Anders Goksøyr, Roger Lille-Langøy, Kit M. Kovacs, Christian Lydersen, Heli Routti & Mark D. Engstrom. 2024. "Population Structure and History of North Atlantic Blue Whales (*Balaenoptera musculus musculus*) Inferred from Whole Genome Sequence Analysis." *Conservation Genetics* 25, 357–371. https://doi.org/10.1007/s10592-023-01584-5.

Kalland, Arne. 2014. *Hval og hvalfangst på Vestlandet 1600–1910*. Oslo, Norway: Novus forlag.

Kelley, Neil P. & Nicholas D. Pyenson. 2015. "Evolutionary Innovation and Ecology in Marine Tetrapods from the Triassic to the Anthropocene." *Science* 348 (6232): aaa3716-3-aaa3716-7. https://doi.org/10.1126/science.aaa3716.

Kyllingstad, Jon R. & Thor Inge Rørvik. 2011. "Zootomisk museum og hvalforskningen." In *Universitetet i Oslo 1870–1911. Vitenskapens universitet*, 297–300. Oslo, Norway: Universitetsforlaget.

Lane, Meredith A. 1996. "Roles of Natural History Collections." *Annals of the Missouri Botanical Garden* 83 (4): 536–45. https://doi.org/10.2307/2399994.

Lindqvist, Charlotte, Anja Probst, Anthony R. Martin, Øystein Wiig & Lutz Bachmann. 2009. "Molecular Species Identification of Historical Whale Remains from South Georgia." *Marine Mammal Science* 25 (1): 229–38. https://doi.org/https://doi.org/10.1111/j.1748-7692.2008.00248.x.

Lister, Adrian M. 2011. "Natural History Collections as Sources of Long-Term Datasets." *Trends in Ecology & Evolution* 26 (4): 153–54. https://doi.org/https://doi.org/10.1016/j.tree.2010.12.009.

McLean, B. S., K. C. Bell, J. L. Dunnum, B. Abrahamson, J. P. Colella, E. R. Deardorff, J. A. Weber, A. K. Jones, F. Salazar-Miralles & J. A. Cook. 2016. "Natural History Collections-Based Research: Progress, Promise, and Best Practices." *Journal of Mammalogy* 97 (1): 287–97. https://doi.org/10.1093/jmammal/gyv178.

Museum of University History. 2012. Hvalmuseet. Accessed 15 January 2023.

Nordvik, Helge W. & Lewis R. Fischer. 1985. "Peter M. Stuwitz and the Newfoundland Inshore Fishery in 1840." *Newfoundland & Labrador Studies* 1 (2). https://journals.lib.unb.ca/index.php/NFLDS/article/view/567.

Pyenson, Nicholas. 2017a. "The Ecological Rise of Whales Chronicled by the Fossil Record." *Current Biology* 27 (11): R558-R564. https://doi.org/10.1016/j.cub.2017.05.001.

Pyenson, Nicholas. 2017b. "Museums and Collections." In *Encyclopedia of Marine Mammals*, edited by Bernd Würsig, J. Thewissen & Kit M. Kovacs, 1190. New York, US: Academic Press.

Pyke, Graham H. & Paul R. Ehrlich. 2010. "Biological Collections and Ecological/Environmental Research: a Review, Some Observations and a Look to the Future." *Biological Reviews* 85: 247–66. https://doi.org/10.1111/j.1469-185X.2009.00098.x.

Rasch, Halvor H. 1845. "Beskrivelse over en i Christiania-Fjorden fanget nye Delphinart." *Nyt Magazin for Naturvidenskaberne* 4: 97–125.

Reeder, DeeAnn M., Kristofer M. Helgen & Don E. Wilson. 2007. "Global Trends and Biases in New Mammal Species Discoveries." *Papers Museum of Texas* 269. https://repository.si.edu/bitstream/handle/10088/3790/VZ_Global_Trends.pdf.

Ride, W. D. L., H. G. Cogger, C. Dupuis, O. Kraus, A. Minelli, F. C. Thompson & P.K. Tubbs. 1999. *International Code of Zoological Nomenclature.* International Trust for Zoological Nomenclature.

Robovsky, Jan & Petr Benda. 2006. "Catalogue of the Cetaceans (Mammalia: Cetacea) in Selected Collections of the Czech Republic, with Special Respect to the Collection of the National Museum, Praha." *Časopis Národního muzea, Řada přírodovědná – Journal of the National Museum, Natural History Series* 175 (3–4): 127–56.

Rocha, Robert C., Phillip J. Clapham & Yulia V. Ivashchenko. 2014. "Emptying the Oceans: A Summary of Industrial Whaling Catches in the 20th Century." *NOAA Marine Fisheries Review* 76 (4): 37–48. https://doi.org/10.7755/MFR.76.4.3. https://www.whalingmuseum.org/wp-content/uploads/2021/04/mfr7643-Emptying-the-Oceans.pdf.

Sars, Michael. 1851. "Beretning om en i Sommeren 1849 foretagen zoologisk Reise i Lofoten og Finnmarken." *Nyt Magazin for Naturvidenskaberne* 6 (2): 121–211.

Simon-Ekeland, Alexandre & Lene Liebe Delsett. 2024. "Whales as a Nordic Specialty: Collecting for Natural History Museums (1860s–1910s)." *Nordic Museology* 6 (1): 28–43. DOI: 10.5617/nm.11591.

Thewissen, J. G. M. & Sunil Bajpai. 2001. "Whale Origins as a Poster Child for Macroevolution." *BioScience* 51 (12): 1037–49. https://doi.org/10.1641/0006-3568(2001)051[1037:Woaapc]2.0.Co;2.

Tønnesen, J. N. & A. O. Johnsen. 1982. *The History of Modern Whaling.* London: C. Hurst & Company.

van Beneden, P.-J. 1883. "Note sur un *Mesoplodon sowebyensis* de la cote de Norwege." *Bulletins de l'Académie royale des sciences, des lettres et des beaux-arts de Belgique* 2 (22): 218–21. https://www.biodiversitylibrary.org/item/111247#page/224/mode/1up.

Vollmer, Nicole L., Erin Ashe, Robert L. Brownell Jr., Frank Cipriano, James G. Mead, Randall R. Reeves, Melissa S. Soldevilla & Rob Williams. 2019. "Taxonomic Revision of the Dolphin Genus *Lagenorhynchus*." *Marine Mammal Science* 35 (3): 957–1057. https://doi.org/https://doi.org/10.1111/mms.12573.

Wiig, Øystein & Lutz Bachmann. 2013. "The Mammal Type Specimens at the Natural History Museum, University of Oslo, Norway." *Zootaxa* 3736 (5). https://doi.org/10.11646/zootaxa.3736.5.9.

Wilson, Don E. & DeeAnn M. Reeder. 2005. *Mammal Species of the World: A Taxonomic and Geographic Reference, Volume 1.* Baltimore, US: John Hopkins University Press.

FIGURE 15.1 A food barrel (2004–66–1), preserved at the Local Museum in Görðum

CHAPTER 15

"Treasures and Useful Things": Everyday Objects in Two Archives

Anna Heiða Baldursdóttir

Abstract

This chapter explores the material culture of nineteenth-century Iceland by comparing everyday objects from two archives: probate inventories in the National Archives of Iceland and a part of the collection at the National Museum of Iceland. It highlights how these archives reflect different societal narratives. While the probate inventories reveal the practical possessions of "ordinary" people, illustrating their daily lives and survival strategies, the museum's collection tends to emphasize artistic and valuable artefacts associated with the upper class, thus presenting a more idealized view of Icelandic heritage.

This study raises questions about the representation of everyday life and how material culture shapes our understanding of historical narrative and Nordic identity. By examining the differences between these collections (and within), this work advocates for a more inclusive interpretation of the past that recognizes the diverse experiences of all societal groups, ultimately enriching our perspective on the past.

Keywords

material culture – everyday history – museum – museum collection – cultural heritage – archives – materiality

The theme of this book is to explore the North through objects from various angles: the objects themselves, such as artefacts and souvenirs, or collections and collection practices. In this chapter the focus will be on the objects in two archives, the *Sarpur* database, which contains information about the museum collection from the National Museum of Iceland, and probate inventories in

 | DOI:10.1163/9789004749665_016

the National Archives of Iceland.[1] Both sources are preserved in official institutions and can be called places where heritage is shaped and performed. In the book *Performing Nordic Heritage*, which examines the performance of Nordic heritage and the shaping of the very idea of being Nordic, the editors, Peter Aronsson and Lizette Gradén, state that: "the performance of heritage takes place in prestigious institutions such as museums and archives" (Aronsson & Gradén 2016). The National Museum and the National Archives in Iceland are two of the largest establishments in the country for preserving and mediating cultural heritage. They are also under legal obligations, leading the way in shaping and creating collections in the country as national centres for management of cultural heritage (Alþingi 2011). It is therefore interesting to study their collections when thinking about material culture in Iceland because of their influence on what is considered as Icelandic heritage, with preservation and representation to a large-scale audience. The authority of these institutions, which consists of their role as cultural policy makers and cultural assets is due to the longevity of their existences. One could say that, due to this power, the museum and the archive have great impact on Iceland's historiography – one of the narratives of the North.[2]

In daily discussion about the value of the National Museum of Iceland, its collection, or at least some of it, has often been described as a national treasure. *Gersemar og þarfaþing* (*Treasures and Useful Things*) is the title of a book published by the National Museum in 1994, in which a part of its collection was described and discussed (Björnsson 1994). From the title it can be concluded that everyday objects are in fact not treasures but rather useful things, which they most certainly were. This instinctive classification between objects leads to contemplations about the collection of the National Museum of Iceland. Are there more treasures than useful things? What can be considered a normal proportion between these? From these thoughts I will focus on everyday objects from the nineteenth century that are preserved at the National Museum of Iceland and in another archive (probate inventories).

There are two main reasons for using these sources. Firstly, because they contain information about material culture from the past. The probate inventories are records of objects listed as being in the possession of people who

1 The subject of this paper is based on the author's doctoral research: Baldursdóttir 2022.

2 In Anna Lísa Rúnarsdóttir's lecture at the conference *Collecting the North* (2022) this issue was discussed with regard to the National Museum of Iceland. See further discussion in: Aronsson 2008, 5–20; Aronsson & Gradén 2016; Smith 2006; Rúnarsdóttir, 2022, 49–62.

have died, and the museum archive is a database of preserved items, often with detailed information about them. Secondly, it can be argued that material culture can give a better insight into everyday life due to how material things can impact human behaviour on a daily basis. Some scholars have even argued that material culture can reveal everyday life better than public textual sources. In their opinion, objects can reveal quite lot of information about individuals and societies (see Baldursdóttir 2022, 15–27; Hamling & Richardsson 2010; Holmes 2024).

By focusing on the everyday objects in two archives, the following questions will be addressed. Are the same things found in both archives? What objects are the most common? The objective of this comparison is to gather information about what kind of material culture is to be found in these different collections, focusing on how everyday life is collected and represented in one of Iceland's influential museums. Before answering these questions, another important one must be raised and considered: what is an everyday object and what is a treasure? According to what can be gathered from the book *Treasures and Useful Things,* "treasures" are objects of more sophisticated value than "useful things", often with regard to artistic matters and valuable materials like silver, silk and gold (Björnsson 1994). The term "everyday object" is certainly a very broad concept and things people nowadays use daily are very different to what was used in the past. It is also vital to understand that, when studying the past, the individuals or societies are not all the same. There were differences between the genders, for example. Women often carried out domestic chores while men tended to physical ones outside, making their use of everyday objects dissimilar. It should also be pointed out that people or households of higher social status had better access to more valuable things because of their favourable financial resources. In other words, they could buy more, and more expensive, things than others.

The everyday life of different individuals was different and therefore they did not use the same objects on a daily basis. Pierre Lemonnier has defined mundane things as something that people often did not take an interest in, but if the objects were not there then everyone would notice, and daily needs and actions would be lacking (Lemonnier 2016, 13). An example would be a household in the nineteenth century that owned no barrels or other large containers. Daily storage of food would not be possible, which would cause inconvenient circumstances for the people in the household. The food they produced would be in danger of decaying, due to the lack of containers to store, for example, meat or fish. Therefore, the daily necessity and action of preserving food for longer periods of time would be unfulfilled.

These are not simple matters, and they are different for everyone in each society and culture, where needs are dependent on resources, tradition, status, gender, hobby, work – in short, the way of life people lead. The subject of this chapter will therefore involve a part of material culture that was important for the everyday life or basic needs of people in the nineteenth century – clothes and domestic equipment. Clothes that shielded and protected people – like trousers, wool sweaters, and socks – that, for example, kept them warm during wintertime. Nourishment or food is vital for humans. Therefore, equipment to make, store, prepare and cook it on a daily basis – like pots, pans or barrels – were equally important. All these were objects that were used in everyday life and served the basic needs of their owners.

1 Probate Inventories in the National Archives of Iceland

The first archive explored consists of probate inventories, preserved by the National Archives of Iceland. There are about 33,000, dating from the eighteenth century to the beginning of the twentieth century.[3] For comparison, the population of Iceland in the year 1801 was 48,587, and at the end of the nineteenth century it was 78,203 (Jónsson and Magnússon 1997, 45–48, 50). The probate inventories are documents that contain inventories and valuations of the estates of deceased individuals, carried out if they left heirs of minor age or died in debt, or in cases where the division of assets among heirs was disputed. They were created from examples in Norwegian inheritance laws from the late eighteenth century and were mostly effected in Iceland until 1860–1870 when it began to deregulate due to changing society in the late nineteenth and beginning of twentieth century (Jónsson 1981, 7–30). The sources commonly listed things in certain categories: books, clothing, bed linen, tools, domestic equipment, livestock, real estate and wood.

This chapter draws its data from the doctoral thesis Things from the Past, by Anna Heiða Baldursdóttir (2022). The research was based on 104 probate inventories from the nineteenth century with individuals who were farm workers, farmers and widows of a certain age. The study sampled people at the beginning, middle and end of a "typical" lifespan, using three age groups, featuring people who were 20–25, 45–50 and 70–75 years old when they died. All of them owned average possessions considering their status and age group and were

3 Database with these records is available at: www.danarbu.skjalasafn.is.

distributed geographically – in other words, from all over the country (Baldursdóttir 2022). The oldest inventory dates from 1804, and belonged to a farm worker, Jón Vilhjálmsson (1758–1804) from the southern part of Iceland. The farmer Guðfreður Guðmundsson (1873–1898) from Westfjords was the owner of the latest estate valued in the study (Baldursdóttir 2022, 109, 140). In total there were 37 farm workers, 34 farmers and 33 widows. The thesis revealed that the probate inventories gave a good perspective on material culture in Iceland during the nineteenth century and gave a rare insight into the everyday life of people (Baldursdóttir 2022, 302–10).

The 104 probate inventories listed 15,287 items, which were analysed and sorted into the same categories as those used by the officials who appraised the deceased's possessions during their valuation. The figure below shows how the belongings were divided into object categories.

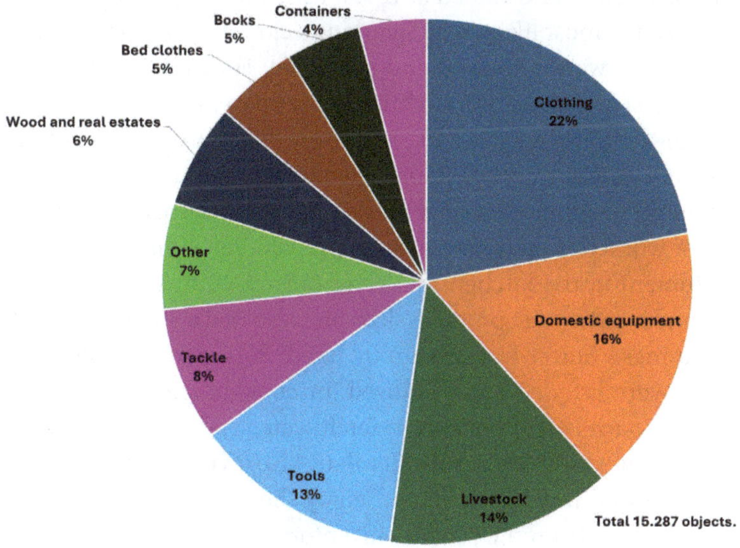

FIGURE 15.2 Object categories from probate inventories
COPYRIGHT: ANNA HEIÐA BALDURSDÓTTIR

Figure 15.2 shows that important everyday objects like *clothing* and *domestic equipment* played a large role in people's lives and material culture at that time. In a conservative and rural society, which depended on agriculture and home production, it is perhaps no surprise that domestic equipment made up so much of people's possessions during the nineteenth century. Objects that were used to make products from, for example, sheep and cows and used to provide food for household members (see Gunnlaugsson 1988; Magnússon 2010). This category consisted of things to prepare, produce, cook, serve and eat food – such as pots, barrels, spoons, bowls, buckets, pans, trays and other useful things in the kitchen and pantry. The next largest category was *livestock* and then *tools* to take care of farm animals or maintain and build things/houses, such as hammers, rakes, saws, shovels, etc. This distribution of people's possessions would look a lot different today, mainly because individuals/homes have become consumers and are not producers anymore. But let us turn to two case studies to quickly demonstrate how probate inventories can be used to study material culture and everyday life.

Sigurlaug Ólafsdóttir from Ketilsstaðir in the western part of Iceland, died as a widow aged 45 in 1866. Her belongings totalled 728 rixdollars (rd.), most of which involved farm animals and real estate. She had quite a lot of clothing but not of much value. Her most valuable piece of clothing was a homespun cloth skirt valued at 4 rd. In comparison, the most valuable piece of clothing in the whole study was a silk dress valued at 12 rd., which belonged to a wealthy widow in Reykjavík. It is most likely that Sigurlaug's clothes were the ones she used every day to keep warm, covered and protected; For example, her *styttuband*, which was a string or tie used to bind skirts up to prevent them from being dragged in the dirt. Her aprons, too, shielded her skirt and sweaters from being covered in food scraps or other stains when working in the kitchen.

Sigurlaug's domestic equipment was rather excessive, comprising 95 pieces, which she and her workers would have used to prepare, serve and make food. Among her equipment in the kitchen was skyrgrind which was a bottomless wooden tray for making the milk product 'skyr'. She also had a strokkur or butter churn, used to make butter from cream by hand. Pots, pans, a coffee mill and other objects were listed. They were used for cooking and making food and drinks. She had barrels and buckets to fetch water and milk or preserve meat with, for example, sour whey or salt. Sigurlaug also owned askur (wooden bowls), spoons, cups and trays to serve and enjoy the food and drink. From these objects (and others from the probate inventory) it can be seen that Sigurlaug's farm produced skyr and butter for trading and commerce – a part of the farm's production. The skyr and the butter was also vital produce for the private consumption of the household. The household was therefore equipped with objects to make, preserve, cook, serve and consume food which, for obvious reasons, must have been used every day to provide and feed the family and others living at Ketilsstaðir.

The farm worker Rannveig Þorsteinsdóttir died at the age of 45 in the year 1810. Her belongings were appraised at a total value of 57 rd. and most of it was her clothes. To put this amount in perspective it has been estimated by Guðmundur Jónsson that the yearly income for farm workers in Iceland was around

3–6 rd (Jónsson 1981, 33–48). This salary was, of course, variable throughout the century, and could vary between regions and genders. Men had better salaries than women, and there are even examples where women only got food and clothes for their work (Jónsson 1981, 33–48). Of 73 entries in Rannveig's probate inventory, 50 involved clothing: good suits and other items of clothing that she wore during her chores at the farm, like underwear, a bodice and a skirt, a "used" sweater and mittens. Her most precious clothes were first to be listed in the probate inventory – for example, a blue bodice with silver lace hooks with a value of 5 rd. What is interesting here is that for a farm worker – a person often without sufficient resources – Rannveig had quite impressive clothing, more valuable than that owned by many other women in her position, and even those who were of higher status, like farmers.[4]

A few things can be deduced from her clothes. First, it is possible that she was trying to "dress up" to catch the eye of an eligible bachelor and improve her status in society. Second, it is an interesting fact that she was working at an upper-class household at Brekka in Reykjavík. That position could have given her better access to more valuable fabrics for clothes. It is also worth mentioning that her employer, Ísleifur Einarsson (1765–1836), showed interest in using clothes to reflect the social status of people, arguing that the higher class and their staff should dress in accordance with their position and so forth. Ísleifur wrote about this opinion in papers and even attended a well-known meeting about eliminating Danish influence on Icelandic clothing (Aspelund 2015, 76–78). Rannveig's household would therefore also explain her valuable clothes, due to her employer's interest in dressing his staff in better clothes than other workers on common farms. Third, and perhaps the simplest reason for her beautiful wardrobe was her good taste for fashion or ability for making clothes. Her handcraft tools would suggest Rannveig's good skills for making fine clothes and her position in a higher-ranking household would have given her access to valuable fabrics.[5]

In Rannveig's probate inventory there are few items of domestic equipment: only four objects. She had two clasp knifes, a spoon and a bottle. This indicates that she had her own things to consume food both at home and for travelling. The reason for so few objects in this category is mostly due to Rannveig's status as a farm worker. Rannveig worked at farms where the farmers were bound by law to provide their servants and workers food, clothes and shelter (Stephensen & Sigurðsson 1854, 683–86; 1853, 600–05; Stephensen 1885, 544–57).

4 *Þí*. Sýsl. Gullbringusýsla. ED2/7,2: 11. 1810.

5 *Þí*. Sýsl. Gullbringusýsla. ED2/7,2: 11. 1810.

It can be assumed that this kind of equipment would have been in place at the household of Brekka where Rannveig died, but for some reason she owned these objects. This is probably for personal reasons; perhaps she preferred to eat with her own things instead of using those owned by her employer. It was possibly her attempt at maintaining personal space within a household that was simultaneously a home and a workplace of many.

Rannveig's few domestic equipment possessions clearly show that she did not produce or make food. As a farm worker, she did not need these kinds of things and only had to take care of herself. In other words, she was not providing for others like the widow Sigurlaug. Brekka's farm provided those kinds of tools for the household's production and consumption. Rannveig therefore gathered and owned things for her own personal use and interest as well as receiving things as salaries or gifts.

Clothes had an obvious and highly practical role in people's daily lives. They were used to shield against harsh working conditions, weather and dirt and to conceal nudity. But it was also a way for individuals to dress up and try to idealize their status, and to impress their fellow countrymen. Single items of domestic equipment were not of much worth in the probate inventories, but their existence and use made people able to produce extensive valuables; not only economically, but also for basic survival needs – food. The clothes and domestic equipment of Sigurlaug and Rannveig were its owner's treasures and highly useful things. Different kinds of things though, because of the women's distinct circumstances. Not only did the possessions indicate their everyday life but also the kinds of objects the appraisers considered worth listing and later sold at auction, which was often the end point of the process of appraising the estate of the deceased in question (Jónsson 2016, 7–30).

It should be mentioned that it cannot be asserted that every single item in people's possession was valued and listed in the probate inventories. Már Jónsson has, for example, discussed this in his work on Icelandic probate inventories. He has pointed out that because of no standardized rule of making a probate inventory, officials had their own ways of doing it. There are also cases where heirs or employers were accused of withholding things from officials (Jónsson 2012, 96–101; see also Baldursdóttir 2022, 210–11). So it cannot be assumed that every single object was listed. The probate inventories are, despite that, a highly valuable source regarding material culture and everyday life. The case studies have shown that using probate inventories can give a valuable perspective on people's everyday lives. The archives can show various ways of life, which adds to current understanding of the past. From these archives a certain view into material culture and everyday life has been attained, but how is everyday life collected and represented at the National Museum of Iceland?

2 Museum Collection of the National Museum of Iceland

In 1863, the Icelandic Antiquities Collection was founded on a base of 15 objects, donated by Helgi Sigurðsson (1815–1888), a priest in the western part of Iceland. This later became the National Museum of Iceland in the year 1911, and is the largest museum in Iceland. Today the museum collection is believed to comprise around 300,000 objects of various kinds and is documented in two ways – in catalogues and in an online digital database called *Sarpur* (Þjóðminjasafn Íslands 2015). For this project, the *Sarpur* database was used to gather data from the museum collections and compare it to that of the probate inventories. The first step was to define the time range – the date was set to 1740–1900, which was in accordance with the period that the 104 individuals who were studied in the probate inventories lived and died in. Their belongings therefore were used most likely during these 160 years and a longer period than the nineteenth century made it possible to include more museum objects into the study (Baldursdóttir 2022, 215–32).

In *Sarpur*, 2,864 items of the museum's collection were dated and documented from 1740–1900. It is safe to say that the collection from this period is much larger than this number, but these objects were the ones where the date was known and registered in the database[6] (Baldursdóttir 2022, 217–20). The next step was to sort the collection into the same objects categories that were used for the possessions in the probate inventories. The distribution can be seen here in Figure 15.3, but with two exceptions. For obvious reasons no living animals are preserved in the National Museum and it includes a whole other category with things from public services like churches, doctors and photographic studios (*official objects*).[7]

The first impression from this data is that these two archives are in most ways alike; everyday life has been collected at the museum and perhaps is well represented here. *Clothing* and *domestic equipment* are the largest categories, others like *tools* and *wood* are also a similar proportion of the collection to that

6 The accurate number is not known but is believed to be around 20,000; see Baldursdóttir 2022: 217–20.

7 The collection was sorted by the author for the doctoral thesis *Things from the Past.*

in the probate inventories. However, examining the material culture in each of the categories quickly reveals that these archives are not the same.

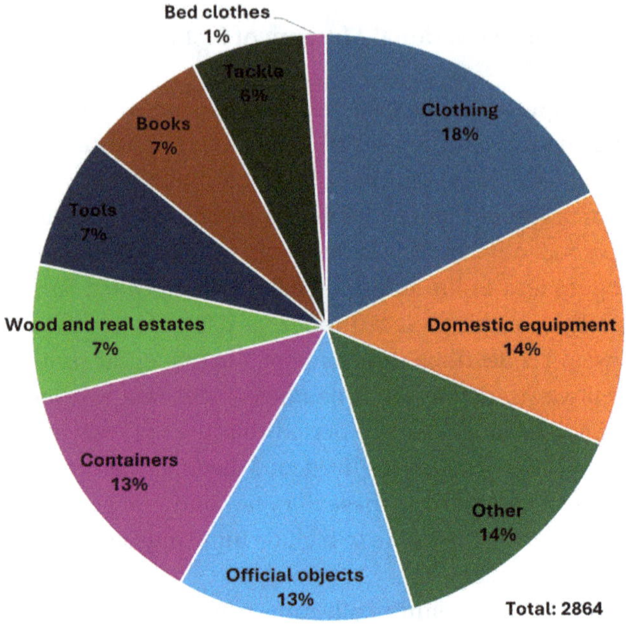

FIGURE 15.3 Object categories from the museum collection of the National Museum of Iceland

This is most obvious in the case of clothes in the museum. The most common clothing in the probate inventories protected its owner and were practical – kept them warm and shielded. Things like socks, bodices, skirts, shirts, trousers, scarfs, mittens and sweaters. (Baldursdóttir 2022, 57–58, 105–06 and 159) The list of the most common clothes in the National Museum of Iceland 1740–1900 are, on the other hand, what could be considered accessories.

Of these 11 kinds of clothing only three are actually clothes: cardigan (from the female national costume), shoe and bodice. The collection mostly comprises buttons and other decoration used on clothes to fasten them or add to their aesthetics and details. All these artefacts are small and easier to preserve than "real" clothes made of fabrics that are delicate. Materials like textile are highly sensitive to factors like humidity or vermin so it is likely that they were thrown away because of their poor condition, or were reused for other purposes – transformed into curtains or used to stuff mattresses, for example (National Park Service 2023). In that process, details like buttons and silver lace hooks were taken off to use again or keep for personal reasons – belongings from a close relative, perhaps.

The collection here shows a great lack of clothes for men. Of 500 garments that were analysed, only 21 were registered as having been used by men. In the museum archives are listed, for example, three skin shoes, three vests, two pairs of trousers, two sweaters, a cardigan, knitted mittens, a balaclava, a belt and a cap. Female clothing, on the other hand, is much more plentiful, as can be seen in the table above plus the addition, for example, of five caps, five aprons and three skirts, all used by women. The national female costume is very noticeable in the museum, both in the collection and the museum's exhibition.

TABLE 15.1 The most common clothing in the National Museum of Iceland

Milla/Ornamental fastenings	137
Skúfhólkur/Tassel tube	31
Kvenbelti/Female belt	22
Kvenkragi/Female collar	17
Skauttreyja/Festive costume cardigan (female)	15
Hnappur/Button	14
Belti/Belt	13
Skinnskór/Shoes made of skin	11
Skautkragi/Festive costume collar (female)	10
Beltispar/Hook for belts	10
Upphlutur/Bodice	10

COPYRIGHT: ANNA HEIÐA BALDURSDÓTTIR

It can be assumed that women did not wear their best clothes daily in the nineteenth century and certainly not the Icelandic festive costume. The costume was designed by Sigurður "the painter" Guðmundsson (1833–1874) and was not introduced until 1864, although it seems to be important for the National Museum (Aspelund 2015, 73–116). It is found both complete or in parts – like buttons, braces, shirts or skirts. In this context it is worth mentioning that the Museum's official key object for the nineteenth century in its exhibitions is the first festive costume Sigurður designed, which was made by his friend, Sigurlaug Gunnarsdóttir (1828–1905). The "key object" is a concept at the exhibition in the museum where one artefact is displayed as a symbol of every century from settlement to the present (Þjóðminjasafn Íslands, n.d.).

Domestic equipment in the probate inventories indicated the need for things that could be used to preserve, produce and prepare food. The most common object was a *trog* (wooden tray), which had various uses such as separating milk or making *skyr*, serving food or to gather rubbish. Buckets, barrels, pots, *keröld* (wooden tubs or barrels), *askar* (wooden bowls) and pints were also quite abundant in people's possessions in this study (Baldursdóttir 2022, 70–71, 117–18 and 177–78). The items included in the National Museum from this part of everyday life in the past is in some ways alike, but still different in nature.

TABLE 15.2 The most common domestic equipment in the National Museum of Iceland

Spónn/Spoon	66
Matskeið/Tablespoon	46
Mataraskur/Wooden bowl for food	26
Askur/Wooden bowl	22
Kanna/Jug	17
Skeið/Spoon	16
Brauðmót/Bread pan	10
Pottur/Pot	9
Kaffikanna/Coffeepot	9
Teskeið/Teaspoon	8

It can be seen from the most common domestic equipment in the collection that the museum preserved all sorts of spoons for no obvious reasons. It is perhaps because, as in the case of all the buttons, they are small and easy to preserve. The spoons are even more plentiful than can be seen here in Table 15.2, because after teaspoons come soup spoons, silver spoons, coffee spoons, and others. Added together, they are by far the most common tool collected from Icelandic kitchens and pantries from the nineteenth century. Wooden bowls, jugs and coffeepots were also used to serve or consume food and other beverages. In fact, the whole collection here is mostly things that were made for that purpose. The material culture of this particular collection in the museum is therefore different to that noted in the probate inventories. This fact merely indicates the dissimilarity between the nature of these two archives, which were made for different purposes. The probate inventories existed to value an individual's belongings to even out debts and divide inheritances. The museum archives are to make and shape a national identity and represent cultural heritage over a long period of time.

Objects used for the production and preservation of food, such as barrels, trays and pots, are also found in the collection – but are not as noticeable as the tools used for the consumption of food. Similar objects to those in the probate inventories are therefore to be found in the collection, although not in the same proportion. There is, though, one other matter which raises questions about the collection's representation. It is notable that the catalogued domestic equipment in the museum is of high quality in both its materials and making. Of 402 registered items of domestic equipment, nearly 200 are made from metal – including 141 of silver. The items under the registration number

2015–62, for example, form part of a beautiful set for serving coffee and tea – all made from silver. One item is a sugar bowl (No 2015–62–3), which is described as follows: "made from sterling silver, English, from the year 1845 […] The bowl has two handles and is very flowery and figured, with flower and leaf decorations."[8]

The materials were not the only sign of high-quality items; the design or making was too. Examples of this are the wooden artefacts from the collection that are mainly carved with decorations or inscriptions. Item No 1921–202 is described as follows: "Wooden bowl made from pine […] the lid is curved from below and is beautifully carved on top with leaf and branch carving."[9] The registration of the collection also shows on a number of occasions, links between objects and their maker/artist, such as Hjálmar Jónsson, 1796–1875 ("Bólu-Hjálmar") and Stefán Jónsson (1832–1868), well-known folk artists (carpenters) in Iceland during the nineteenth century. In the cataloguing of item No 2008–5–225 is an added comment to the objects description from Freyja Hlíðkvist a specialist at the National Museum: "the wooden bowl is very similar to other preserved bowls made by Stefán Jónsson from Malland and is without a doubt carved by him."[10] Given this, one could wonder whether the collection in the National Museum is more suited to representing works of art than everyday life in the past.

The collection from the National Museum of Iceland regarding everyday life in clothing and domestic equipment shows material of higher quality and is perhaps more artistic than that which the belongings in the probate inventories hinted at. From this it can be said that there are more "treasures" than "useful things" preserved at the museum. The comparison with the belongings in the probate inventories, and the nature of the collection in these object categories, raises doubts about how, and even *if*, everyday life in the North is collected and represented. The collection presents, however, the everyday life of a small group in Iceland – people of higher social class. A manifestation that may be seen more as an idealisation of Icelandic materiality in the nineteenth century. But why is there this difference between the archives?

8 *Þmjs.* 2015–62–3. Author's translation – originally: "úr sterlingsilfri, enskt, frá árinu 1845 … Karið er með tveimur eyrum/höldum og það er mikið flúrað og mynstrað, með blóma og laufaskreyti."

9 *Þmjs.* 1921–202. Author's translation – originally: "Lokið er hvelft að neðan og fallega útskorið að ofan, með laufaskurði og greinum."

10 *Þmjs.* 2008–5–225. Author's translation – originally: "Askurinn er keimlíkur varðveittum öskum eftir Stefán Jónsson á Mallandi og er án efa skorinn af honum."

3 Material Culture in Two Archives

Material culture encompasses not only objects or the material itself, but also the relationship between people and things, which is constantly changing. From the object's maker to its owner and user – and perhaps later observer or a viewer at museums. The information from the collection in the National Museum often shows this process. The wooden bowl (No 2008–5–225) made by Stefán Jónsson in the nineteenth century was his product, given or sold to its owner and user (who is unknown). The owner/user in question used the bowl to consume food such as porridge and took good care of it, given its good condition. Then the bowl (for unknown reasons) ended up at the *Fr. Möller* store in Eskifjörður (in the eastern part of Iceland). From there it was sold to a collector in 1887, who sold or gave it to Nordiska Museet in Sweden in the beginning of the twentieth century. There it was preserved until 2008 when it was given to the National Museum of Iceland.[11]

From this series of events in the object's biography a different use and meaning has been attached to this wooden bowl throughout history.[12] From practical and economic uses to the status of a curiosity or status symbol, as souvenir from a foreign country in the North Atlantic. Then later given to a museum, which collected and preserved it as a part of Nordic heritage. Finally, it was brought to another museum, which is more focused on Icelandic traditions, while also a Nordic narrator, where it is now represented as a relic from the past indicating, for example, food culture and folk art. The relationship between the bowl and the people changed. In other words – the material culture of the bowl changed.

This process is the first indicator of why the material culture is different in these archives. The belongings in the probate inventories are those used for the personal and practical needs of their owners/users and other people in their lives. To provide and be able to participate daily in a rural society driven by agriculture (Magnússon 2010, 18–32, 238–57). The collection in the National Museum is not collected for the same purposes, and therefore is not the same. From the registration or the information about the museum objects, a few factors can be used to shed more light on this matter.

11 *Þmjs.* 2008–5–225.

12 The information in the object biography is based on the registration of the wooden bowl. There are still blanks, such as who owned it between Stefán and the store, or whether Stefán sold it to the store. It is also possible that Nordiska Museet bought the item from *Fr. Möller.* That part is not obvious.

First is the fact that most of the clothing and domestic equipment was collected at the beginning of the twentieth century. To put this into perspective, 89% of the clothes here were already in the museum by 1950, as was 68% of the domestic equipment. At that time, it is safe to say that the discourse in Iceland was very much under the influence of nationalistic views, considering that sovereignty was claimed in 1918 and independence in 1944. This discourse gave little room for the daily life of the common public and focused more on creating (and praising) national identity based on a "golden age" when Iceland was independent (Hálfdanarson 2006, 237–54; Karlsson 1980, 77–89).

Matthías Þórðarson, director of the National Museum of Iceland during 1908–1947, was the person who amassed most of the collection at issue here (Baldursdóttir 2022, 217–92). He turned the museum to a more modern take on museology and was an influential figure in cultural decisions in Iceland at that time. Anna Lísa Rúnarsdóttir and Þorgerður Þórhallsdóttir have studied Matthías's ways of acquisition and his emphasis during his time as a director – and, it is noted, the only staff member for a long time. Their conclusions are that part of his emphasis in collecting was on feminine handiwork and objects of great aesthetic value (Rúnarsdóttir 2022, 49–62; Þórhallsdóttir 2022, 221–34). The clothes and domestic equipment that was discussed earlier here is also an example of that; Icelandic female festive costume, its accessories, silver spoons and carved wooden artefacts are very noticeable. Museum personnel can therefore affect what the institution collects and displays.

Secondly, it must be taken into consideration that the public is also a factor in heritage policy making. The museum relies on donations from the public, which was not immune to political discourse at the turn of the twentieth century. Þorsteinn Þorsteinsson (1825–1912), a farmer and carpenter from northern Iceland, donated over 100 buttons and other decorative items related to clothing around 1900 (Baldursdóttir 2022, 222–23). This generous contribution has most certainly had an impact on what the collection consists of – the clothes are more accessories than clothing. Public opinion on what constitutes a "museum artefact", therefore, matters in regard to what ends up in a museum's collection. In the case of clothing at the National Museum, the public did not consider old everyday clothes from "grandpa and grandma" to be museum objects and tossed them aside, only keeping the ones that were considered appealing and "Icelandic". As a demonstration of how people think about things, valuables like silver, decorated clothing or artistically carved objects form the basis of the collection. Societies interest in everyday history came later in the twentieth century, but by then it was too late to obtain those kinds of objects (Baldursdóttir 2022, 303–05; see also Iggers 1997, 101–17).

Other factors do play a part in how the National Museum of Iceland collects and represents everyday life. Here in this chapter, a certain part of the museum's collection has been described. The acquisitions policy has been shaped from the beginning and is still evolving. It is a "conversation" between participants like museums, the public, educational institutions, scholars, students, media, politicians and the tourism industry (Aronsson & Gradén 2016; Smith 2006). When the National Museum was founded in the nineteenth century, its role was seen as being to nurture and promote the Icelanders' sense of national identity, spurring on the campaign for self-determination after centuries of Danish rule. Hence the museum's acquisitions policy was understandable in the context of the time, and its influence is seen in the bulk of the artistic and valuable objects in the collection.

The museum's policy on acquisitions went on to evolve during the twentieth century, but nineteenth century everyday objects naturally tended to be lost over the decades, so when the museum and the public turned their attention to it, nineteenth century relics proved hard to source. In other words, by the time the public and the museum became interested in the less obvious aspects of social history, it was already too late to acquire that "heritage", which had been lost over the years. A similar development occurred at the National Archives, which originally only collected public records, but later expanded its interest to other documents like personal archives from common people (Baldursdóttir 2022, 217–92; Aronsson & Gradén 2016). This course of collecting practices led to the fact that the everyday life of people of higher status (a class that was only a small part of the nation) are more and better represented in the museum's collection. The belongings in the probate inventories, however, gave a better insight into the everyday life of the Icelandic majority – the common people.

4 Everyday Life of the Nordic

At the beginning of this chapter, a question was raised about the proportions between treasures and useful things in the collection of the National Museum of Iceland and the probate inventories in the National Archives. This seems an unfair question now that the features of these archives have been unfolded. It was evident, when comparing the belongings in the probate inventories to the collection in the National Museum of Iceland, that these sources were different. The archives are made for dissimilar purposes, which gave the appearance of two distinct material cultures, making the comparison an impossible task – like comparing "apples and oranges" – but nevertheless an important one. With this comparison it is possible to gain a better understanding of how powerful tool collection practices can be when it comes to making and shaping Nordic heritage. Given this, it is necessary for historians (and others) to understand the nature of various sources – how they were made and why. The comparison and use of probate inventories and a collection has given a diverse insight into the past which would not have been the case if only one of the archives had been studied.

One thing is certain, we will never be able to know for sure how everyday life was (and is) for everyone. Daily life is varied because people have different obligations, interests, status and routines. What was a "normal" everyday experience and usage of materials for one person was not the same for the next person. Seasonal differences are also a part of how individuals spent their day. During summer when collecting hay, tools like rakes and scythes would have been everyday objects, but not during wintertime. Domestic equipment and clothing in the possession of common people did not consist of the valuable objects the upper class had access to. All of this, therefore, means that the probate inventories are perhaps a better means to study the everyday life of common people, while the collection in the National Museum of Iceland is better suited to focus on the daily life of the upper class.

Some might say that the collection represents, therefore, a glossy image of the past, but that was its purpose from the beginning: to make Iceland a nation to be reckoned with and making it comparable to other nations in the North. Consequently, a big part of materiality in the ordinary life of the public fell between the cracks when collecting and creating heritage. This is also relevant when it comes to collecting and representing the North (Aronsson & Gradén 2016). On the other hand, the advantage of the museum collection is preservation and research on handcrafts, art, artists, fabrics and jewellery making, carpentry techniques and silver smithery. The museum objects themselves contain and reflect on how the materiality of the past would have looked. The objects themselves and their existence gives later generations ideas of how matter was used, made and assembled. The collection brings its audience, researchers and other people, closer to a visual encounter of the nineteenth century, but with certain gaps that are important to acknowledge. That is not the case in the probate inventories, where only the verbal concepts of objects or matter are included.

The method of comparing these two archives is not perfect. First, the data in the probate inventories was much smaller than the whole collection – the probate inventories amount to 33,000 and it is believed that there are around 15,000–20,000 objects from the nineteenth century preserved at the National Museum of Iceland, but not dated. Second, the origin and features of these two archives are very different. Third it is worth mentioning that neither of these archives are flawless; there are gaps in both. The collection lacked, for example, clothes for men, everyday clothing and domestic equipment for food production and preparation. In the probate inventories there is a tendency with the appraisal to not list worthless objects in the case of a farmer or others who owned other more valuable possessions. An example of this are practical things that can be assumed to have been a part of a farmers' inventory, such as tools for handcraft such as needles, wool combs and knitting needles (Baldursdóttir 2022, 208–14). With that in mind, the collection and the probate inventories (and other archives) can make up for each other's shortcomings.

The reality of individuals in the nineteenth century was to go about their daily routines, interests, specializations, farm production, raising children, education, farm duties and so on, which required diverse materiality to survive. The collection in the National Museum of Iceland is more in relation to contemporary interest – what the society at a current time finds to be of interest from the past. At the beginning of the twentieth century, Icelandic society was not interested in the everyday life of common people such as farm workers or farmers, but today their belongings and history are part of the National Museum acquisition policy[13] (Aronsson & Elgenius 2015; Aronsson & Gradén 2016).

The representation of useful things and treasures in each archive was revealed differently in regard of everyday life. Some may find this a negative result for the archives, but what it brings to the table is of much more value for scholars. The comparison gives a more varied picture of the past and also from different parts of society. It also shows the pros and cons of using these sources and confirms that society and the past is diverse. When discussing how the North has been collected, it is therefore important to approach the subject from various angles to get a better understanding of it, including how the collection came about and how it was collected. Then the treasures will definitely become useful things for the present and later generations of the North.

The treasures found in this study were not only "shiny" objects, but also everyday objects. In the case of people's possessions, this meant practical things

13 Regarding similar processes throughout Europe, see Aronsson & Elgenius 2015, and Aronsson & Gradén 2016. Regarding the policy of the National Museum of Iceland, see Rúnarsdóttir 2017.

like clothes and domestic equipment. They were valuable because of their use, not their appearance, and played a big role in people's lives. If it was not there then surely daily life was impacted. People with few or damaged clothes would be cold, and with a limited range of domestic equipment the farm's production would be more complicated. The treasures in the museum collection were more artistic and delicate, but also everyday objects – just for another type of people (the upper class). Everyday life is/was adjustable and depended on a lot of factors such as gender, status, time, region and season. This was evident in the material culture of the probate inventories and the museum collection.

Bibliography

Alþingi. 2011. "Lög um Þjóðminjasafn Ísland. 2011 nr. 140, 28 September." https://www.althingi.is/lagas/151c/2011140.html.

Aronsson, Peter. 2008. "Comparing National Museums. Methodological Reflections." In *Comparing: National Museums, Territories, Nation Building and Change,* edited by P. Aronsson & A. Nyblom, 5–20. Linköping: Linköping University Electronic Press.

Aronsson, Peter & Gabriella Elgenius, eds. 2015. *National Museums and Nation-building in Europe 1750–2010: Mobilization and Legitimacy, Continuity and Change.* Abingdon: Routledge.

Aronsson, Peter & Lizette Gradén, eds. 2013. *Performing Nordic Heritage; Everyday Practices and Institutional Culture.* Farnham: Ashgate.

Aspelund, Karl. 2015. "Breytileg merking menningararfs. Skautbúningurinn, konurnar, landið og þjóðin." In *Menningararfur á Íslandi. Gagnrýni og greining*, edited by Ólafur Rastrick & Valdimar Tr. Hafstein, 73–116. Reykjavík: Háskólaútgáfan.

Baldursdóttir, Anna Heiða. 2022. *Things from the Past: People's Possessions and Museum Collection from 19th Century.* PhD. diss., University of Iceland.

Björnsson, Árni, ed. 1994. *Gersemar og þarfaþing. Úr 130 ára sögu Þjóðminjasafns Íslands.* Reykjavík: Þjóðminjasafn Íslands.

Gunnlaugsson, Gísli Ágúst. 1988. *Family and Household in Iceland 1801–1930: Studies in the Relationship Between Demographic and Socio-Economic Development, Social Legislation and Family and Household Structures.* Uppsala: Uppsala Universitet.

Hamling, Tara & Catherine Richardsson, eds. 2010. *Everyday Object: Medieval and Early Modern Material Culture and its Meaning.* Abingdon: Routledge.

Hálfdanarson, Guðmundur. 2006. "Severing the Ties. Iceland's Journey from a Union with Denmark to a Nation-State." *Scandinavian Journal of History* 31 (3) (September): 237–54.

Holmes, Helen. 2024. *The Materiality of Nothing. Exploring our Everyday Relationships with Objects Absent and Present.* Abingdon: Routledge.

Iggers, Georg. 1997. *Historiography in the Twentieth Century. From Scientific Objectivity to the Postmodern Challenge.* Hanover, NH: Wesleyan University Press.

Jónsson, Guðmundur. 1981. *Vinnuhjú á 19. öld.* Ritsafn Sagnfræðistofnunar 5. Reykjavík: Sagnfræðistofnun Háskóla Íslands og Ísafoldarprentsmiðjan.

Jónsson, Guðmundur & Magnús S. Magnússon, eds. 1997. *Hagskinna. Sögulegar hagtölur um Ísland.* Reykjavík: Hagstofa Íslands.

Jónsson, Már. 2016. "Securing Inheritance. Probate Proceedings in the Nordic Countries, 1600–1800." *Sjuttonhundratal: Sällskapet för 1700–talsstudier* 13: 7–30.

Jónsson, Már. 2012. "Skiptabækur og dánarbú. 1740–1900." *Saga* 50 (1): 78–103.

Karlsson, Gunnar. 1980. "Icelandic Nationalism and the Inspiration of History". In *The Roots of Nationalism: Studies in Northern Europe*, 77–89. Edinburgh: Donald.

Magnússon, Sigurður Gylfi. 2010. *Wasteland with Words: A Social History of Iceland.* London: Reaktion Books.

National Park Service. *Museum Handbook, Part 1.* Acknowledgements 2023. https://www.nps.gov/museum/publications/handbook.html.

Lemonnier, Pierre. 2016. *Mundane Objects: Materiality and Non-verbal Communication.* London: Routledge.

Rúnarsdóttir, Anna Lísa. 2022. "'Hver er heiður þjóðarinnar?' Söfnun Matthíasar Þórðarsonar 1908–1947." In *Heimsins hnoss. Söfn efnismenningar, menningararfur og merking,* edited by Davíð Ólafsson & Kristján Mímisson. Sýnisbók íslenskrar alþýðumenningar 30, 49–62. Reykjavík: Háskólaútgáfan.

Rúnarsdóttir, Anna Lísa, ed. 2017. *Safnastefna á sviði menningarminja. Skýrslur Þjóðminjasafns Íslands 2017.* Reykjavík: Þjóðminjasafn Íslands.

Smith, Laurajane. 2006. *Uses of Heritage.* London: Routledge.

Stephensen, Oddgeir & Jón Sigurðsson. 1853–1854. *Lovsamling for Island,* II, IV. Copenhagen: Höst.

Stephensen, Oddgeir. 1885. *Lovsamling for Island,* XVIIII. Copenhagen: Höst.

Þjóðminjasafn Íslands. 2015. "Saga safnsins. Frá 19. öld til nútímans." Last modified 21 September 2015. https://www.thjodminjasafn.is/stofnunin/um-safnid/saga-safnsins/.

Þjóðminjasafn Íslands. N.d. "Þjóðin og þéttbýlið 1800–1900." Accessed 16 December 2023. https://www.thjodminjasafn.is/syningar-vidburdir/grunnsyning/thjod-verdur-til/timabil/nr/205.

Þórhallsdóttir, Þorgerður. 2022. "Skúfhólkar og skotthúfur." In *Heimsins hnoss. Söfn efnismenningar, menningararfur og merking,* edited by Davíð Ólafsson & Kristján Mímisson. Sýnisbók íslenskrar alþýðumenningar 30, 221–34. Reykjavík: Háskólaútgáfan.

Archival Material

ÞÍ. Þjóðskjalasafn Íslands

Sýslumannasafn (Sýsl.)

Dalasýsla. ED2/2,9: 39. Fylgiskjöl, dánarbú 1866.

Gullbringusýsla. ED2/7,2: 11. Fylgiskjöl dánarbú 1810.

Reykjavík. ED2/2,11. Fylgiskjöl, dánarbú 1811.

Þmjs. Þjóðminjasafn Íslands

1921–202.

2008–5–225.

2015–62–3.

FIGURE 16.1 "Daily life at the end of the 20th century" vignette from "Into the Modern World"

REPRODUCED WITH PERMISSION FROM NATIONAL MUSEUM OF ICELAND. ALL PHOTOS BY AUTHOR

CHAPTER 16

Northern Vernaculars: National and Sundry Objects

John Bodinger de Uriarte

Abstract

The sheer density of museum-object-to-site-to-nation in Iceland is remarkable. This chapter critically considers two stand-out institutions: the National Museum of Iceland in Reykjavík (in particular, its "Into the Modern World" gallery), and the Museum of Sundry Objects, not far from Akureyri.

Both the Sundry and National Museums engage the vernacular as a focus for collecting and exhibition. The Sundry's collection showcases one person's extended engagement with changing circumstances in the North and its relationship to modernity – shifts in traditional building practices, handwork and touch in everyday objects, and both the recession of the vernacular as an element of daily life and work and a celebration of it in the face of its fading. In the National, the vernacular surfaces and expands its object density across the breadth of the twentieth century, again reflecting changes in traditional to modern objects (from handmade woolen slippers of the early twentieth century to the mass-manufactured plastic geegaws of the late twentieth century). These changes are reflected as its object vignette labels shift over time; the Sundry as a site for the wear of objects – especially hand tools – illustrates a different engagement with the vernacular and a shifting imagination of what actually constitutes the everyday.

Keywords

museums – vernacular objects – modernity – Iceland – the lives of objects – collections – exhibitions

Given the number of museums in Iceland, the sheer density of museum collections in the small island nation makes it a singular site for critically considering museums and museum practice, both Nordic and other. With a population of

 | DOI:10.1163/9789004749665_017

about 375,000, Iceland has 266 museums; the range is remarkable, from large and formal museums of history and art and Viking-era settlement to smaller museums dedicated to birds, turf house sites, fish, penises, rocks, and vernacular objects. This chapter specifically explores the use of vernacular objects in the more mainstream National Museum of Iceland in Reykjavík (in particular, elements from its "Into the Modern World" gallery), and the more idiosyncratic Museum of Sundry Objects, a small museum not far from Akureyri, Iceland's "Capital of the North". The National Museum receives about 100,000 visitors a year; the Sundry Museum about 3,000. I see these two museums as representing institutions from both ends of the museum spectrum. The National Museum is a large museum in the nation's capital; it focuses on representing Icelandic history and the processes to contemporary nationhood, it enjoys governmental support and serves as one introduction to Iceland for its increasingly large number of foreign tourists (approximately 2 million in 2023). The Sundry hosts an extensive collection of vernacular objects – the outcome of one man's collecting practices – it is far from the nation's capital, and it is housed in a much humbler edifice (a repurposed schoolhouse and small community center).

Both the Sundry and National museums use vernacular objects as key components of their overall narratives, and both engage with different aspects of the vernacular as a focus for collecting and exhibition. For the purposes of this chapter, vernacular objects are those designed for everyday use, in common circulation, often without embellishment (like hand tools). Vernacular objects

FIGURE 16.2 Central element from "Into the Modern World"
REPRODUCED WITH PERMISSION FROM NATIONAL MUSEUM OF ICELAND

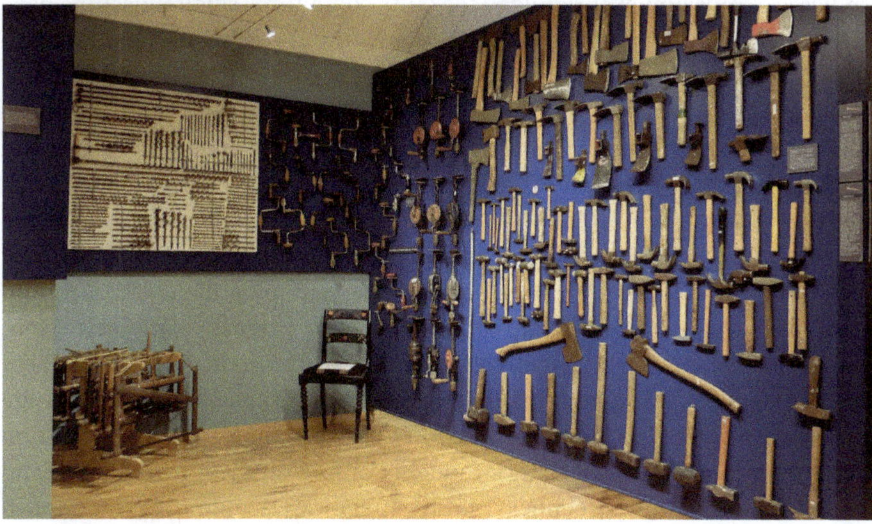

FIGURE 16.3 Gallery from the Museum of Sundry Objects
REPRODUCED WITH PERMISSION FROM MUSEUM OF SUNDRY OBJECTS

also include mass-produced and mass-circulated objects, those that may reveal a traffic in international or global commodities. The objects from the National Museum in focus for this chapter – those featured in "Into the Modern World" – are connected to Iceland's history over the course of the twentieth century. The Sundry Museum's collections were also primarily established during the twentieth century.

The organized collection that forms the Sundry Museum was gifted to the district of Eyjafjarðarsveit in 2003 by Sverrir Hermannson, a local master carpenter, on condition of its public display. The collection is the outcome of his extended engagement with the changing circumstances of the North – shifts in building practices; the roles of handwork, carpentry, and touch; the recession of the vernacular as expressed in hand tools and hardware, and a celebration of such objects in the face of their fading. While the primary focus of the display areas in the Sundry feature carefully arrayed tools and hardware, the museum also exhibits collections of ephemera, like matchbooks and give-away pens from banks and businesses.

While the museum's collection is fairly extensive, I suggest that it fits Fiona Candlin's definition of a small museum: the collection is the result of a single collector's efforts; it is housed in a relatively modest structure; it is well off the beaten path; and it exists outside of some of the main categories for large and often government-supported museums, like art, history, or natural history. While national museums have been the subject of close attention for museum studies and museum anthropology for quite some time, small museums have only recently received increased attention in those fields (see Candlin 2016; Candlin, Butler & Watts 2022; or Robertson & Gregory 2018, for example).

National museums are built to affirm and validate the authenticity of the nation – its claims to sovereignty, which are particularly salient to Iceland and its histories of Norwegian and Danish colonization, and military occupation by Great Britain and the United States – and to articulate the story of the nation to its citizens and other visitors. Critical literature focused on the role of national museums, especially "new museological" work critiquing the colonial mono-narratives of national belonging and identities, is extensive and well established (see Marstine 2006; Lonetree 2012; or Karp et al. 2006, for example). For this chapter, I will be drawing from Arjun Appadurai's consideration of the life of objects, Walter Benjamin's distinctions between concentration/attention and distraction, Michael Taussig's thoughts on tactility and distraction, Paul Basu's concept of "inbetweenness", Kathleen Stewart's contemplation on nostalgia, and some careful thinking on the role of object singularity and density as elements of museum exhibition strategies. My aim is to add to the literature in English of Icelandic museums, with particular attention paid to collection objects.

1 Coming to Ground with the National Museum of Iceland

Collecting in the North, specifically in Iceland, is inextricably tied to the declaration and affirmation of a specific and unique national identity. A settlement history on unpopulated land, *terra nullius,* is an integral element of Iceland's history. It both afforded the opportunity for unique identity construction – how or who is an Icelander, and how is a unique Icelandic identity established over time – but also the absence of some sort of autochthonous narrative of identity and belonging.[1] Iceland's history cannot be tied to the annihilation, displacement, or incorporation of an indigenous population. In this sense, the affirmation of a coherent national identity and history is very much an element of the national museum's main narrative. Somewhat unique for Iceland

1 In here I am making a distinction between museums in settler colonial states with Indigenous populations, like the United States, that work to establish an ur-narrative of belonging through an incorporation of indigeneity, even if it is only to provide a justification for the eradication – culturally and physically – of said Indigenous population. Indeed, a good deal of new museology as applied to US institutions critiqued this sort of presence-becoming-absence of indigenes in the landscape of political and territorial history.

is the connection of a discrete national identity to the declaration of sovereignty and the ending of colonial control in 1944, which allowed Iceland to further enter and participate in the challenges and affirmations of modernity. The National Museum was closed in 2000 for repairs and for the installation of "Into the Modern World" on the top floor; the Museum reopened in 2004. While the lower floors of the National Museum focus on elements of a premodern history, "Into the Modern World" offers a narrative of Iceland embracing modernity over the course of the twentieth century.

Most national museums display some collection of vernacular objects – after all, many of the everyday elements of daily lives, no matter how distantly located in the past, are often featured in national museum narratives and their presented arcs of history. The two museums at the focus of this chapter use contemporary (or recently contemporary) objects to build different stories about the more recent past. In "Into the Modern World" at the National Museum, vernacular objects surface and expand their material presence, collecting greater object density across the breadth of the twentieth century (from handmade woolen slippers in 1906 to the mass-manufactured plastic geegaws of the late 1990s). The National reflects this passage of time through its object vignette labels, in particular those deployed in its central exhibitionary object: a large, oval luggage carousel salvaged from the Icelandic airport. Of course, the expansion and modernization of the airport in the face of the growth in the tourist industry toward the end of the twentieth century is also an important element in understanding contemporary Iceland. In this sense, the carousel itself serves as both a nostalgic and vernacular object, firmly embedded in the modernity narrative of Iceland-confirming-nationhood and coming increasingly into a global view.

The large luggage carousel is segmented into a series of twenty-five mini vignettes, each using a collection of objects to illustrate a particular moment in Iceland's history and its progress "into modernity". As might be expected for a small island nation, many of these vignettes use objects to illustrate key firsts in technologies (the first national phone service and first national postal service, for example), while others later in the series present an array of commodities made increasingly available through global trade and increased access to global marketplaces, like cell phones and sneakers. The initial vignettes include panel texts that identify each displayed object; by the end of the series there are far too many objects to identify singly. The increasing density of objects is an important element of the luggage carousel display. The sequence of the vignettes blurs the hard and fast boundaries between their temporal singularity and the arrangements of those elements that precede and follow. They jostle between illustrating specific moments: the establishment of the telephone

FIGURE 16.4 1906 vignette element from "Into the Modern World"
REPRODUCED WITH PERMISSION FROM NATIONAL MUSEUM OF ICELAND

company, for example, or compulsory education; decades ("Daily Life in the 1960s"); or broader themes, like "Commerce and Industry" or "Safety". The gallery thus stresses the horizontalness of national identity-making, to find it in the telephone and radio networks as well as the declaration of the Republic. The organization and conceit of the luggage carousel may also reflect comparative strengths in the collections (there may simply be more small objects in the collection from the late twentieth century than earlier eras). The twentieth century is a long and complicated time span to cover comprehensively in any museum. The carousel presents both "voiced" and "unvoiced" objects as exhibition elements – not every object on display is referenced in the corresponding vignette's text panel – and the likelihood of an object being "unvoiced" increases as the objects accrete and tumble toward the end of the century, to a display that is almost overwhelming.[2]

Object density is one strategy of the carousel, connected to a sense of the visitor being overwhelmed by objects by its end vignette. Walter Benjamin's consideration of the "distracted gaze" as providing opportunities for "shocks of recognition", moments where an object slips from being part of a general aggregate and demands attention, is also a useful way to consider the work

2 In this they resemble Walter Benjamin's Angel of History where he sees a "single catastrophe, which keeps piling wreckage upon wreckage" we see a chain of events—the temporal narrative illustrated by the carousel's object collections.

of the carousel. In this capacity, the crowding elements of the display offer a number of "ways in" to recognizing and connecting with modern Iceland, different potential moments for recognition to break through the crowd noise of the assembled and entangled things (Basu 2017). Some of these avenues work within the space for imagining the seductive and productive powers of nostalgia, of a longing to connect with the past. Susan Stewart (1993) identifies a key element in considering nostalgia: a seduction of the viewer based on a feeling of recognition, one that comes from a bridging of the gap between signifier and signified. In applying this thinking to exhibitions, museum objects serve as signifiers for museum narratives – they physically embody or "mark" points in the overall museum story about the past, about Iceland ascending to nationhood and the formation of national identity(ies).

In addition to the carousel elements, the gallery also features "heroic objects" – those single, stand-alone objects that museums often use as synecdoches to indicate much larger stories. These objects, and their attendant wall texts, identify larger historical moments or innovations to connect them clearly to their role in "making" the Icelandic nation – a fire hydrant used to indicate the advent of municipal water systems, for example, or a trawler line-cutter used during the Cod Wars (1958–1961, 1972–1973 and 1975–1976). Distinct from the carousel texts, here the wall panels reflect a more extensive and nuanced engagement with the object(s) on display. They connect these objects to larger historical/nation-making themes – the explanatory text for the Trawl Wire Cutters, for example, connects them to the struggle for economic and political independence following the declaration of the Republic, and to the challenges of international militarized and economic conflicts. In their singularity the heroic objects in this gallery gesture toward an imagined whole while they confirm their own authority to "speak" for the general (once there were many hydrants like this) (Bodinger de Uriarte 2021).

The carousel offers both a compelling strategy and a difficult one – while there is provenience provided for some of the objects, and quite a number of them are fairly specific, there are also a number that surface without a clear, associative history. Like commodities, they represent one among many, often story-less, providing "context clusters" without clear grounding. In this sense, they provide an experience of distraction (Benjamin), a display strategy that both challenges some of the contemplative practices we might imagine in a museum, in part reflecting the overwhelming press of capitalism and mass-produced objects as we careen into the end of the twentieth century. Additionally, we can connect the visitor experience to Benjamin's consideration of the optical unconscious, or what may be revealed through the distracted gaze partially confirmed through the exhibit's sheer object density. While these theoretical concepts are not directly linked in his writing – one is offered as a way to consider the power of film and photography and its ability to "fix" an entire scene in place for more careful consideration, often discovering things that the eye certainly saw, but did not register; much like the state of apperception that is an element of consciousness, they are useful in thinking through the exhibition's impact. Distraction, however, offers a way to understand distinctions between those objects that we contemplate, that we fix under our concentrated gaze – as in the way the heroic objects are presented and viewed in the museum – and those that we catch from the corner of our eye, embedded in a mixed multitude of attention possibilities – as in the sheer density of the objects offered by the carousel's last vignette, devoted to the 1990s. The density and materiality of the vignettes thicken, congeal, contain even more objects than can be interpreted by their panel texts. The popular cultural circulation of commodity objects is almost overwhelming by the end of the century, and the carousel is mimetic of this growing circulation density. The number of things that can be pointed to as particularly or specifically Icelandic also changes over time as Iceland's transnational and global trade increases, an expanding entangled network of connection and reference. The nostalgia of the nation and nationhood offered by the first vignettes slowly becomes the nostalgia of internationalism and participation in global commerce.

2 Exploring the Museum of Sundry Objects

The Museum of Sundry Objects also offers an exhibitionary space dense with the display of miscellaneous things and vernacular objects, the most compelling of which may be the collection and careful, orderly display of various hand tools. Sverrir Hermannsson's collection reflected his specialization in the restoration of local old houses and other structures primarily from the nineteenth and earlier twentieth century. His work led him to a variety of objects – those salvaged from houses and other buildings, discovered in crawl and storage spaces, old nails and fixtures, and things recovered from dumpsters and other sites. As the museum's brochure states: "Where others saw trash, Sverrir spotted treasures" (The Museum of Sundry Objects & Þráinsdóttir 2017). He was an insatiable collector, adding over a thousand objects to his collection each year, sometimes absorbing other collections into his own. He arranged his collected objects painstakingly in specially built cabinets and on custom-built display boards of various sizes. In one section of the museum a disordered array of objects is displayed, offering a glimpse into Sverrir's process between collecting and organizing for display.

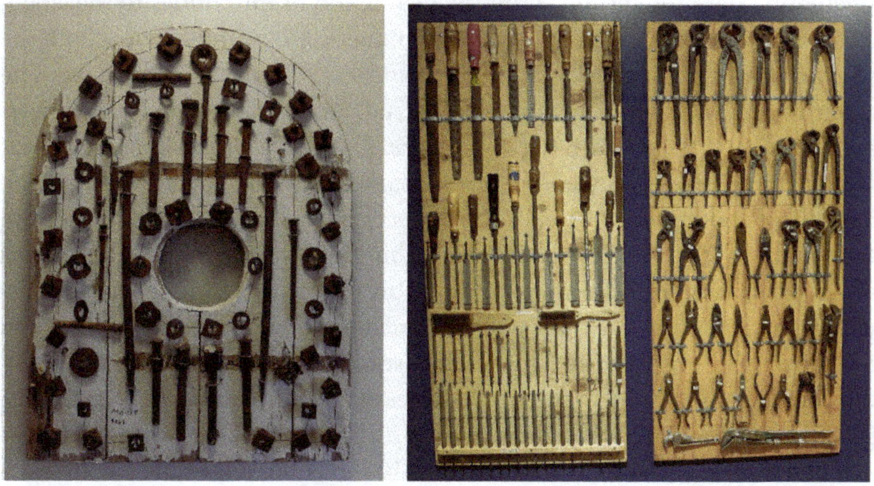

FIGURE 16.5 Exhibition details from the Museum of Sundry Objects
REPRODUCED WITH PERMISSION FROM MUSEUM OF SUNDRY OBJECTS

One compelling element of the Sundry's display strategy is its use of arrays of similar objects rather than singular "heroes". Its claim to the singularity of objects, however, is little different – displaying multiple "like" objects calls attention to the unique attributes of each. The arrays re-tune the visitor perspective to seek difference in the similar, and to recognize the networks that give them meaning. The majority of the objects in the Sundry show the marks of their own singular passage through time and use, "worn with the hand's obligations, as by acids, steeped in sweat and in smoke" (Neruda 1974, XXI). The lived history of the objects reveals their singularity through their own constellations of wear. The marks of wear shift implications over time. During the COVID pandemic, for example, the touch implied by handwear presented a point of potential contamination and danger – the life course of the object was shadowed by different suspicions, different enacted cautions (Bodinger de Uriarte 2021b). In other considerations they mark the transition (or surrender) from human-powered tools to electric and pneumatic tools.

The exhibition spaces in the Sundry are thick with object density. While pathways through the galleries are clear and uncluttered, the walls feature panels offering arrays of hammers, for example, augers, or collected keys or lock-plates. Each object bears a set of handwritten initials indicating its origin – those that Sverrir collected during his work, those that were donated, those that came from construction-site dumpsters or the Salvation Army. Here the trace of the object is also inscribed through its collection. The orderliness of Sverrir's objects can be read as their disorder, or at least their coming to singularity. Here the move is from the distraction of multitudes to the singular focus of contemplation. An array of hammers, for example, raises both the question of design (collective) and the histories of their use (individual). The collections of nails offer examples of past nail design, for example, the hinges and locks past designs of hinges and locks. The utility of the objects, in part arrested from their discovery site, is still marked with the potential of use. The hum of this potentiality is part of the objects' backstory, their life course, and their eddying in the display boards and cabinets of the museum. In a very real sense, the objects exist in a liminal state, on the threshold of permanent disuse, but making bare the marks of their once-busy lives. They are drawn from the teeming of everyday life, marked by use and nostalgic projections, forgetfulness and remembrance (the twin engines of any museum). The grouped displays impart agency for each object, in part through making the human wear and handling of each of them more accessible and more of the life path of each object. Where the heroic museum object wants one to move from the individual to the imagined array, the array conceit of the Sundry opens a possibility for recognizing the singularity of objects. The museum also serves to deliver objects from the liminality of abandonment, ruin, and disuse to the liminality of the collection, the exhibition, and the museum display. In many ways, this reflects Basu's concept of inbetweenness: "a middle ground, a contact zone, a borderland" (Basu 2017, 8).

The familiarity of tools is an important element of the Sundry's main exhibition strategy. The handedness of them, how they hold trace elements of touch,

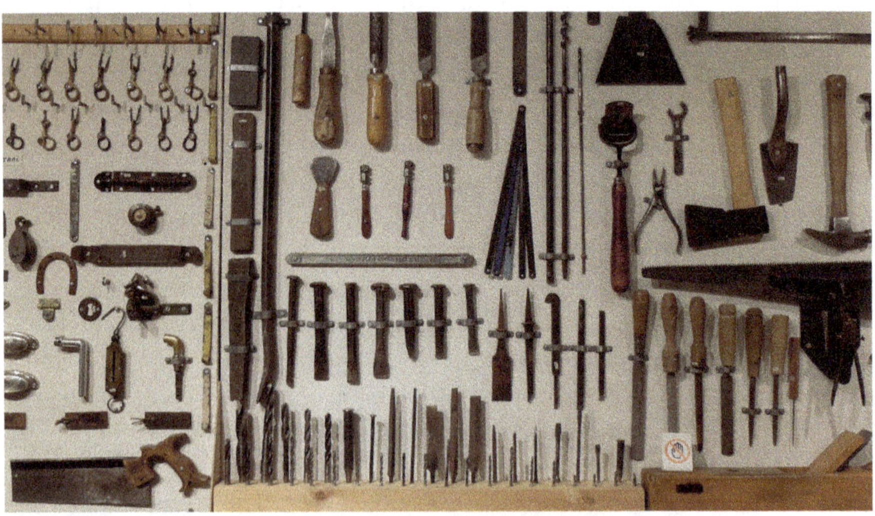

FIGURE 16.6 Exhibition detail from the Museum of Sundry Objects
REPRODUCED WITH PERMISSION MUSEUM OF SUNDRY OBJECTS

how they are shaped to respond, to be directed, the stains of hand oils and accidents and experiences all embedded on the surfaces or dented down in deep, relaying histories. It is not hard to imagine the lives of these tools, nor is it hard to think that they now live in a state of arrest, a sort of limbo of practical usefulness, changed from one utility to a different set of utilities, held in place.

The tools in the Sundry illustrate how "the vernacular" is a kind of liminal status, moving from new object to one used in the everyday, in the negotiation of work and non-work, between object singularity and a collection of "fellows" or like kind. Exhibitions of the everyday in museums large and small help make plain how exhibition designs work to establish distinction and difference at the same time they may encourage recognition ("my dad or mom had one of those"). Vernacular collections potentially accentuate the porousness of museums, a negotiation between the rarified and the recognizable. It is in this space that these museums operate, extending the authority of museum display. Vernacular collections open a door to recognition or question the validity of the objects displayed, or their institution ("why am I looking at this? I have one of these at home"), or their potential market value ("I never realized that it was 'museum worthy'). It is in this space that museums operate, between nostalgia and discovery. The arrangement/design is integral in keeping a dialogue with objects active. But nostalgia is charged, is tricky. As Kathleen Stewart suggests: "Nostalgia, like the economy that it runs with, is everywhere. But it is a cultural practice, not a given content; its forms, its meanings, and effects shift with the context – it depends on where the speaker stands in the landscape of the present" (Stewart 1992, 252). The force and experience of nostalgia are elements in the exhibition of contemporary vernacular objects. There is often likely a connection between these exhibitionary objects and the life experiences of the viewer(s).

3 Conclusion

Both museums under discussion use a pressing multitude of vernacular objects to create a space of potential distraction (Benjamin) as an element of the visitor experience. "Distraction" challenges some of the contemplative practices we might imagine in a museum, a response, in part, reflecting the overwhelming press of capitalism and mass-produced objects. In the National Museum, the visitor passes through the lower galleries, often focused on heroic objects firmly lodged in a projected and imagined past – a long boat, for example, or a turf house interior – to a closing gallery that is almost overwhelming in its object density. The objects in the Sundry do not completely throw off the elements of mass production, but the patina and hand-oil staining of the objects locates them in a different kind of use, marked by labor and personalized through application, the hand tools offer a clearer human/bodily connection, the ghost of their tactile lives. Modern vernacular collections are often informed by their imagined history of touch. Not the possibility of touch – these are museum objects, after all – but the objects present in a space nostalgic of touch, or one becoming nostalgic as visitors are offered different moments and opportunities to identify with the objects through their own personal histories with useful things.

The distraction offered in both museums reflects "the type of flitting and barely conscious peripheral-vision perception unleashed with great vigor by modern life at the crossroads of the city, the capitalist market, and modern technology" (Taussig 2009, 148). I do not want to make the argument that these exhibits – both the Sundry's and the National's "Into the Modern World" galleries – only operate at the level of peripheral experience. Indeed, the National's stand-alone objects demand contemplation both by their singularity and their accompanying and developed panel texts. But I will suggest that the objects on the carousel and the objects in both ordered and disordered display spaces in the Sundry make some of their impact through their sheer density, through the kind of directed and distracted gaze afforded by their multitude. The National's carousel objects accrete as the depicted decades advance; the Sundry's objects demand a first appreciation of their distracting aggregation, then beckon for closer inspection of different singular elements in the offered arrays.

When thinking about collections, one direction for contemplation is to consider how the objects come together – what logics do they reveal, what efforts cohered, what stories narrated? What a focus on the vernacular makes possible for the museum visitor is an oscillation between the contemplation of the exhibition object and its recognition as an element of a recognized everyday. Here the performative space of the museum – a space that brings objects into potential close attention or veneration – also questions itself. "Why is this object brought to my attention *here* and how might that shape how I see similar objects in non-museum spaces?" Here the connective tissue between the museum object and its vernacular fellow in non-museum circulations may be a threshold of re-engagement, what Sandra Dudley calls incorporation, something that moves the object from their "home" in the museum world to "affective engagement" (Dudley 2017). What Dudley identifies seems similar to Benjamin's notion of the distracted gaze that ends up fixing on something, that responds to something that rises up from the surround. Indeed, Benjamin's "shock of recognition" that awakens attention brings to mind Roland Barthes's formulation of *punctum* – the piercing revelation that is sometimes felt through an image. Although Barthes uses this as a way to talk about photography, the sheer density of display objects can also be seen in this light, as offering a plane of objects embedded in a sea of potential triggers of connection or recognition. In Sandra Dudley's words:

> How can objects become once more fully incorporated and active within social life? How, from their own viewpoints, can they come to be "at home" in their museum world? My contention is that this occurs in rare, unpredictable and fleeting, intensely affective engagements between the object and the museum visitor: moments when something catches someone's eye and surprises, delights, perplexes or appalls [sic], drawing them to stop and stare. (Dudley 2010, 51)

Another element consistent in both collections and exhibitions is an engagement with modernity which offers a fruitful intersection between the Sundry and the National. Sverrir was working to preserve vernacular things from the past, but also elements of construction detail and tools that were losing ground in the face of the modern. The National's collection – as do most national museum collections – asserts its historical narrative as reflecting a "glide into the limitless future" (Anderson 1991, 11–12). Not only does the museum visitor "land" in a present made inevitable by the progression of the museum's narratives, but the future is also presented as a potentially untroubled territory, a continuation of the clear steps of progress made visible throughout the museum's galleries and stories. Institutional collections are not discrete networks of collected and ordered (finite) groupings, but entangled assemblages (Tsing 2017) that cross back and forth between different understandings and contexts. In thinking about collections, one direction for contemplation is how the objects come together – what logics do they reveal, what efforts cohered, what stories narrated, what nostalgias invoked? However, as Kathleen Stewart (1992, 227) suggests above, "the experience of nostalgia depends on where one stands". In other words, nostalgia is a positioned and positioning strategy, a way to consider the powerful feelings invoked by objects and narratives and their pasts, but not as an encapsulating answer for how they work – a way to understand process, not closure.

Bibliography

Anderson, Benedict. 1991. *Imagined Communities: Reflections on the Origin and Spread of Nationalism.* NY: Verso Press.

Appadurai, A. 1986. "Introduction: Commodities and the Politics of Value." In *The Social Life of Things*, edited by Arjun Appadurai, 3–63. Cambridge, UK: Cambridge University Press.

Basu, Paul (ed). 2017. *The Inbetweenness of Things: Materializing Mediation and Movement between Worlds.* New York: Bloomsbury Academic Press.

Bodinger de Uriarte, John. 2021. "Selfies, Sovereignty, and the Nation-State: 'Into the Modern World' at the National Museum of Iceland." In *Mobility and Transnational Iceland: Current Transformations and Global Entanglements*, edited by Kristín Loftsdóttir & Unnur Dís Skaptadóttir, 193–208. Reykjavík: University of Iceland Press.

Bodinger de Uriarte, John. 2021. "Time, Contamination, and the Handedness of Things." Part of the panel "Small Museums: Large Truths" for annual meetings of the American Anthropological Association.

Bodinger de Uriarte, John. 2020. "At Large in the Empire of Things: The Museum of Sundry Objects." *Museum and Society* 18 (2): 280–89.

Candlin, Fiona, Toby Butler & Jake Watts. 2022. *Stories From Small Museums.* Manchester, GB: Manchester University Press.

Candlin, Fiona. 2016. *Micromuseology*, New York: Bloomsbury Academic.

Crane, Susan. 2000. "Curious Cabinets." In *Museums and Memory*, edited by Susan Crane, 60–80. Stanford, CA: Stanford University Press.

Desilvery, Caitlin. 2017. *Curated Decay: Heritage Beyond Saving.* Minneapolis, MN: University of Minnesota Press.

Dudley, Sandra H. 2017. "The Buzz of Displacement: Liminality among Burmese Court Objects in Oxford, London and Yangon." In *The Inbetweenness of Things: Materializing Mediation and Movement between Worlds*, edited by Paul Basu, 38–56. New York: Bloomsbury Academic Press.

Dudley, Sandra H. 2010. "Museum Materialities: Objects, Sense, Feeling." In *Museum Materialities: Objects, Engagements, Interpretations*, edited by Sandra H. Dudley, 1–17. New York: Routledge Press.

Stewart, Kathleen. 1992. "Nostalgia – A Polemic." In *Rereading Cultural Anthropology*, edited by George Marcus, 252–66. Durham, NC: Duke University Press.

Stewart, Susan. 1993. *On Longing*. Durham, NC: Duke University Press.

Taussig, Michael. 2009. "Tactility and Distraction." *Cultural Anthropology* 6 (2): 147–53.

The Museum of Sundry Objects & Þráinsdóttir, H. 2017. "Sverrir Hermannsson's Saundry [sic] Collection," translated by Frances Jane Milne McQueen & Rafn Kjartansson [museum brochure].

Tsing, Anna. 2017. *The Mushroom at the End of the World.* Princeton, NJ: Princeton University Press.

Turkle, Sherry. 2011. "Introduction: The Things that Matter." In *Evocative Objects,* edited by Sherry Turkle, 1–10. Cambridge, MA: The MIT Press.

van de Wetering, Ernst. 2012. "The Surface of Objects and Museum Style." In *Museum Objects: Experiencing the Properties of Things*, edited by Sandra H. Dudley, 103–08. New York, NY: Routledge Press.

OBJECT: Island

FIGURE 17.1 A modified photo of polar explorer Robert Peary with a fictitious island, Crocker Land, in the background which was generated by the author using AI
ORIGINAL PHOTO OF PEARY "SELF-PORTRAIT, CAPE SHERIDAN, CANADA, 1909" ACCESSED FROM WIKIPEDIA

CHAPTER 17

False Islands of the Arctic: The Enigmatic and Enduring Arctic Landmass, Crockerland

Grace Shephard

Abstract

Explorers have long been driven by the desire to discover new continents, stake claim to resources, and collect a piece of history. The promising sight of distant land, however, had sometimes been revealed to be a falsity. Concepts like mysterious islands, lost cities, sunken landmasses, and *Fata Morgana* have fascinated sailors, societies, and scientists. Are these purely imaginary, or do they have roots in reality? With harsh environmental conditions, including sea ice, glaciers and strong climatic seasonality, the highest latitudes are ideal locations for finding such enigmatic landmasses. The late nineteenth and early twentieth centuries saw the race to the North Pole and the supposed discovery of unchartered Arctic islands. One famous example is 'Crocker Land', sighted by explorer Robert E. Peary in 1906 near Ellesmere Island. The ill-fated 1913 "Crocker Land Expedition" aimed to prove its existence, but land was never reached. Erased from maps and minds in its original form, the idea of Crocker Land has instead evolved into a material scientific concept in the field of geology adopting the same name. An ancient landmass, termed 'Crockerland' by Canadian geologist Ashton F. Embry, was once located at the centre of the Arctic and has since been fragmented by geological processes. Crocker Land might therefore not be purely imaginary; Peary was just 200 million years too late in its sighting. This article explores links between Arctic exploration and geology, the concept of lost or false islands, and how society and scientific discoveries (and un-discoveries) are intertwined.

Keywords

geology – exploration – mirage – Arctic Ocean – Crockerland

1 Reaching the Unmapped

Achieving exploration firsts, discovering resources, collecting objects, facilitating territorial expansion, and gaining notoriety amongst peers have variously driven explorers throughout the centuries. It follows that the sighting of a previously undiscovered continent or landmass was somewhat of a holy grail for explorers. The polar regions, in particular, have long been considered a final frontier for exploration – and can still be considered so in modern times, especially in the scientific sphere. Considering the navigational difficulties and climatic conditions, it is not surprising that the physical nature of the high Arctic was not known until recent times. Today, we know that an ocean basin occupies the center of the Arctic, and many parts of the Arctic Ocean's seafloor are still being mapped at high resolution (Jakobsson et al. 2012). However, the idea of a northernmost land persisted for many centuries, dating back to the ancient Greek's idea of *Hyperborea*. With the North Pole remaining unreached by the turn of the nineteenth to twentieth centuries, it therefore held the potential for new territorial discoveries. Indeed, around the early twentieth century the tangible promise of a new landmass lay on the horizon, the "Lost Atlantis of the North" – and it was a landmass named Crocker Land.

Crocker Land's original sighting (Figure 17.2a) is attributed to Robert Edwin Peary (1856–1920), who was an American polar explorer, Navy admiral, and engineer. Peary is most famous for his 1909 claim to being the first to reach the North Pole. He has several namesakes around the world including Peary Land, -Glacier, -Nunatek, -Cape, -Bay, -Channel, -Mount, -Crater, USS-, Camp-, and has featured on stamps, and many other memorabilia. To explain the story of

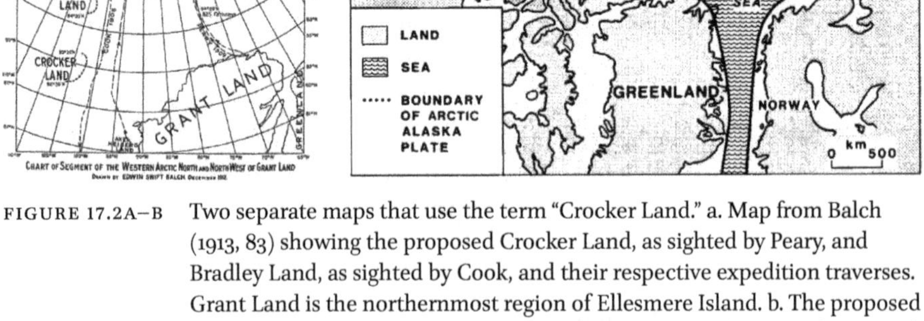

FIGURE 17.2A–B Two separate maps that use the term "Crocker Land." a. Map from Balch (1913, 83) showing the proposed Crocker Land, as sighted by Peary, and Bradley Land, as sighted by Cook, and their respective expedition traverses. Grant Land is the northernmost region of Ellesmere Island. b. The proposed geological landmass, Crockerland (Embry, 1993), located in a similar central Arctic region but at an ancient geological time (around 200 million years ago). 17.2b is reproduced with permission from Ashton Embry

Crocker Land, one must first turn to the succession of Peary's Arctic exploits, which included several expeditions between 1886 and 1909. These included the 1886 dogsled trip to Greenland, the 1891–1892 expedition to determine if Greenland was an island or a peninsula extending to the North Pole, the 1898–1902 expedition to Cape Morris Jessup (northernmost point of mainland Greenland), the 1905–1906 "Farthest North" journey to 86 °30' latitude (which becomes crucial for this chapter here), and the famous 1908–1909 North Pole claim. These expeditions had several financial and scientific backers, including the American Geographic Society, the National Geographic Society, and the Peary Arctic Club (established 1898), amongst others.

The 1908–1909 North Pole expedition included 22 explorers, with the final leg to the pole comprising six men: Peary, Matthew Henson, and four Indigenous guides; Ootak, Egigingwah, Seegloo, and Ooqueah (Peary 1910). According to their records, the North Pole was reached on 6 April 1909 and this was communicated via telegram on 6 September 1909. The feat was certified later in 1909 by the National Geographic Society and Naval Affairs Subcommittee of the U.S. House of Representatives. However, the claim was not without significant controversy (Bravo 2019). Focusing on the Peary expedition itself, there was a lack of additional independent and navigational validation, for example, with longitudinal, speed, and camera lens records.[1] The National Geographic Society apparently limited access to diaries and records from the expedition, and only made Peary's diary available in 1986 where it was re-examined in 1989 by polar explorer Wally Herbert (1934–2007) who cast doubt on Peary's claim (Herbert 1989). The American Geographic Society and Scandinavian Society also never formally recognized Peary's claim, and has been the topic of numerous historical investigations (e.g. Henderson 2005).

An additional major dispute was only revealed once Peary had returned from his expedition, namely, that American Dr. Frederick Cook (1865–1940) had reached the North Pole before him. Cook was a member of Peary's earlier expeditions including the 1891–1894 North Greenland voyage. For Cook's own North Pole claim (1907–1909 expedition), he was accompanied by two Inuit men, Ahwelah and Etukishook, and claimed to have reached the pole on 21 April 1908 (a full year before Peary). The telegram communicating this was sent on 1 September 1909 from the Shetland Islands, five days before Peary's own telegram. The delay from Cook was caused by adverse conditions on their return journey and the need to over-winter on Devon Island in the Canadian Arctic Islands. According to reports, the navigational details of Cook's expedition were sparse

1 The Peary controversy has generated a variety of discussions. For further information see: Henderson 2009; Tierney 2009.

and the claim was rejected by a Danish panel. Debate about the two competing claims continued for several years, the saga featured in international newspapers of the time and included appeals to national agencies.

In an attempt to disentangle truth from fiction, the parties involved and their supporters returned to past expedition records and Peary's publications. Crucially, in Peary's 1907 book, *Nearest the Pole,* he details that during his 1905–1906 expedition he spotted a distant land to the northwest from Cape Thomas Hubbard on Axel Heiberg Island.

> North stretched the well-known ragged surface of the polar pack, and northwest it was with a thrill that my glasses revealed the faint white summits of a distant land which my Eskimos claimed to have seen as we came along from the last camp. (Peary 1907, 202)

This distant land was estimated to be around 210 km away, located at 83 °N and 100°W. Maps from this book (Peary 1907, 443–48) include this mark as a dashed line and "Crocker Land". Crocker refers to George Crocker (1856–1909) who was a banker from San Francisco, a member of the "Peary Arctic Club", and financial backer of Peary's 1905–1906 expedition during which this feature was sighted. It is not inconceivable that Crocker may have been seen to hold the purse strings to future expedition financing, however, the San Francisco earthquake of 1906 and Crocker's death in 1909 changed that. This apparent sighting of Crocker Land is important for Peary's North Pole claim in 1909. This is because Cook (who traversed nearby the area a year prior to Peary) claimed that there was no evidence of Crocker Land. It has been suggested that if Peary, or his supporters, could prove that Crocker Land existed it may provide the smoking gun to invalidate Cook's North Pole claim, because he must have lied about passing through that area (and therefore the Pole). Furthermore, it would also suggest that Nansen's (1907a, 470) statement that "There is, in fact, no known land in the north whose northern coasts are not now explored" was also erroneous.

It is worth noting, however, that Cook claimed to have seen a different land. This island was located further north than Crocker Land (Figure 17.2a), at around 84–88°N, and was named Bradley Land, also after an expedition financer. Bradley Land was documented in Cook's 1911 book *My attainment of the Pole*:

> Soon the western heavens, ever a blank mystery, cleared. Under it, to my surprise, lay a new land. I think I felt a thrill such as Columbus must have

> felt when the first green vision of America loomed before his eye [...] But it was real land with all the sense of security solid earth can offer [...] This new land was never clearly seen. A low mist, seemingly from open water, hid the shoreline [...] there were two distinct land masses [...] the lower coast resembled Heiberg Island. (Cook 1911, 243–46)

The account also included photographs of the proposed land, however, they have since been reported to have been taken whilst looking southwards towards Axel Heiberg Island, and not in the purported direction of Bradley Land (Bryce 2008). Both Crocker and Bradley lands appeared in several other map iterations from the time, including in Balch (1913) *The North Pole and Bradley Land* (Figure 17.2a). It is worth commenting here that given the imprecision of the sightings, resolution of maps at the time, and their locational similarities, it seems ambiguous as to how one would clearly distinguish between Crocker Land and Bradley Land, in any case.

Today, we know that there are no islands or lands in the central Arctic Ocean. The Arctic is centered by a deep ocean (around 3000 m depth; Figure 17.3c) which is variably covered by seasonal or multi-year sea-ice. Surrounding this central deep area are continental shelves (typically 300 m depth on average) which include islands and archipelagos, such as Svalbard or the De Long Islands. Even today, many parts of the Arctic Ocean remain impenetrable to icebreaking ships, particularly in the regions of the thickest sea-ice north of Greenland. However, we know that a central Arctic landmass does not exist (or other deep marine islands for that matter). So, what happened to the concept of Crocker Land? To discuss further, we return to a brief summary of northern polar exploration and mapping followed by alternative explanations for Crocker Land's apparition.

2 Mapping the Unreached

Perhaps more so than anywhere else in the world, cartographic maps depicting the Arctic region have significantly evolved over recent centuries (Figure 17.3). Coastlines, typically indicating the location of sea-level at the time of mapping, are key reference markers and have been the physical and metaphorical departure and arrival points for exploration throughout time. With population centers globally concentrated at coasts, those with highly traded routes, fishing and whaling hotspots, and/or easily navigable regions became more resolved with time (and territorial claims along with it). The push for mapping final frontiers

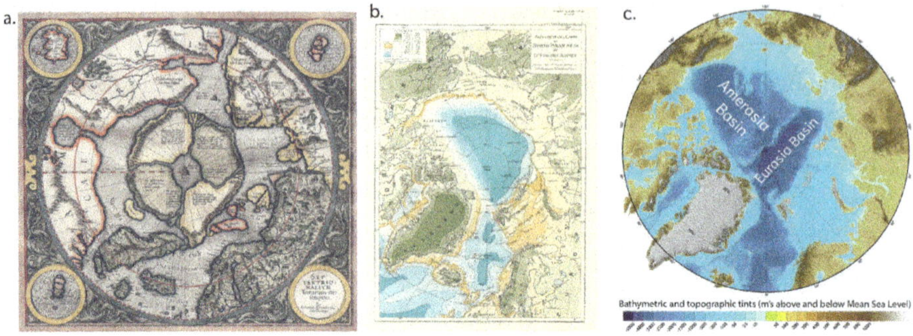

FIGURE 17.3A–C Historical and modern maps of the Arctic, from left to right: a. Mercator map republished by Hondius in 1607, b. Fridtjof Nansen's map in 1907, and the IBCAO map (International Bathymetric Chart of the Arctic Ocean; Jakobsson et al., 2012). The two major deep basins of the Arctic Ocean, the Amerasia and Eurasia basins, are labelled in c

and claiming any "new" lands naturally expanded to lesser known or seldom reached areas of the globe. This expansion corresponded to locations farther away from Europe and those with challenging climatic conditions, such as the poles.

A survey of historical maps of the Arctic showcases navigational advancements and scientific progress as well as imagination and artistic flair. By projecting a spherical globe (the Earth) onto a 2-D surface, any given map has inherent distortions of certain regions, such as the relative size or shape of a continent. Historically, maps showing the world were often rectangular in shape and centered on the Equator, leading to distortion of the polar regions. The development of polar-centered projections was therefore essential for high latitude mapping and exploration. One of the first and most famous historical maps to have a polar-centered projection was the Mercator-Hondius map of 1607 (Figure 17.3a).[2] This map depicted the centuries-old idea that there was a large magnetic rock at the North Pole, which was surrounded with a quadrant of lands with rivers and mountains. Encircling this was a circum-polar ocean, bordered by coastlines of the Eurasian and American continents, though with very uncertain detail and control. In a letter from Mercator in 1577, the Arctic is described as:

> In the midst of the four countries is a Whirl-pool [...] into which there empty these four indrawing Seas which divide the North. And the water rushes round and descends into the earth just as if one were pouring it

2 Atlas minor Gerardi Mercatoris a I. 1628.

> through a filter funnel. It is four degrees wide on every side of the Pole, that is to say eight degrees altogether. Except that right under the Pole there lies a bare rock in the midst of the Sea. Its circumference is almost 33 French miles, and it is all of magnetic stone. (Taylor 1956, 60)

Three centuries later, by the time the 1907 map of Fridtjof Nansen was made (Figure 17.3b),[3] first-order details of the coastlines surrounding the Arctic were featured, including the islands of Svalbard, Frans Josef Land, the De Long Islands, and the Canadian Arctic Islands. Mainly based on depth soundings from marine expeditions, Nansen's map was one of the first to propose that the Arctic Ocean was a deep ocean basin, with no central landmasses or islands. Nonetheless, as of the turn of the twentieth century no-one had laid claim to reaching the North Pole, and thus the comprehensive nature of lands or oceans was not verified.

The first International Polar Year (1882–1883) heralded a focus on scientific exploration, and the turn of the twentieth century corresponded to a number of major polar expeditions. For the Arctic, whilst Indigenous knowledge and exploration occurred for centuries prior, the scale and descriptions of polar exploration during this period is very much from a perspective of scientific colonialism. Referring to the Western expeditions, the majority of the largest and most expensive expeditions during this period aimed at addressing one, or more, of these major topics; reaching the North Pole, crossing Greenland, discovering the fate of lost expeditions, such as the 1845 Franklin expedition, and navigating the Northwest and Northeast passages.

The financial costs associated with expeditions, both for equipment and team members, created a competitive scenario for attaining financial backing. Both government-funded and private-supported expeditions occurred, sometimes financed by a single wealthy patron or organization. It follows that a centuries-old tradition is to name a newly discovered feature or place after someone or something (e.g. city or place name). In many instances, these were named after a Western male in high social, cultural and/or economic standing, especially someone who may have financed this expedition and/or the city or places they were from. This is a classic exploration tactic and example of colonialist perspectives i.e. a "newly" discovered land (whether an Indigenous population existed, or not) would become the objective (and subsequent object) of a Western/Imperialist country or person's possession. Naming the discoveries after the name of their patrons (if not the explorer themselves) was one important incentive to not only honour the expedition's supporters but, crucially, to pre-empt further exploration financing. Geographic features

3 Nansen 1907b.

around the world are testament to this tactic and can be considered a form of tribalism i.e. imposing a name, culture, and/or geographic demarcations. Another potential and distinct element of tribalism in early polar exploration is that of groups, teams, and rivalry – to undermine the claims of others and to use the findings for their own ambitions. To disprove the existence of a newly discovered feature would undoubtedly create an element of conflict and competitiveness amongst those who it was named after, and amongst those who opposed it. It follows that the Arctic was subject to this side of exploration as well.

3 The Hunt for Crocker Land

Returning to Crocker Land, in order to establish whether it existed or not, several members of the Peary Club set out on what is known as the Crocker Land Expedition from 1913–1917 (Welky 2016; Solberg 2019). Crocker Land was once described as the "last geological problem of the North" by Donald Baxter MacMillan (1874–1970), who led this campaign. MacMillan wrote:

> In June 1906, Commander Peary, from the summit of Cape Thomas Hubbard, at about latitude 83 degrees N, longitude 100 degrees W, reported seeing land glimmering in the northwest, approximately 130 miles [210 km] away across the Polar Sea. He [Peary] did not go there, but he gave it a name in honor of the late George Crocker of the Peary Arctic Club. That is Crocker Land. Its boundaries and extent can only be guessed at, but I am certain that strange animals will be found there, and I hope to discover a new race of men. (MacMillian 1918)

A central Arctic landmass had been hinted at, even before Peary's sighting and publication. Tidal charts, sea-ice patterns, bird and animal migration patterns, reports by whalers, and Indigenous stories (Welky 2016), were used as evidence for a northern land. For example, in 1904 and onwards, American oceanographer and mathematician Rollin A Harris (1863–1918) looked at ocean current and tidal information and suggested a large "North Polar Land". This landmass also featured in later government-produced publications including US Coast and Geodetic Survey maps (Harris 1904).

A main objective of the Crocker Land expedition was therefore to investigate the existence of Crocker Land but also "geology, geography, glaciology, meteorology, terrestrial magnetism, electrical phenomena, seismology, zoology, botany, oceanography, exthnology, and archaeology". The expedition was backed by the American Museum of Natural History, the American Geographical Society, the University of Illinois, and others. A large collection of items from this expedition are now located at The Peary-MacMillan Museum of Bowdoin College in Maine and the Spurlock Museum of the University of Illinois Urbana-Champaign. The expedition is also documented in the 2016 book *A Wretched and Precarious Situation* by David Welky, amongst other literature. It is worth noting that a Canadian Arctic Expedition (1913–1916) led by Viljalmur Stefansson was also established at this time to search for new land.

The Crocker Land Expedition included MacMillan and 12 men, including at least seven Inuit guides. According to records, the challenging conditions meant that most members turned back early on in the trip. MacMillian and three men; Green, Piugaattoq and Ittukusuk, continued.

On 21 April 1913, Crocker Land once again was hinted at, and it was detailed: "Great heavens! What a land! Hills, valleys, snow-capped peaks extending through at least one hundred and twenty degrees of the horizon" (MacMillian 1918, 87). One of the Indigenous guides, Piugattoq, however, stated that it was an illusion or "poo-juk" (mist):

> I turned to Pee-a-wah-to anxiously and asked him toward which point we had better lay our course. After critically examining the supposed landfall for a few minutes, he astounded me by replying that he thought it was poo-jok (mist). E-took-a-shoo offered no encouragement, saying "Perhaps it is." Green was still convinced that it must be land. (MacMillian 1918, 80)

Later, however, on the fifth day of the expedition it was written:

> The day was exceptionally clear, not a cloud or trace of mist; if land could be seen, now was our time. Yes, there it was! It could even be seen without a glass, extending from southwest true to north-northeast. Our powerful glasses, however, brought out more clearly the dark background in contrast with the white, the whole resembling hills, valleys and snow-capped peaks to such a degree that, had we not been out on the frozen sea for 150 miles, we would have staked our lives upon its reality. Our judgment then, as now, is that this was a mirage or loom of the sea ice. As we watched it more narrowly its appearance slowly changed from time to time so we were forced to the conclusion that it was a mirage of the sea ice. (MacMillian 1918, 87–88)

It was thus proven that Crocker Land (and perhaps Bradley Land in turn) did not exist. At this point, the expedition changed objectives and took a murderous turn. Two members of the party, Piugaattoq and Green, were subsequently sent to explore to the west. It is documented that the pair argued and Piugaattoq was fatally shot in the back, despite Green claiming he died in an avalanche. Green was never prosecuted.[4] The remainder of the expedition was marred by poor weather and delays of rescue ships, and the party returned in 1917.

4 False Islands

It has also been proposed that Peary's original Crocker Land sighting may have been fabricated from the start: a distinct lack of expedition-contemporaneous diary entries by Peary referring to Crocker Land lends strongly to this (Welky 2016). Perhaps it was a last-minute attempt to boost sales of Peary's book, or a ploy to finance a future expedition by Crocker (or other patrons). However, maybe it was a genuinely false sighting and type of mirage. In any case, mysterious objects or shapes on horizons are not an uncommon phenomenon and featured prominently in past exploration accounts and in folk-lore alike. Mirages are often synonymous with, for example, mythological lands, lost continents or lands, sunken islands, *Fata Morgana*, ghost or phantom islands, and water worlds. Definitions and use of these terms widely varies but can be somewhat separated into two scenarios. The first scenario is if the feature was once real and subsequently lost, for example to sea-level rise, landslides, shifting tides, or other mechanism of destruction. Doggerland, now submerged in the North Sea, is one such example of an ancient landscape that was once in existence (Cohen et al. 2017). The second scenario is if there was a false or mistaken observation i.e. the object or feature identified was never in existence. The human reasons for this later scenario could include navigational error, sea or travel-sickness, false memories or recollections, hallucinations, mistaken identification (e.g. from poor visibility or an optical illusions), or more pessimistically, pure fabrication.

A special category of mystery lands includes those that once featured on published maps or charts but were subsequently removed, or "un-discovered". One of the most famous cases is Plato's Atlantis. Another example is a

4 This incident is considered in *Murder in the Arctic* by Kim Schmidt. Accessed from https://storied.illinois.edu/crocker-land/#!

longitudinally aligned string of islands in the Southern Ocean, to the south and east of Australia, variously named Royal Company's Islands, Emerald Island, the Nimrod group, and Dougherty Island. These islands persisted in multiple iterations of hydrographic charts, including in British, German, French, and Dutch maps. Another example was Sandy Island, near New Caledonia in the western Pacific, which persisted in hydrographic maps for nearly two centuries after initially reported by British Captain James Cook in 1774. It was removed from maps by the French Hydrographic Service in 1974 and Australian Hydrographic Service in 1985 but remained in other datasets such as Google Maps until 2012 when a geological voyage to the region sailed directly over it for the first time (Seton et al. 2013). Sandy Island's initial sighting (an example of a false or mistaken observation) may have been a floating raft of pumice rock from a nearby volcanic eruption, or an intentional cartographer's mark for copyright.

In addition to Crocker Land, several other erroneously mapped or described features in the high north have also been documented. These include Sannikov Land, which Russian cartographers Yakov Sannikov and Matveu Gedenschtrom claimed to have seen in an 1809–1810 expedition to the New Siberian Islands. It is possible that this island was shallow shifting sands, or a mirage from Bennett Island of the New Siberian Islands archipelago. Another false Arctic island was Giles Land, which first noted by Dutch explorer Cornelis Giles in 1707. Subsequent searches did not find the land, and its confusion was compounded by erroneous mapping which placed land too far to the north; Giles Land was likely the island known today as Kvitøya within the archipelago of Svalbard (Lantz 2019). Another purported landmass was known as Croker's Mountains (unrelated to Crocker Land) which was apparently spotted in an 1818 expedition by John Ross to Lancaster Sound in search of the NW Passage (Ross 1819). It was named for the First Secretary to the Admiralty, John Wilson Croker. These mountains were deemed to completely impede passage through Baffin Bay and thus Ross did not attempt to sail through. This choice cost Ross drastically, because the next year, William Parry, of Ross's earlier crew, sailed straight through this area. Whether Crocker Mountains were caused by a mirage (though atmospheric conditions were reported to not have been conductive to such a phenomenon) or due to other reason has not clearly been established (Giaimo 2018).

5 Optical Phenomena and the Polar Regions

Mirages, which are naturally occurring optical phenomena, are frequently cited as the cause for many of these lost islands, and can be produced by a range of environmental and atmospheric conditions. Returning to Arctic exploration, there are a number of unique geographic elements worth considering in the context of mirages. As mentioned, one peculiarity is that at the turn of the twentieth century it was unclear how deep the Arctic Ocean was. It was therefore possible that isolated islands or archipelagos had not yet been discovered. At these times, many modern geological concepts which would have helped explain the first-order geography of ocean basins had not yet been proposed, such as the theory of plate tectonics, which was developed from the 1960s onwards. Another consideration is that sea-ice is thickest and oldest near the northern margin of Greenland, which was a launching site for many polar expeditions. Sea-ice can generate relatively significant topographic variation across a few meters, for example via ice ridges and icebergs, or the creation of fog banks. Related terms such as 'Peary-Land-Ice' and 'Cook-Land-Ice' are documented in polar expedition diaries. Sea-ice is also typically associated with stable atmospheric temperatures (as opposed to colder and more variable temperatures on land), though temperature inversions can occur leading to mirages of these topographic undulations. Another significant consideration is in navigation and mapping; at the poles, the meridians (lines of longitude) converge. One consequence is that an obliquely deviating path can lead to very different true directions. It also renders the commonly used Mercator projection to be distorting at high latitudes. Furthermore, the geographic and magnetic poles do not coincide, with the latter migrating (currently at a rate of around 50 km per year). Modern navigation, including, the use of satellites, radar, and inertial navigation systems, did not exist in the era of Peary and Cook. Early navigation techniques using solar compasses and shadows were also dependent on weather conditions, which could be very variable in the polar regions and was also complicated by moving sea-ice. With these considerations in mind, coupled with the physical and mental exertion of polar exploration, it is perhaps not surprising that many mirages occur in the polar regions.

To understand what a mirage is, requires a brief review of optical phenomena (Rees et al. 1991). Light doesn't always travel in a straight line; if a ray of light encounters a density contrast, such as a change in temperature or shift in medium (e.g. air to water), it will refract (Figure 17.4) and appear to travel in a new line. The human brain cannot correct for this redirection, leading to optical phenomena i.e. a mirage is a real optical effect, which can also be captured on camera, and it rests with the viewer to choose how to interpret the image. Notably, these types of optical phenomena are not termed 'illusions' because they are not an invention of the mind, such as for a hallucination.

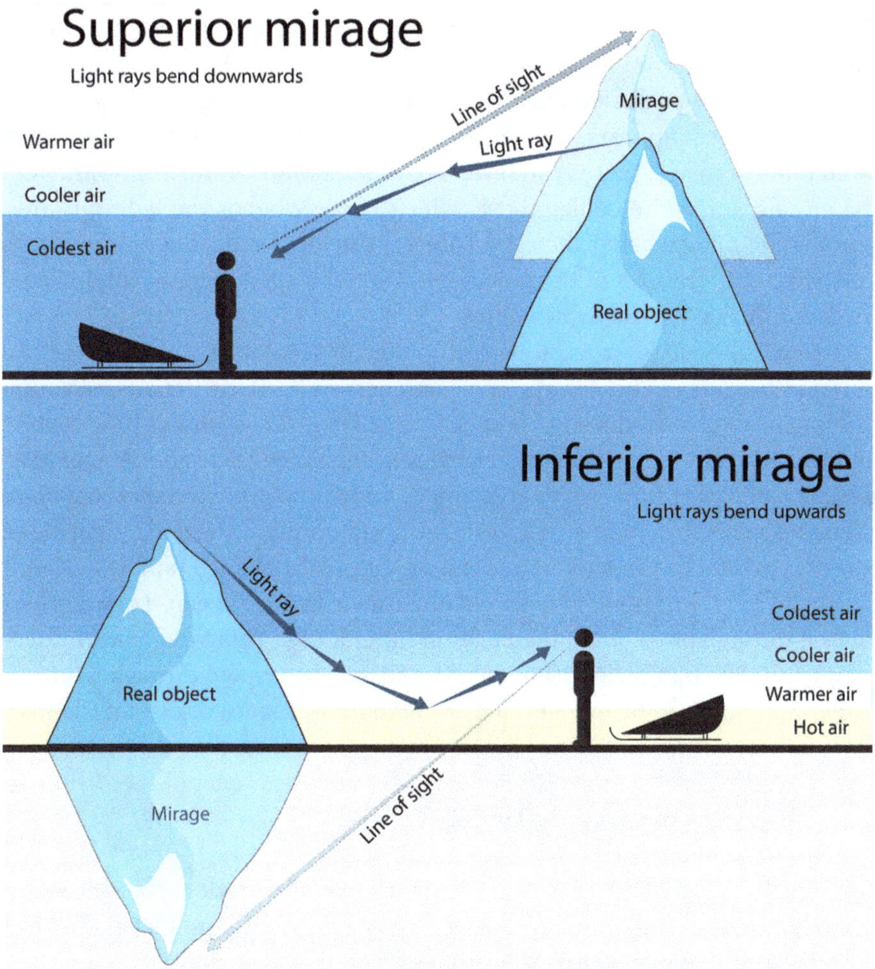

FIGURE 17.4 Simplified scenario of the atmospheric conditions that can refract light and might produce a superior-type mirage (top) or inferior-type mirage (bottom)

There are generally two types of mirages; a superior mirage and an inferior mirage e.g. Greenler (1987) (Figure 17.4). A superior mirage is where an image appears above its actual location, such as a floating ship on the horizon, and are relatively common in the polar regions. By contrast, an inferior mirage renders an image to appear below its location, such as water in the desert. In the case of a superior image, the atmosphere at the surface (e.g. where the viewer is standing) is cold, dense, and has a greater refractive index. Above this are layers of warmer air, which have a lower refractive index. A viewer assumes there is a straight path of light between the object and our eyes, but the temperature gradient refracts the light ray downwards. The resultant effect is that the object appears to be elevated, and a mirage is created. In the case of an inferior image, there is an inversion in the temperature, from warmer at the ground surface to cooler above, which refracts/bends the light upwards, rending the mirage below the actual object. The local atmospheric conditions and geometric features of the object may render a slightly more complicated resultant image than described above.

A *Fata Morgana* is a special type of mirage or optical phenomena that can occur on land, seas, and deserts. The name derives from the Italian translation of Morgan the Fairy who was an enchantress from the legend of King Arthur, and was thought to have conjured castles in the air or false land. A *Fata Morgana* is a complex form of superior image which distorts, stretches and compresses objects and may lead to several upside down (inverted), right-sided (erect), and stacked images. These mirages typically change rapidly and may appear to be continuous with the landscape or include a gap. They are best viewed through binoculars, telescopes or other lenses. Fata Morgana are documented in the polar regions (Rees 1998), and may be created through temperature inversions and the topographic undulations of sea-ice or distant islands.

6 Defining the Arctic in Time and Space

Geology, literally and figuratively, underlies human exploration. The mountains climbed, the seas sailed, and the deserts traversed, are all a product of the long-term evolution of our planet. The surface of Earth is the result of interactions occurring across a range of timescales from the instantaneous to millions and billions of years. The most challenging and distant landscapes which explorers long sought to reach, conquer and/or collect are so because of their geological evolution. To shift from human timescales described in the sections above to those of geology it is convenient to review the concept of a 'frame of reference'.

The 'Arctic' can be defined in many different ways, including from social, indigenous, geo-political, celestial, environmental, and/or climatic perspectives. Within the natural sciences, in particular, there are many definitions for the Arctic, including those based on temperature, sea-ice, permafrost, tree-limit, latitude (e.g. the Arctic Circle (Figure 17.5)) and ecological range. From a geological perspective of deep time, i.e. one that covers tens and hundreds of millions of years, the Arctic is a geologically "recent" and geographic-restricted concept. The Arctic we are familiar with today, with its

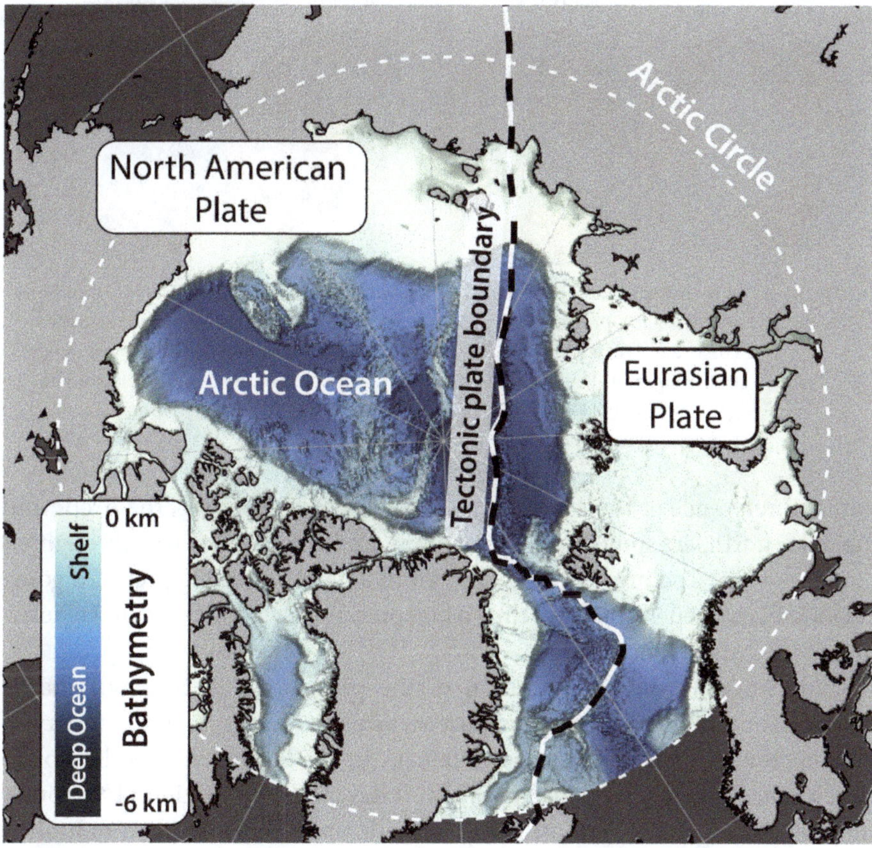

FIGURE 17.5 Overview of the polar ocean located north of the Arctic Circle (astronomically-derived dashed white line located at around 66.5° N), coloured with bathymetry (depth to the seafloor, from IBCAO, Jakobsson et al. 2012). The Arctic Ocean today is located across two tectonic plates (the North American and Eurasian plates) with the location of the tectonic plate boundary shown by the thick white and black dashed line
FIGURE MADE BY THE AUTHOR

forests, tundra, mountains, coasts, sea-ice, glaciers, continental shelves and central deep ocean, did not always exist. Indeed, the geological fragments of the Arctic were once located in low latitudes, including in the southern hemipshere, and did not have the same configuration as today (Figure 17.6). Thus, strictly speaking, the northern 'polar' regions (as defined by a latitudinal-based domain) and the 'Arctic' (the region that occupies these high norther latitudes today) are not the same thing. A similar story applies to the southern polar regions and the continent of Antarctica, which was once located further north and even covered by forests. Naturally, the

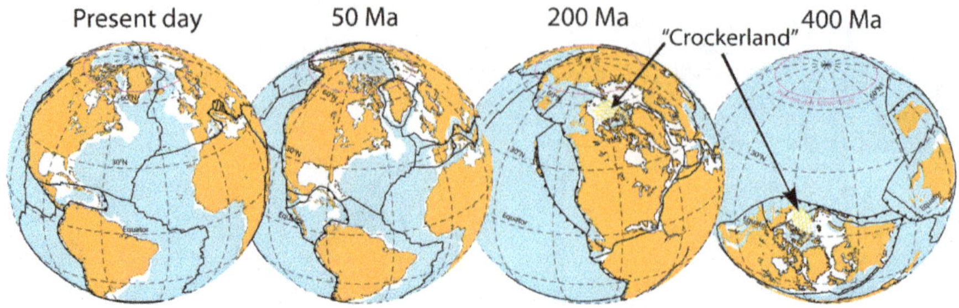

FIGURE 17.6 A reconstruction of the world at four different times in the past (Ma = millions of years ago), present-day, 50 Ma (before the Eurasia Basin opened) 200 Ma (before the Amerasia Basin opened), and 400 Ma (when the Arctic was close to the equator. The black lines are the boundaries of the tectonic plates. The geological Crockerland may have been located in the hatched area
FIGURE MADE BY THE AUTHOR

frame of reference of the Arctic is that of today; the two areas, the Arctic and the polar north, are somewhat synonymous, and will have been so throughout human history. The present-day is therefore a natural starting point for going backwards in deeper time, and thus using present-day nomenclature i.e. referring to the Arctic, is convenient.

Understanding the configuration of the present-day Arctic and what it looked like in the deep geological past turns us to the fundamental theory of plate tectonics. Plate tectonics is a concept in the Earth Sciences developed largely since seminal work in the 1960's. It describes the motion of the outer shell of our Earth (the lithosphere) through time. Tectonic plate motion is the direct response to the gradual cooling of the planet over geological eons, which has set up a pattern of movement in the interior of the Earth referred to as mantle convection – the Earth is in fact a solid "rock" down to the outer core, but behaves like a fluid over millions of years. As the interior of the Earth moves, so does its outer shell of tectonic plates, and the continents and oceans that are located atop of them. Tectonic plates (such as the Eurasia, Pacific, Australia, and Antarctic plates) are separated by tectonic plate boundaries, which incorporate the majority of deformation near their edges, which is why most earthquakes and volcanoes occur at these locations. The various tectonic plates on which we stand today are moving at the rates of a few centimeters per year (similar to finger nail growth), which means that in the course of 1,000,000 (one million) years, a plate may have shifted by around 10–100 kilometers. As a side note, the debunked 'Expanding Earth' theory, in which Earth is purported to be expanding in size, was in part fueled by the geology of the Arctic and a so-called 'Arctic Paradox,' which erroneously invoked generally northward-moving tectonic plates as an argument for an expanding Earth (Carey 1976).

Today, the Arctic domain is located across two large tectonic plates, the Eurasia Plate and the North America Plate (Figure 17.5 & Figure 17.6). The plate tectonic boundary between them runs from the middle of the North Atlantic Ocean, northwards through the Eurasia Basin, then transitions onshore at the Laptev Sea, before heading east towards the Pacific. The tectonic boundary accommodates relative motion along its length; including areas of extension as well as compression or strike-slip motion. Over millions of years, this movement means that the Eurasia and North American tectonic plates have shifted location and shape. Looking at the Arctic, there have been many episodes of the plates moving, with the mountains and oceans all shifting along with them. Figure 17.6 shows the configuration of the Arctic region at different times in the deep geological past, at 50, 200 and 400 million years ago. Two major changes include an overall northward motion of the Arctic away from the equator (when going forward time), as well as changes in the more local configurations. One important observation is that the central Arctic Ocean, as comprised by the deep Amerasia and Eurasia oceanic basins today (Figure 17.3c), did not always exist. The Eurasia Basin started to form around 55 million years ago and the Amerasia Basin started to form around 120 million years ago; before these times the continental regions of the Arctic were closer together (e.g. Shephard et al. 2013). This process is analogous to the assembly and dispersion of supercontinents, such as Pangea. It is this reassembled landmass located in the Arctic region that we here focus on for the concept of Crockerland.

7 The Deep-Time Return of Crockerland

What was located in the 'Arctic' long before the Arctic Ocean formed is a discussion point amongst geologists. Nonetheless there was thought to be some continental rocks (i.e. a landmass) in the central region. Canadian Geologist Ashton F. Embry (b. 1946) of the Canadian Geological Survey is credited with the alternative use of Crocker Land and adding it to the vernacular of Arctic geologists. Crockerland (typically spelled in this context as one word) was proposed in Embry (1993) for an ancient landmass that was located in the central Arctic during the Devonian and mid-Jurassic times (*ca.*400–120 million years ago). The existence of such a landmass at these past times explains several other geological observations around the Arctic. These include deformed rocks related to an ancient collision in the Canadian Arctic when Crockerland itself was formed, and that Crockerland served as the source of sediments which were later eroded (for example, due to running water), transported (for example, via rivers) and deposited to form the sedimentary rocks that are now located across the Canadian Arctic Islands and the Barents Shelf (some of these sedimentary rocks are in turn related to oil and gas formation) (e.g. Anfinson et al. 2012; Embry & Anfinson 2013). Embry directly acknowledges Peary as the origins for his naming choice, stating in Embry (1993) that "The northern landmass is named Crockerland after the mythical land which Robert E. Peary, the polar explorer, claims to have sighted north of Axel Heiberg Island following his failure to reach the north pole in 1906" (Peary 1907). Perhaps one could say that Peary was not wrong with his claim of land to the northwest of the Arctic Islands, but rather he was just late. Embry (pers. comm.) also expanded this, explaining that one member of the Crocker Land Expedition was the geologist and botanist Elmer Ekblaw. Like many other explorers of the time, Ekblaw named various geological features after notable American geologists including Embry's own hero, Joseph Barrell (1869–1919). Embry writes:

> When it comes to Crockerland, I often paraphrase the great last line of Barrell's (1913) paper on Devonian clastics of the Appalachians [mountain chain]. "Crockerland is gone and where once it rose from the northern sea, it now lies fragmented, scattered, and buried. But, in the remnants of formations, born of destruction, we can read the epitaph that recorded its history." That nicely sums it up. (Embry, pers. comm.)

The geological Crockerland, or parts thereof, also has other names, including Arctida, which is often used by the Russian geological community. While the geology today hints at the existence of this ancient landmass, Crockerland cannot be reached (there is an ocean at the centre of the Arctic today, not a landmass). Crockerland gradually broke up due to plate tectonics, specifically via the process of continental rifting and seafloor spreading within the Amerasia Basin i.e. one of the deep oceanic basins that make up the Arctic Ocean today (Figure 17.3c). As such, the rocks that once formed Crockerland are thought to now be dispersed and comprise elements of the northern Canadian Arctic Islands, Chukotka and East Siberia, the Chukchi Borderland, and potentially part of the submarine Amerasia Basin (Embry 1993; 2009).

8 Concluding Remarks

In a symbolic way, perhaps Peary's sighting (and that of Cook and others) was not just the proverbial crock; his vision of northern land was just a few hundred million years off the mark (also suggested in Embry 1993). It is worth noting that the geological concept of Crockerland is also subject to definition, and just like the original exploration inception of Crocker Land, is not without its own controversy. Recently, a geological study by Gilmullina et al. (2022) looked at ancient river and delta drainage systems and stated "we propose that the Crockerland concept should be abandoned". In any case, and without giving more geological details in this chapter, there is still geological evidence for an ancient landmass during the Arctic's earlier geological history (Anfinson et al. 2016; Embry, pers. comms.), and research is ongoing.

Similarly to Peary's Crocker Land, the geological Crockerland is a divisive concept. In both eponymous uses (geological and exploration) it is a somewhat intangible "object" that requires detailed and iterative scientific verification. Nonetheless, the use of naming is an interesting intersection of society and the natural sciences, and the continuing legacy of Arctic exploration i.e. George Crocker to Crocker Land to Crockerland. It also highlights that there are still geographic features, both on present-day maps and paleo-maps, in which scientists are able to propose the name of the feature. Underwater features found today, for example, are submitted to the General Bathymetric Chart of the Oceans (GEBCO) sub-committee on Undersea Features Names (SCUFN).[5]

Crocker Land also offers a cross-over between science and the arts. At the time of writing, several bands and songs released with eponymous nods to Crocker Land were found online. These include the ambient-electronic song "Crocker Land" by Nordic band Northaunt,[6] the song "Crocker Land" by alt-rock Spanish band Los Spam,[7] and the Canadian folk-rock band "The Crocker Land" whose title image includes sea-ice and a mirage.[8]

The collection of objects or items from the northern regions is typically considered in a physical sense. The example of Crocker Land detailed here also serves as an example of collecting something that is sensory (including visual or imaginary). In a sense it is also unobtainable and uncollectable; for the polar exploration version of Crocker Land – it was a vision of something on the distance and that was later proven to not exist (i.e. a false island), and for the geological version of Crockerland it is a feature that once did exist but

5 GEBCO Sub-Committee on Undersea Feature Names (SCUFN) https://www.gebco.net/about_us/committees_and_groups/scufn/ Accessed March 2024.

6 Song title: Crocker Land, Album: Cryosphere [ALBUM] by Band: Northaunt. Digitally released 25th Feb 2022 by Glacial Movements label, originally released in 2013. Accessed on YouTube https://www.youtube.com/watch?v=J9KNnGga2zo.

7 Song title: Crocker Land, Album: Roses by Band: Los Spam! (The Spam!). Released September 2021. Accessed on SoundCloud https://soundcloud.com/losspam.

8 "Music from The Crocker Land" Accessed from https://soundcloud.com/thecrockerland.

no-longer exists today (i.e. it is a dispersed geological feature). Crocker Land serves as a unique tribute to connections of human history and exploration, the ever-changing physical environment, and the ongoing scientific process in disciplines including the geosciences.

Acknowledgments

The 'Collecting Norden' group at the University of Oslo, led by Ulrike Spring and Janicke S. Kaasa, is thanked for instigating this paper and for related discussions about objects collected in the high north. Ashton Embry is thanked for providing valuable additional information behind the Crockerland concept and naming. Thanks also to Lene Liebe Delsett and Owen Anfinson for helpful feedback. G.E.S. acknowledges support from the Research Council of Norway for 'POLARIS – Evolution of the Arctic in Deep Time', Project Number 326238, and its Centres of Excellence Scheme, Project Number 332523. While best efforts have been made, there is the possibility of historical inaccuracies or interpretations, so the reader is referred to original and additional sources.

Bibliography

Anfinson, Owen. A., Andrew L. Leier, Rich Gaschnig, Ashton.F. Embry & Keith Dewing. 2012. "U–Pb and Hf Isotopic Data from Franklinian Basin Strata: Insights into the Nature of Crockerland and the Timing of Accretion, Canadian Arctic Islands." *Canadian Journal of Earth Sciences* 49 (11): 1316–28. https://doi.org/10.1139/e2012-067.

Atlas minor Gerardi Mercatoris a I. 1628. *Hondio plurimis aeneis tabulis auctus et illustratus [cartographic material]: denuo recognit[io], additisque novis delineationibus emendatus*. Accessed from https://archive.org/details/atlasminorgerardoomerc/page/n5/mode/2up.

Balch, Edwin Swift. 1913. *The North Pole and Bradley Land*. Kessinger Publishing.

Barrell, Joseph 1913. *Part III The Relations of the Delta to Appalachia. Ar. In Article XVII The Upper Devonian Delta of the Appalachian Geosyncline*. Accessed from https://zenodo.org/record/1521782/files/article.pdf.

Bravo, Michael. 2019. *North Pole: Nature and Culture*. London: Reaktion Books.

Bryce, Robert M. 2008. *Fredrick A. Cook: From Hero to Humbug*. http://www.humbug.polarhist.com/.

Cohen, Kim M., Kieran Westley, Gilles Erkens, Marc P. Hijma & Henk K.T. Weerts. 2017. "The North Sea." In *Submerged Landscapes of the European Continental Shelf*, edited by N.C. Flemming, J. Harff, D. Moura, A. Burgess & G.N.Bailey. https://doi.org/10.1002/9781118927823.ch7.

Carey, Samuel W. 1976. *The Expanding Earth*. Amsterdam: Elsevier.

Cook, Frederick Albert. 1911. *My Attainment of the Pole; Being the Record of the Expedition That First Reached the Boreal Center, 1907–1909, With the Final Summary of the Polar Controversy*. New York: The Polar Publishing Co.

Embry, Ashton F. 1993. "Crockerland. The Northern Source Area for the Sverdrup Basin, Canadian Arctic Archipelago." In *Arctic Geology and Petroleum Potential*, edited by T. O. Vorren, E. Bergsager, Ø. A. Dahl-Stamnes, E. Holter, B. Johansen, E. Lie & T. B. Lund. Norwegian Petroleum Society, Special Publication 2: 205–16.

Embry, Ashton F. 2009. "Crockerland – The Source Area for the Triassic to Middle Jurassic Strata of Northern Axel Heiberg Island, Canadian Arctic Islands." *Bulletin of Canadian Petroleum Geology* 57 (2): 129–40.

Embry, Ashton F. & Owen Anfinson. 2013. *A History of Crockerland – The Little Arctic Terrane that Could*. 3P Arctic Conference and Exhibition, Stavanger, Norway. (Abstract).

Giaimo, Cara. 2018. "How a Fake Mountain Range Slowed Down Arctic Exploration." *Atlas Obscura*. https://www.atlasobscura.com/articles/john-ross-arctic-mountain-fake.

Gilmulina, Albina et al. 2022. "Arctic Sediment Routing During the Triassic: Sinking the Arctic Atlantis." *Journal of the Geological Society* 18. https://doi.org/10.1144/jgs2022-018.

Greenler, Robert G. 1987. "Laboratory Simulation of Inferior and Superior Mirages." *Journal of the Optical Society of America A* 4: 589–90. https://doi.org/10.1364/JOSAA.4.000589.

Harris, Rollin A. 1904. "Some Indications of Land in the Vicinity of the North Pole." *National Geographic* 15 (June): 255–61.

Henderson, Bruce. 2005. *True North: Peary, Cook, and the Race to the Pole*. W.W. Norton & Company.

Henderson, Bruce. 2009. "Who discovered the North Pole?" Accessed from https://www.smithsonianmag.com/history/who-discovered-the-north-pole-116633746/.

Herbert, Wally. 1989. *The Noose of Laurels: Robert E. Peary and the Race to the North Pole*. New York: Atheneum Books.

Jakobsson, Martin et al. 2012. "The International Bathymetric Chart of the Arctic Ocean (IBCAO) Version 3.0." *Geophysical Research Letters* 39, L12609: doi:10.1029/2012GL052219.

Lantz, Björn. 2019. "Giles Land – a Mystery for S.A. Andrée and Other Early Arctic Explorers." *Polar Research* 38 (July). https://doi.org/10.33265/polar.v38.3433.

MacMillan, Donald Baxter. 1918. *Four Years in the White North*. New York: Harper & Brothers.

Nansen, Fridtjof. 1907a. "On North Polar Problems." *The Geographical Journal* 30 (5): 469–87. https://doi.org/10.2307/1778052.

Nansen, Fridtjof. 1907b. "Bathymetrical Chart of Northern Polar Seas." *The Geographical Journal* 30 (5): 584.

Peary, Robert. 1907. *Nearest the Pole: A Narrative of the Polar Expedition of the Peary Arctic Club in the S.S. Roosevelt, 1905–1906*. Doubleday.

Peary, Robert. 1910. *The North Pole: Its Discovery in 1909 Under the Auspices of the Peary Arctic Club*. New York: Frederick A. Stokes Company.

Seton, Maria, Simon Williams, Sabin Zahirovic & Steven Micklethwaite. 2013. "Obituary: Sandy Island (1876–2012)." *Eos Trans. AGU* 94 (15): 141. https://doi.org/10.1002/2013EO150001.

Rees, W. G., C. M. Roach & C. H. F Glover. 1991. "Inversion of Atmospheric Refraction Data." *Journal of the Optical Society of America A* 8: 330–38.

Rees, W. G. 1988. "Polar Mirages." *Polar Record* 24 (150): 193–98. doi:10.1017/S0032247400009141.

Ross, John. 1819. *Voyage of Discovery*. London: John Murray.

Shephard, Grace, R. Dietmar Müller & Maria Seton. 2013. "The Tectonic Evolution of the Arctic Since Pangea Breakup: Integrating Constraints from Surface Geology and Geophysics with Mantle Structure." *Earth-Science Reviews* 124: 148–83. https://doi.org/10.1016/j.earscirev.2013.05.012.

Solberg, Winton U. 2019. *Arctic Mirage: The 1913–1917 Expedition in Search of Crocker Land*. Jefferson, N.C.: McFarland.

Taylor, E. G. R. 1956. "A Letter Dated 1577 from Mercator to John Dee." *Imago Mundi* 13: 56–68.

Tierney, John. 2009. "Who Was First at the North Pole?" Accessed from https://archive.nytimes.com/tierneylab.blogs.nytimes.com/2009/09/07/who-was-first-at-the-north-pole/.

Welky, David. 2016. *A Wretched and Precarious Situation: In Search of the Last Arctic Frontier*. W.W. Norton & Co.

Index

www.ingramcontent.com/pod-product-compliance
Lightning Source LLC
Chambersburg PA
CBHW071248230926
51849CB00001B/273
9789004749672